Humanizing Healthcare – Human Factors for Medical Device Design

Russell J. Branaghan • Joseph S. O'Brian
Emily A. Hildebrand • L. Bryant Foster

Humanizing Healthcare – Human Factors for Medical Device Design

 Springer

Russell J. Branaghan
Research Collective
Tempe, AZ, USA

Emily A. Hildebrand
Research Collective
Tempe, AZ, USA

Joseph S. O'Brian
Research Collective
Tempe, AZ, USA

L. Bryant Foster
Research Collective
Tempe, AZ, USA

ISBN 978-3-030-64435-2   ISBN 978-3-030-64433-8   (eBook)
https://doi.org/10.1007/978-3-030-64433-8

This Springer imprint is published by the registered company Springer Nature Switzerland AG
The registered company address is: Gewerbestrasse 11, 6330 Cham, Switzerland

Russ dedicates this book to Michael G. Egleston for his inspiration in handling challenges, great and small, medical and otherwise. I am proud to call you family! Joe dedicates this book to his daughter, Taylor. The two most powerful words in learning are "how" and "why." Never stop wondering how the world works!

Emily dedicates this book to all the wonderful teachers and mentors she has had, especially Russ, without whom I would not be where I am today, working in the field of human factors engineering. Bryant dedicates this book to his parents, Larry and Wendy Foster. Thank you for teaching me the value of hard work and that life is meant to be enjoyed.

Preface

Like most human factors engineers, I learned about the field completely by accident. As an undergraduate interested in neuroscience, I was pursuing majors in psychology and biology when I took a job as a research assistant in the psychobiology lab. Just prior to that, one of the professors in the department passed away, and his wife donated his entire library to our school. As the assistant, I was tasked with shelving all his books, and one book, *Human Engineering Guide to Equipment Design*, edited by Harold P. Van Cott and Robert G. Kinkade, caught my eye. As I paged through, I discovered all kinds of facts, figures, and rules about human vision, hearing, memory, attention, and decision making. These weren't just musings or guesses about how people behaved; they were real honest to goodness data compiled from hundreds of scientific studies. It then showed how to apply these scientific facts to design. It combined my interests in psychology and physiology perfectly and, more than that, proved that some lucky people actually did this for a living. I decided immediately to search for graduate programs in human factors.

Back then, there were only a few PhD programs in human factors, and they were housed in either psychology (cognitive psychology, engineering psychology, industrial psychology, experimental psychology) or industrial engineering. Interestingly, they taught largely the same courses: Research methods, statistics, sensation and perception, cognition, biomechanics, and of course, human factors, which usually combined the other topics.

All four of us have stories somewhat similar to this. We were studying something related, learned about human factors engineering (HFE) by chance, and recognized we had a real affinity for it. In recent years, device manufacturers, hospitals, and regulatory entities have recognized the perils of medical device use error and the need for human factors engineering. Because devices failed to accommodate well-known human capabilities and limitations, patients, providers, and caregivers were injured or died. This has led more people to discover the field and recognize their affinity for it, as well.

Rather than human factors engineering degrees, however, practitioners often have backgrounds in mechanical engineering, quality engineering, medicine, technical communications, industrial design, user experience design, or service design, to name a few. As a result, many have come to us to learn about the subjects we took in graduate school. They can take courses and read books about risk analysis, formative and validation usability testing, and preparing documents for submission to regulatory industries and there are a few good edited volumes about human factors in medical device design (e.g., Privitera, 2019; Sethumadhavan & Sasangohar, 2020; Weinger, Wiklund, & Gardner-Bonneau, 2011). Also, there are good human factors texts (e.g., Lee, Wickens, Liu, & Boyle, 2017; Proctor & Van Zandt, 2018). Unfortunately, however, there were no single authored (or in our case, team authored) books that taught the fundamental human factors engineering topics, and these are important. This book is our way to share them with you. It is our hope that you will integrate the material into your own work to make the world in general, and medical devices in particular, more useful, usable, pleasant, and safe.

References

Lee, J. D., Wickens, C. D., Liu, Y., & Boyle, L. N. (2017). *Designing for people: An introduction to human factors engineering*. Scotts Valley, CA: CreateSpace.
Privitera, M. B. (Ed.). (2019). *Applied human factors in Medical Device design*. Cambridge, MA: Academic Press.
Proctor, R. W., & Van Zandt, T. (2018). *Human factors in simple and complex systems*. Boca Raton, FL: CRC Press.
Sethumadhavan, A., & Sasangohar, F. (2020). *Design for health: Applications of human factors*. Amsterdam: Elsevier.
Weinger, M. B., Wiklund, M., & Gardner-Bonneau, D. (2011). *Handbook of human factors in medical device design*. Boca Raton, FL: CRC Press.

Tempe, AZ, USA

Russell J. Branaghan
Joseph S. O'Brian
Emily A. Hildebrand
L. Bryant Foster

There are some patients whom we cannot help: there are none whom we cannot harm.

—Lambert (1978)

Acknowledgments

We would like to start by acknowledging our families for their support and inspiration: Tonya, Kevin, and McKenna Branaghan, Jamie and Taylor O'Brian, Devin, Elaine, and Jackson Steward, and Chelsea, Braclyn, Pratt, and Sutton Foster. Writing this book kept the authors really busy and away from family time, which would have been more fun.

Most original images in the book were created by Natalie Sheehan. All we can say is, wow, your graphics look a lot better than ours! Stephanie McNicol also provided outstanding images and design consulting that improved the document immensely. Natalie and Stephanie represent a new kind of human factors engineer, with exceedingly strong backgrounds in experimental psychology and design.

Michael Sheehan is a medical student and photographer. Somehow, among his clinical rotations, board exams, and other medical school rigors, he conducted literature reviews, located statistics, summarized medical and popular press articles, and took photographs. He also fielded numerous phone calls to patiently explain procedures, devices, and challenges. His assistance improved the document and gave us confidence.

Then there was the editing: Several talented colleagues donated their time and talents to edit the chapters. Tonya Branaghan edited the Cognition chapter, Sarai Westbrook edited the chapters on Research Methods—Qualitative, Quantitative, and Usability Evaluation. Stephanie McNicol edited the Displays and Human-Computer Interaction chapters, and Anders Orn volunteered to edit the whole darn thing! In doing so, he provided bold advice and much needed camaraderie.

Greta Bowman was the conductor; she made sure all chapters, headings, figures, and tables were numbered correctly, organized all images and permissions to submit to the publisher, and generally kept us from dropping the ball. Tonya Branaghan kept the company running while we wrote—which is no small feat.

The content and organization of the book were sculpted by questions and discussions with Russ' students at Arizona State University and Northwestern University, as well as colleagues at several design and research companies, especially Mark Palmer at Lextant Corporation, Walter and Scot Herbst at Herbst Produkt, and Bradley Peacock at Peacock 9. Many of the ideas were refined when Russ served as a Visiting Scientist at Mayo Clinic, working alongside Susan Hallbeck, Katie Law, Renaldo Blocker, and Bethany Lowndes.

We thank Mike McCabe and Arun Pandian from SpringerNature for their hard work and dedication. Mike was kind enough to contact us about the need for this book, and Arun guided us through the production process.

Finally, dozens of clients have enlisted our help for literally hundreds of HFE and usability projects. Each project teaches us more, challenges us more, and reminds us why we chose this as our life's work. Thank you to all of them for placing their trust in us!

Abbreviations

AAMI	Association for the Advancement of Medical Instrumentation
ADA	American Disabilities Act
AE	Adverse events
AMD	Age-related macular degeneration
ANOVA	Analysis of variance
ANSI	American National Standards Institute
APA	American Psychological Association
APD	Auditory processing disorder
ATM	Automated teller machines
AU	Action units
BBFG	Bulletin board focus group
CDRH	Centers for Devices and Radiological Health
CGM	Continuous glucose monitors
CHL	Conductive hearing loss
CMC	Control movement compatibility
CPAP	Continuous positive air pressure
CRT	Cathode ray tube display
dB	Decibel
ECG	Electrocardiography
ECRI	Emergency Care Research Institute
EEG	Electroencephalography
EHR	Electronic health records
EMR	Emergency medical records
EU	European Union
FACS	Facial action coding system
FDA	Food and Drug Administration (U.S.)
FEA	Facial expression analysis
FMEA	Failure Modes and Effects Analysis
FOV	Field of view
GSR	Galvanic skin response
GUI	Graphical user interface

HAI	Healthcare-associated infections
HCD	Human-centered design
HCP	Healthcare providers
HF	Human factors
HFE	Human factors engineering
HFE/UE	Human factors and usability engineering
HTA	Hierarchical task analysis
HVAC	Heating, ventilation, and air conditioning
HZ	Hertz
ICU	Intensive care unit
IFU	Instructions for use
ILD	Interaural level difference
IRB	Institutional review boards
ITD	Interaural time difference
IVD	In vitro diagnostic medical device
JIT	Just in time
LCD	Liquid crystal display
LED	Light emitting diode
LTM	Long-term memory
LVAD	Left ventricular assist device
MAA	Minimal audible angle
MHRA	Medicines and Healthcare Products Regulatory Agency (EU)
MRI	Magnetic resonance imaging
NICU	Neonatal intensive care units
NIHL	Noise-induced hearing loss
NIOSH	National Institute for Occupational Safety and Health (U.S.)
NNR	Noise reduction rating
OR	Operating room
OTC	Over the counter
PACU	Post-anesthesia care unit
PPE	Personal protective equipment
PRP	Platelet-rich plasma
PTZ	Pan-tilt-zoom
RaS	Robotic assisted surgery
RCA	Root cause analysis
RME	Reusable medical equipment
RN	Registered nurse
ROM	Range of motion
RSI	Repetitive strain injury
SAW	Surface acoustic wave
SME	Subject matter expert
SNHL	Sensorineural hearing loss
SOP	Standard operating procedure
SPL	Sound pressure level
sRGB	Standard red green blue

SUD	Single use devices
ToT	Time on task
UE	Use-error
UI	User interface
URA	Use risk analysis
UX	User experience
WCAG	Web content accessibility guidelines
WM	Working memory
μg	Microgram
μs	Microsecond

Mnemonics

AEIOU	Activities, Environments, Interactions, Objects, and Users
BASIC	Breakdowns, Anecdotes, Scenarios, Insights, Custom Tools
FACES	Flow, Artifacts, Context, Environment, Sequence
MAUDE	Manufacturer and User Facility Device Experience
RIMS	Redundancy, Immediacy, Modality, Specificity
ROYGBIV	Red, Orange, Yellow, Green, Blue, Indigo, Violet
SOAP	Subjective, Objective, Assessment, and Plan
WEIRD	Western, Educated, and from Industrialized, Rich, and Democratic Countries

Contents

About the Authors

Russell J. Branaghan Russ is Co-founder and President of Research Collective, a Human Factors consultancy and Usability Laboratory in the Phoenix metropolitan area. He partners with medical device and pharmaceutical companies all over the world for user research, human factors analysis, usability testing, and regulatory consulting. Russ has more than 30 years of human factors experience in industry, consulting, teaching, and research. He says that industry was OK, but he likes consulting, teaching, and research a lot more. In addition to his consulting, Russ holds appointments at Arizona State University and Northwestern University and has won teaching awards at both. An avid writer, Russ has published over 100 peer-reviewed articles, book chapters, and proceedings papers. In his spare time, Russ plays guitar (poorly) and runs (slowly).

Joseph S. O'Brian Joe has been active in the world of Human Factors since 2013, contributing to the community by speaking at conferences, as well as participating on various standards and review committees. He was drawn to Human Factors through the unconventional means of learning how to fix a broken guitar as a Psychology undergraduate student. That research revealed an area of ergonomics and design that stuck with him, and instilled a passion to help make products more comfortable, easier to learn, and enjoyable to use. After completing a Master's degree in Applied Psychology at Arizona State University, Joe joined Research Collective and is currently the company's Senior Human Factors Scientist.

Emily A. Hildebrand Emily is the Director of Human Factors at Research Collective, a Human Factors consultancy and Usability Laboratory in the Phoenix metropolitan area. She leads usability, product design, and user experience-related projects for Fortune 100 and Fortune 500 clients across a variety of fields, but has over a decade of experience performing human factors (HF) studies to support the development of medical devices. In addition to consulting, she has performed research internally on medical devices and workflow processes at the VA and Mayo Clinic. She has extensive experience in product failure analysis and expert witness litigation support for medical devices. Emily has a PhD in Applied Cognitive

Science from Arizona State University and participates in the larger HF community as a member of various AAMI human factors committees, a reviewer for the HFES Healthcare Symposium, and as a speaker and attendee at HF and medical device conferences.

L. Bryant Foster Bryant is Co-founder and VP of Human Factors at Research Collective, a HF consultancy and Usability laboratory in the Phoenix metro area, where he has performed human factors and usability research for dozens of medical devices including surgical instruments, point-of-care devices, diagnostics, combination products, home-use devices, OTC products, and more. He serves as an active member of the Human Factors Engineering committee within the Association for the Advancement of Medical Instrumentation (AAMI) and also teaches a Human Factors and Design Controls course for the Regulatory Affairs Professional Society (RAPS). Bryant is also a member of the Healthcare and Product Design technical groups within the Human Factors and Ergonomics Society (HFES) and regularly speaks at both the annual HFES meeting and the HFES Healthcare Symposium. Bryant has written numerous articles about human factors, usability, and human-centered design for several publications.

Chapter 1
Introduction

1.1 Medical Error

While caring for her patient, a nurse attempted to program an infusion pump to deliver 130.1 mL/h of a particular medication. She pressed all the right keys, "1 - 3 - 0 - . - 1," but unfortunately, on this model of infusion pump, the decimal point did not work for numbers over 99.9. As a result, the pump ignored the decimal point key press and was programmed to deliver 1301 mL/h, a ten times overdose (Zhang, Patel, Johnson, & Shortliffe, 2004).

In another hospital two nurses cared for a 15-day-old baby with a congenital heart defect, breathing problems, and a rapid heart rate. The nurses gave the baby digoxin, a common drug for slowing heartbeats. Tragically, they made a mathematical mistake and administered 220 µg of digoxin rather than the intended 22 µg. The massive dose caused the baby to go into cardiac arrest, and he died a few days later (BBC, 2005).

This problem, called "death by decimal," illustrates some of the dangers of medical error in our healthcare environment. Errors in medicine are common. One recent study (Makary & Daniel, 2016) concluded that medical error kills 251,000 Americans per year, making it the third leading cause of death, behind heart disease and cancer (Fig. 1.1). According to this estimate, medical error accounts for 9.5% of all US deaths, the equivalent of two 747 jumbo jets (loaded with 364 passengers each) crashing every day, just in the United States (US). This death rate is comparable to one September 11 attack every 4 days. Even more troubling, this estimate only accounts for inpatient deaths. Many people die from errors in ambulatory settings, clinics, therapy, and home.

Medical error happens in a variety of circumstances—in hospitals, in surgery, when delivering medications, when using a medical device, and so on. Let us start by discussing medical errors in hospitals. To do that we need to understand the notion of an adverse event (AE). Adverse events (AEs), also known as harms, are injuries resulting from medical care rather than from illnesses themselves (Wachter,

© Springer Nature Switzerland AG 2021

R. J. Branaghan et al., *Humanizing Healthcare – Human Factors for Medical Device Design*, https://doi.org/10.1007/978-3-030-64433-8_1

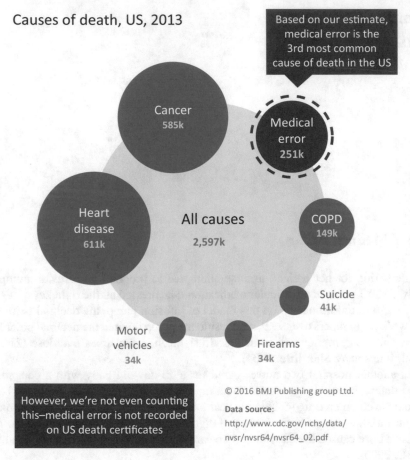

Causes of death, US, 2013

Based on our estimate, medical error is the 3rd most common cause of death in the US

Cancer
585k

Medical error
251k

Heart disease
611k

All causes
2,597k

COPD
149k

Suicide
41k

Motor vehicles
34k

Firearms
34k

However, we're not even counting this—medical error is not recorded on US death certificates

© 2016 BMJ Publishing group Ltd.

Data Source:
http://www.cdc.gov/nchs/data/
nvsr/nvsr64/nvsr64_02.pdf

Fig. 1.1 Causes of death in the United States in 2013 (BMJ Publishing group, Ltd. is licensed under CC BY 4)

2012). Some AEs are not preventable, but those that can be prevented usually involve some type of error: either acts of omission (failing to do something) or acts of commission (doing something wrong). Approximately one-third of hospitalized patients experience some type of AE (Classen et al., 2011). While roughly two-thirds of AEs cause little-to-no harm, the remaining third unfortunately do cause harm. This is not only dangerous, but also expensive; the cost of preventable AEs is estimated to be between 17 billion and 29 billion dollars per year just in United States hospitals (Wachter, 2012). These costs are even higher when considering preventable AEs in ambulatory clinics, nursing homes, assisted living facilities, and other settings.

Problems can occur during bedside procedures as well. Several procedures related to insulin pumps, ablation systems, automated external defibrillators, duodenoscope reprocessing, and many more (FDA, 2016) have complication rates exceeding 15%. For example, patients undergoing central venous catheter placement are at risk of arterial laceration, pneumothorax, thrombosis, and infection, each potentially deadly.

Many medical errors occur in the surgical suite. More than 20 million patients undergo surgery every year in the US. Although surgeries have become safer in recent years, many safety issues remain. For example, approximately 3% of patients who undergo operations suffer an AE and half of these are preventable (Lindenauer et al., 2007). These include anesthesia-related complications, wrong site and wrong patient surgery, medication errors, retained foreign objects, and surgical fires (Wachter, 2012). These are referred to as "never errors" because they should never happen, under any circumstances. They would be similar to a commercial jet taking off on an overseas flight without any fuel. And yet, never errors occur all the time.

One type of never error, retained objects, involves leaving surgical instruments, sponges, or other objects behind in the body after surgery. Gawande, Studdert, Orav, Brennan, and Zinner (2003) reviewed 54 patients with retained foreign bodies over 16 years, and found that about two-thirds of the items left behind were sponges or pieces of gauze used to soak up blood. The remaining one-third were surgical instruments. The rate of retained objects is about 1 in 1000, roughly equivalent to one case per year for a typical large hospital in the US (Wachter, 2012). On the other hand, this estimate is probably low because it is derived from an analysis of malpractice cases. Many, if not most, retained object errors never lead to malpractice claims, since it often takes years to discover that a surgical sponge has been left behind (Wan, Le, Riskin, & Macario, 2009). Now radio-frequency (RF) surgical sponge detection devices are used at the end of each case. The device detects RF chips placed in most sponges.

Another challenge is wrong site surgery. For example, due to diabetes and circulatory disease, a 51-year-old retired construction worker needed to have his left leg removed below the knee. Appropriately, the operating room (OR) schedule, surgical suite blackboard, and hospital computer system all indicated that the patient was to have his *left* leg amputated. Unfortunately, the patient accidentally signed a consent form to amputate his *right* leg. And, that is exactly what the surgeon did (Lieber, 2015).

One study of 1000 hand surgeons showed that 20% of them admitted to having operated on the wrong site at least once in their career. An additional 16% had prepared to operate on the wrong site but caught themselves before cutting (Meinberg & Stern, 2003). Simple solutions to this include "sign your site," in which the surgeon marks the surgical site in indelible ink (Fig. 1.2). However, even the "sign your site" strategy presented its own problems: some surgeons placed an "X" on the surgical site (as in "X marks the spot") whereas others placed an "X" on the opposite limb, meaning "Do not cut here."

Time outs as required by the joint commission have also been implemented. The time out is performed in the OR once the patient is prepped and before incision. It confirms patient identity, correct site, and correct procedure. The operating surgeon has to be present and agree to the time out.

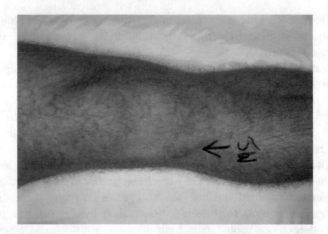

Fig. 1.2 Sign your site

Fig. 1.3 Comparison of
adult and child dosage
vials of heparin (Image
courtesy of ISMP www.
ismp.org)

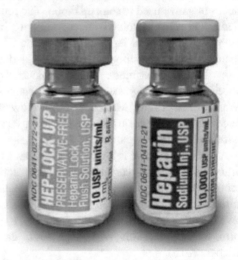

Many medical errors are more mundane than cutting off the wrong leg, but potentially more fatal, like administering the wrong dose of a common medication. Consider the following story. Dennis Quaid, the actor, and his wife Kimberly Buffington brought their newborn twins to Cedars-Sinai Hospital to be treated for staph infections. To prevent clots around intravenous catheter sites, the babies were prescribed a baby-friendly 10 unit-per-mL-dose of the anticoagulant, heparin (shown on the left in Fig. 1.3). Instead, however, they were accidentally administered the adult dosage on the bottle on the right, 10,000 units per mL. Worse, this happened twice, once at 11:30 AM and again at 5:34 PM (Ornstein, 2014). This was a 1000 times overdose of anticoagulant. The error was identified when one of the babies started oozing blood from the puncture site, and blood tests confirmed the problem. We are pleased to report that despite the potentially fatal medical error, the infants survived.

Investigating the event, Cedars-Sinai identified three issues that led to the over-doses. First, the pharmacy technician retrieved the heparin from supply without having a second technician verify the drug's concentration. Second, when delivered to a satellite pharmacy, a different technician failed to verify the concentration. Third, the nurses who administered the heparin failed to verify that it was the correct medication and dose.

When we present this case to undergraduate students, their first reaction is outrage. How could trained medical professionals be so careless? Fire the nurses immediately! Bring them up on legal charges! At the very least, students insist that the nurses and pharmacy technicians should go through training. Cedars-Sinai had a similar reaction. The employees were relieved of their duties during the investigation and "appropriate disciplinary actions were taken."

We do not agree with this reaction, however. In this case, we side with our human factors engineering (HFE) graduate students rather than the undergraduates. Because our graduate students study human performance, cognition, and design, they reach a very different conclusion. They immediately note the similar color, size, shape, font, and words on the bottles. Sure, the labels are different shades of blue, but they are clearly in the same color family, as effective brand guidelines dictate. Now imagine busy pharmacy technicians and nurses trying to care for sick babies, managing numerous medications, pieces of equipment, parents, physicians, and who knows what else. Now remember that these professionals have the same attention span, working memory, and judgment limitations as you or I. Perhaps design is part of this problem; and perhaps HFE could help.

The manufacturer reached the same conclusion as our graduate students. To reduce future errors, they changed the label on the higher concentration vials, modifying the background color, increasing font size, and adding an "alert" tear-off label.

It should be no surprise that medication errors are common, simply because there are over 10,000 prescription drugs and biologicals and 300,000 over-the-counter medications available in the United States (Aspden, 2007). An average hospitalized patient can expect one medication error per day. At least 5% of hospital patients experience some adverse drug event during their hospital stay (Wachter, 2012). And, 5–10% of the patients almost received the wrong medicine or the wrong dose, but the problem was caught in time (this is often called a "near miss").

Patients on numerous medications, as well as older patients, are most likely to be harmed because medication errors are especially common when patients are on high-risk medications, such as warfarin, insulin, or heparin. Classen, Jaser, and Budnitz (2010) found that one in seven patients receiving heparin experienced an adverse drug event. As with many errors, these are expensive. The cost of preventable medication errors in the United States hospitals is approximately 16.4 billion dollars per year (Wachter, 2012). Moreover, nearly 5% of hospital admissions can be traced to problems with medications, many of which are preventable.

1.2 Medical Devices

Now that we have described the problem of medical error in general, let us turn our attention to error in the use of medical devices, and why this has become so problematic. It helps to start by considering how technology has changed in such a short time. If you were a physician during the first decade of the 1900s, new medical knowledge would be revolutionizing your practice. Unlike before, you would now wash your hands, sterilize your instruments, and wear a face mask during surgery. Thanks to the miracles of ether and chloroform, patients would no longer need to remain awake during procedures. You could now utilize cutting-edge diagnostic devices like stethoscopes to diagnose heart and lung problems, ophthalmoscopes to inspect the eyes, and laryngoscopes to inspect the throat. Particularly useful would be a recently invented 5-min procedure using a mercury thermometer to measure body temperature. Depending on your sophistication and financial resources, you could now use microscopes to test for tuberculosis, cholera, typhoid, and diphtheria. Sadly however, you would have no method to cure these diseases, since the first antibiotic, penicillin, had not yet been discovered.

Medical devices at this time were unsophisticated, mostly limited to hospital beds, wheelchairs, crutches, bandages, splints, canes, and crude prosthetics. This is a far cry from medical practice today. In little more than a century, medical technology has led to considerably longer lives (about twice as long!). The world's population of people aged 65 years and older increases by approximately 850,000 every month (Kinsella & Phillips, 2005), and half of the people who have ever reached the age of 65 are alive today (Rowe & Kahn, 2015). And it is not just longer lifespans; technology has also led to longer health spans, low infant mortality, reduced pain, same-day surgery, and generally a safer life.

Although improvements in sanitation, nutrition, antibiotics, pharmaceuticals, and anesthesia starred in this revolution, advances in medical devices played a strong supporting role. Due to advancements in medical devices, modern healthcare providers benefit from thousands of diagnostic blood tests, and dozens of imaging techniques.

The World Health Organization (2020) defines medical devices, writing that medical devices diagnose, prevent, monitor, treat, alleviate, or compensate for disease or injury. The list of devices includes everything from tongue depressors, to robotic surgical systems, and artificial hearts. Indeed, there are more than 1700 different types of medical devices available for orthopedics, cardiovascular and diagnostic imaging, minimally invasive surgery, wound management, ophthalmology, diabetes care, dental devices, nephrology, and many other fields (FDA, 2019). These are used in every environment, from intensive care to patient bedrooms. Even public spaces like airports and train stations house automatic external defibrillators and various devices included in first aid kits. Table 1.1 lists examples of common medical devices categorized into 11 physiological systems. The list is not comprehensive, but provides a sense of the enormous variety in the field.

Table 1.1 Sample medical devices and the physiological systems they serve

Physiological system	Example medical devices
Cardiovascular	Left-ventricular assist device (LVAD), artificial heart, stent, pacemaker, defibrillator, cardiac catheterization, mechanical heart valves, cardiac loop recorder, electrocardiogram, Holter monitor, intravenous/intra-arterial catheter, stethoscope
Digestive	Anastomotic stapler, bipolar electrosurgical device, electrocautery, laparoscopic surgery, endoscope, tongue depressor, ultrasound
Endocrine	Insulin pump, continuous glucose monitor, artificial pancreas device system, autoinjector
Immune	Autoinjector, syringe, genomic testing assay, thermometer, nasal allergy spray
Integumentary	Band aid, suture, stapler, Steri-Strip, wound vac, zipline closure device
Muscular	Recovery device (percussive therapy, ice/heat device), electric stimulation devices (transcutaneous electrical nerve stimulation [TENS] units), pelvic floor therapy devices, patient transfer devices (wheelchairs, scooters, lifts, walkers), braces, splints, orthotics
Nervous	Deep brain stimulator, peripheral electrical nerve stimulator, electromyography, electroencephalogram, cochlear implant, ophthalmoscope, reflex hammer
Renal	Hemodialysis machine, ureteral stents, cystoscopy, urine dipstick
Reproductive	Uterine ablation device, intrauterine device, ultrasound, pregnancy test
Respiratory	Ventilators, asthma inhaler, bronchoscope, intubation devices, continuous positive airway pressure (CPAP), inspiratory spirometer, speculum

According to a recent market study (Fortune Business Insights, 2019), the medical device industry is projected to grow to 612.7 billion dollars per year by 2025. The importance of medical devices is rising due to advances in technology, increases in lifestyle-associated disease (Menotti, Puddu, Maiani, & Catasta, 2015; Weisburger, 2002), and an aging population.

The FDA (2019) classifies devices as Class I, II, or III based on their risks and the regulatory controls necessary to provide assurances of safety and effectiveness. Class I devices pose the lowest risk to the patient and/or user whereas Class III devices pose the highest risk.

- Class I devices are not intended to help support or sustain life or be substantially important in preventing impairment to human health, and may not present an unreasonable risk of illness or injury. Examples include bandaids, tongue depressors, and dental floss.
- Class II devices are subject to special labeling requirements, mandatory performance standards, and post-market surveillance. Examples include acupuncture needles, powered wheelchairs, infusion pumps, air purifiers, surgical drapes, stereotaxic navigation systems, and surgical robots.
- Class III devices usually support or sustain human life, are of substantial importance in preventing impairment of human health, or present a potential, unreasonable risk of illness or injury and require premarket approval. Examples include implantable pacemakers and HIV diagnostic tests.

Reducing the volume and severity of medical error is one of design's greatest challenges. And, good design requires scientific knowledge about people, including their anatomy, physiology, sensory and perceptual systems, cognition, emotion, social behavior, and motor control. This is especially true when designing medical devices. When designed well, medical devices can help patients, healthcare providers, and caregivers. Conversely, when designed poorly, they can cause harm or death.

This book introduces principles, guidelines, research, and design methods in the human factors of medical devices. We provide an overview and reference that is applicable to human factors engineers, product designers, biomedical engineers, and regulatory affairs practitioners, among others. Due to space constraints, we do not provide exhaustive specifications and standards, but instead point you to appropriate recent resources. You can read this book from beginning to end, but it may be more helpful to read specific chapters as you need them. In either sense, we hope it serves as a valuable resource.

1.3   What Is Human Factors Engineering?

There are several definitions of human factors (Association for the Advancement of Medical Instrumentation, 2009; International Ergonomics Association, 2000; Lee, Wickens, Liu, & Boyle, 2017; Wachter, 2012). In this book, we use the following definition, which combines components provided by Chapanis (1985), Sanders & McCormick (1987), and Wickens, Lee, Liu, and Gordon Becker (2004).

HFE studies and applies knowledge from all human sciences to improve the match between people and the world through the design of products, processes, and environments. This includes knowledge of human capabilities, limitations, and behavior.

There are many specializations within engineering, and each applies specific scientific disciplines. For example, mechanical engineering relies on physics, and chemical engineering applies chemistry. Because HFE focuses on human performance and satisfaction with systems and devices, it relies on social and biological sciences such as psychology (cognitive science), anthropology, physiology, sociology, and medicine, to name a few. It also relies on engineering and design disciplines: industrial engineering, biomedical engineering, industrial design, and mechanical engineering. Figure 1.4 illustrates various academic disciplines involved in human factors engineering.

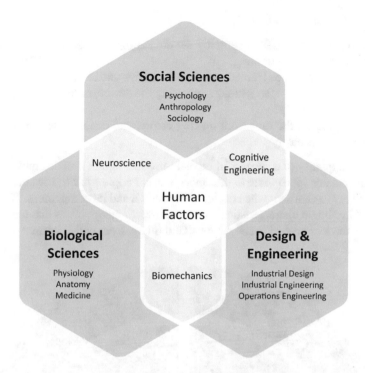

Fig. 1.4 Academic disciplines that frequently contribute to human factors engineering

Fig. 1.5 The qualities of successful medical devices

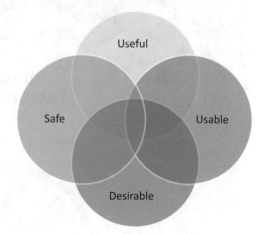

Goals of Human Factors Engineering

Figure 1.5 illustrates that successful medical devices have four interrelated qualities. First, they are useful, meaning they enable the user to do something not easily accomplished without the device. Second, they are usable, providing the following attributes (Nielsen, 1994):

- Easy to learn—the user can employ the device for its intended use quickly and easily without undue training
- Efficient to use—once learned, the device has few extraneous steps or time lags
- Easy to remember—after a break from using the device, users can get up to speed with it again quickly
- Safe—it protects the user from making errors, and enables the user to recover from any errors easily

The third quality of a successful medical product is that it is desirable; it elicits emotions that are appropriate to the device's use. People *want* to use such a device and are likely to choose it when deciding between it and its competitors. The fourth, and most important characteristic from a healthcare standpoint, is that the device is safe. Safe devices protect users, patients, and others from undue harm.

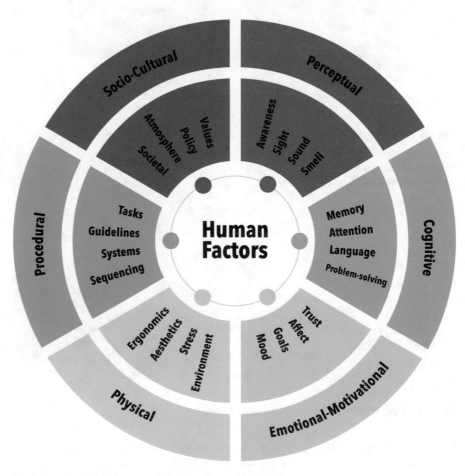

Fig.1.6 A simplified framework for human factors analysis

The connection between the user and the device is the user interface (UI), defined as everything the user comes into contact with physically, perceptually, or cognitively. These components usually come in the form of hardware, software, packaging, websites, communications, labeling, instructions for use (IFU), and so on.

As you can see, that is rather a lot to cover. Over the years, as we have engaged in hundreds of HFE projects, the framework illustrated in Fig. 1.6 has been helpful. When we are faced with a new project, we first analyze the situation through each of the six lenses discussed below and illustrated in Fig. 1.6:

- Perception—vision, audition, tactile senses, olfaction, and proprioception.
- Cognition—attention, memory, learning, judgment, and decision-making.
- Physical—size, shape, strength, flexibility, and endurance, among others.
- Emotion and motivation—characteristics of a device that make people want to use it, or alternatively want to avoid it. It is also likely that positive emotion, such as a lack of stress and anxiety, is likely to improve user interaction.
- Socio-cultural—factors relevant to fit within an environment or group. For example, certain words and colors have different meanings and connotations for different cultures.

What Human Factors Engineering Is Not

Now that we have discussed what human factors engineering is, it is equally important to clarify what human factors engineering is not. For one, HFE is not just applying checklists and guidelines to device design (Lee et al., 2017). Although checklists and guidelines can be useful, they are insufficient for good design. People are wildly variable: capabilities, limitations, and performance vary greatly from one person to another. One person is short, the other tall. One is technologically sophisticated, the other a Luddite. One is a great reader, the other uneducated. To add to this complexity, people vary in their capabilities from moment to moment. Early in the day you might be on the ball, whereas in the late afternoon you might feel sluggish. Some days you are sleep deprived. Other days your mind wanders. The list of variables is endless. Expecting a checklist alone to accommodate this variety is unrealistic.

Second, HFE is not using yourself as a model of the user. Unfortunately, in the absence of data about human performance, capabilities, and limitations, this is exactly what many designers and engineers do. This is not out of laziness or lack of concern. Instead, we are most familiar with ourselves, so we default to ourselves when considering the users of the device. We think that a certain font, color, text size, or contrast looks good, so it should work for everyone else. Unfortunately, due to the wide variety of human capabilities and limitations, this just does not work.

Benefits of Human Factors Engineering

Clearly, integrating HFE requires changes in design approach and process. So, what are the benefits of doing this? A few are listed below:

- Improved sales: Employing HFE data, guidelines, and methods can improve sales. People prefer devices that are easy to use, which translates into an increased willingness to repurchase. Once people become familiar with your easy-to-use device, it erects a barrier to competition. This is advantageous because it is easier to keep customers than to attract them in the first place.
- Improved product reviews: HFE also leads to better product reviews. An increasing portion of technical and professional publications is dedicated to product ease of use. Better usability yields better reviews.
- Improved brand image: Product reviews are repeated less formally through word of mouth discussions. People talk, and in healthcare they have preferred instruments, devices, and products. The good ones are discussed positively while the bad ones, not so much. This is the grass roots reflection of your brand. Positive reviews and discussion bolster the brand, whereas negative ones do just the opposite.
- Improved task performance: Easy-to-use devices improve task performance, making the task easier to execute, faster, and more consistent. These devices require less training time for users to become proficient, translating into improved financial performance. Efficiency increases profit.
- Improved patient outcomes: HFE can help improve patient outcomes and reduce product liability risk. Users are more likely to be compliant when using these products because they require less of a cognitive burden. In other words, HFE can facilitate correct and frequent use. Further, usable devices reduce the likelihood, frequency, and severity of use-error. And when use-errors do occur, usable devices facilitate recovery.
- Facilitates regulatory approval: Finally, good HFE facilitates the regulatory approval process. Regulatory bodies such as the Food and Drug Administration (FDA) recognize that error involving the misuse of medical devices is a considerable source of harm among patients and users. As a result, they require that many Class II and all Class III devices include Human Factors Engineering activities, including use risk analysis, identification of critical tasks, and usability testing (FDA, 2016).

Of course, the opposite occurs when HFE is ignored. Lack of HFE can lead to reduced sales, decreased adoption of the device by patients and healthcare providers, reduced satisfaction, decreased willingness to repurchase, poor product reviews, poor word of mouth reputation, and more frequent and more dangerous use-errors.

Poor HFE processes can even slow down the development process itself. When HFE is not included early in the design process, some usability problems are not identified until late in that process. This is costly. For example, in software, Lederer and Prasad (1993) found that the cost of making a change to the product is 1 unit in

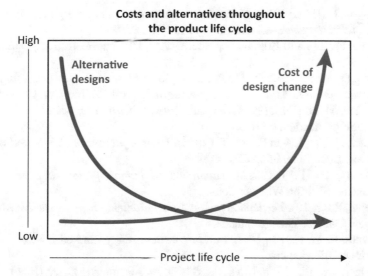

Fig. 1.7 Cost and alternatives by product development phase

definition phase, 1.5–6 units in development, and 60–100 units after delivery. Figure 1.7 provides an illustration of this. Reading from left to right, you can see that during the early stages of development (e.g., product definition), the design team can entertain numerous alternative designs. Making changes at this early stage is inexpensive. This changes, however, as development proceeds. Over time, the number of alternatives available decreases while the cost of making a change increases precipitously. Making a change late in the design process is exceedingly expensive, oftentimes prohibitively so. This is why it is so important to discover (and remedy) usability problems early in the design process.

Frustratingly, when HFE has been implemented appropriately, it is hard to notice (even if it has saved your life). When HFE is done well, nothing newsworthy happens; only when things go wrong do people notice. As Donald Norman (2013) pointed out:

> Good design is actually a lot harder to notice than poor design, in part because good designs fit our needs so well that the design is invisible, serving us without drawing attention to itself. Bad design, on the other hand, screams out its inadequacies, making itself very noticeable (p. xi).

Resources

- AAMI. (2018). *ANSI/AAMI HE75:2009/(R)2018 human factors engineering—Design of medical devices*. Fairfax VA: Association for the Advancement of Medical Instrumentation.

- Carayon, P. (Ed.). (2016). *Handbook of human factors and ergonomics in health care and patient safety*. Boca Raton, FL: CRC Press.
- Human Factors and Ergonomics Society Annual Healthcare Symposium. Hfes. org.
- Lee, J. D., Wickens, C. D., Liu, Y., & Boyle, L. N. (2017). *Designing for people: An introduction to human factors engineering*. Scotts Valley, CA: CreateSpace.
- Privitera, M. B. (Ed.). (2019). *Applied human factors in medical device design*. Cambridge: Academic Press.
- Proctor, R. W., & Van Zandt, T. (2018). *Human factors in simple and complex systems*. Boca Raton, FL: CRC Press.
- Salvendy, G. (Ed.). (2012). *Handbook of human factors and ergonomics*. Hoboken, NJ: John Wiley & Sons.
- Vicente, K. (2010). *The human factor: Revolutionizing the way we live with technology*. Canada: Vintage Canada.
- Wachter, R. M. (2012). *Understanding patient safety* (2nd ed.). New York, NY: McGraw-Hill Medical.
- Weinger, M. B., Wilund, M. E., & Gardner-Bonneau, D. J. (Eds.). (2011). *Handbook of human factors in medical device design*. Boca Raton, FL: CRC Press.

References

Aspden, P. (2007). Medication errors: Prevention strategies. In P. Aspden, J. Wolcott, J. L. Bootman, & L. R. Cronenwett (Eds.), *Preventing medication errors* (pp. 409–446). Washington, DC: National Academies Press.

Association for the Advancement of Medical Instrumentation. (2009). *ANSI/AAMI HE75–2009: Human factors engineering—Design of medical devices*. Arlington, VA: Association for the Advancement of Medical Instrumentation.

British Broadcasting Corporation. (2005). Baby died after 'decimal' error. Retrieved April 29, 2020, from http://news.bbc.co.uk/2/hi/uk_news/england/leicestershire/4566427.stm.

Classen, D. C., Jaser, L., & Budnitz, D. S. (2010). Adverse drug events among hospitalized Medicare patients: Epidemiology and national estimates from a new approach to surveillance. *The Joint Commission Journal on Quality and Patient Safety, 36*(1), 12–AP9.

Classen, D. C., Resar, R., Griffin, F., Federico, F., Frankel, T., Kimmel, N., et al. (2011). Global trigger tool shows that adverse events in hospitals may be ten times greater than previously measured. *Health Affairs, 30*(4), 581–589.

Fortune Business Insights. (2019). Medical devices market size, share and industry analysis by type (orthopedic devices, cardiovascular devices, diagnostic imaging, IVD, MIS, wound management, diabetes care, ophthalmic devices, dental & nephrology), end user (Hospitals & Ambulatory Surgical Centers and Clinics) and regional forecast, 2019–2025. Retrieved July 21, 2020, from https://www.fortunebusinessinsights.com/industry-reports/medical-devices-market-100085.

Gawande, A. A., Studdert, D. M., Orav, E. J., Brennan, T. A., & Zinner, M. J. (2003). Risk factors for retained instruments and sponges after surgery. *New England Journal of Medicine, 348*(3), 229–235.

International Ergonomics Association. (2000). *Human Factors/Ergonomics (HF/E): Definition and applications.* Retrieved July 21, 2000, from https://iea.cc/what-is-ergonomics.

Kinsella, K. G., & Phillips, D. R. (2005). *Global aging: The challenge of success* (Vol. 60, No. 1, p. 3). Population Reference Bureau.

Lambert, E. C. (1978). *Modern medical mistakes.* Bloomington: Indiana University Press.

Lederer, A. L., & Prasad, J. (1993). Information systems software cost estimating: A current assessment. *Journal of Information Technology, 8*(1), 22–33.

Lee, J. D., Wickens, C. D., Liu, Y., & Boyle, L. N. (2017). *Designing for people: An introduction to human factors engineering.* Scotts Valley, CA: CreateSpace.

Lieber, J. B. (2015). *Killer care: How medical error became America's third largest cause of death, and what can be done about it.* OR Books.

Lindenauer, P. K., Rothberg, M. B., Pekow, P. S., Kenwood, C., Benjamin, E. M., & Auerbach, A. D. (2007). Outcomes of care by hospitalists, general internists, and family physicians. *New England Journal of Medicine, 357*(25), 2589–2600.

Makary, M. A., & Daniel, M. (2016). Medical error—The third leading cause of death in the US. *BMJ, 353.*

Meinberg, E. G., & Stern, P. J. (2003). Incidence of wrong-site surgery among hand surgeons. *JBJS, 85*(2), 193–197.

Menotti, A., Puddu, P. E., Maiani, G., & Catasta, G. (2015). Lifestyle behaviour and lifetime incidence of heart diseases. *International Journal of Cardiology, 201,* 293–299.

Nielsen, J. (1994). *Usability engineering.* Burlington: Morgan Kaufmann.

Norman, D. (2013). *The design of everyday things.* New York: Basic Books.

Ornstein, C. (2014). Dennis Quaid files suit over drug mishap. Los Angeles Times, 9/16, 2014. Retrieved June 16, 2020, from https://www.latimes.com/entertainment/gossip/la-me-quaid5dec05-story.html.

Rowe, J. W., & Kahn, R. L. (2015). Successful aging 2.0: Conceptual expansions for the 21st century. *The Journals of Gerontology: Series B, 70*(4), 593–596.

Sanders, M. S., & McCormick, E. J. (1987). *Human factors in engineering and design* (6th ed.). New York: McGraw-Hill.

U.S. Food and Drug Administration. (2016). *Applying human factors and usability engineering to medical devices: Guidance for Industry and Food and Drug Administration Staff.* Washington, DC: U.S. Department of Health and Human Services Food and Drug Administration, Center for Devices and Radiological Health, Office of Device Evaluation.

U.S. Food and Drug Administration. (2019). *How to determine if your product is a medical device.* Retrieved March 30, 2019, from https://www.fda.gov/medicaldevices/deviceregulationand-guidance/overview/classifyyourdevice/ucm051512.htm.

Wachter, R. M. (2012). *Understanding patient safety* (2nd ed.). New York: McGraw-Hill Medical.

Wan, W., Le, T., Riskin, L., & Macario, A. (2009). Improving safety in the operating room: A systematic literature review of retained surgical sponges. *Current Opinion in Anesthesiology, 22*(2), 207–214.

Weisburger, J. H. (2002). Lifestyle, health and disease prevention: The underlying mechanisms. *European Journal of Cancer Prevention: The Official Journal of the European Cancer Prevention Organisation (ECP), 11,* S1–S7.

Wickens, C., Lee, J. D., Liu, Y., & Gordon Becker, S. E. (2004). *An introduction to human factors engineering.* Upper Saddle River, NJ: Pearson Prentice Hall.

World Health Organization. (2020). *Medical device—Full definition.* Retrieved March 30, 2020, from http://www.who.int/medical_devices/full_deffinition/en/.

Zhang, J., Patel, V. L., Johnson, T. R., & Shortliffe, E. H. (2004). A cognitive taxonomy of medical errors. *Journal of Biomedical Informatics, 37*(3), 193–204.

Chapter 2
Qualitative Human Factors Research Methods

2.1 Human-Centered Design

As the name implies, human-centered design (HCD) is a framework for placing user goals, needs, capabilities, and limitations at the center of the product design process. Individual manufacturers will follow their own proprietary design processes, so HCD will be implemented differently in each case, but the following three tenets will be the same (Gould & Lewis, 1985):

1. Early and constant focus on the users and their tasks
2. Reliance on human–system performance data to guide design
3. Iteration

This HCD approach is distributed throughout the design process—not just during initial stages of need finding and requirements gathering but also through iterative refinement and post-market research and surveillance (see Fig. 2.1). The user, rather than the product, takes center stage in all parts of the process, requiring us to research the user's job, tasks, workflow, needs, and preferences.

Focusing on the user throughout the whole design process requires a great deal of research. At the beginning you need to investigate and analyze the user's needs, tasks, environments of use, and other factors. With each design iteration you need to inspect the user interface and conduct usability tests to identify usability problems. Even after the product is released, you should monitor it for use-error, and conduct competitive usability research, to learn how your product stands against the competition.

This chapter discusses the research, usability inspection, and data analysis methods typically employed during HCD. Not surprisingly, the list is long, and the trick for the researcher is to choose the right methods at the right time. As a result, these methods should be thought of as a toolkit rather than a recipe. Depending on the questions you need to answer, you will choose different research methods, and different data analysis techniques.

© Springer Nature Switzerland AG 2021
R. J. Branaghan et al., *Humanizing Healthcare – Human Factors for Medical Device Design*, https://doi.org/10.1007/978-3-030-64433-8_2

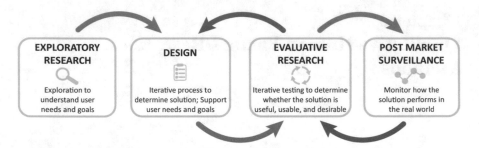

Fig. 2.1 Simplified human-centered design (HCD) process

2.2 Human Factors Research

Human factors engineering (HFE) research involves gathering, analyzing, and interpreting data. Of course, people gather and interpret data every day, but we usually do so intuitively, and non-systematically. Research, on the other hand, answers questions in ways that are objective, systematic, and repeatable.

By "objective," we mean that conclusions are based on data rather than intuition or opinion. Intuitions vary from person to person, and even moment to moment. To be sure, intuition does have a place in design, but intuition improves with increasing knowledge and information provided by research. It is risky to the point of foolishness to base multimillion-dollar design budgets solely on intuition.

Unlike intuition, which happens naturally and with little effort, objectivity is difficult, unnatural, and takes discipline. So, to facilitate objectivity we employ specific research methods designed to answer questions while reducing bias (Cacioppo & Freberg, 2013). We will discuss many of those methods in this chapter.

One note of caution is appropriate here. The primary purpose of research is to answer questions. We are often surprised by clients and students who begin research projects without specific research questions in mind. This can be an expensive use of time, money, frustration, and even professional reputation. Identifying, in detail, the questions you will address makes planning, conducting, and analyzing the study much easier. Undoubtedly, you will serendipitously answer other unexpected questions along the way, but identifying your questions ahead of time is critical. If you have no clearly defined research question, you are not doing research!

2.3 Reliability and Validity

Let us turn our attention to two important characteristics of research measurement—reliability and validity. Imagine you stood on your bathroom scale and looked at your weight. Now imagine doing it again, and then yet a third time. You would expect the weight to be just about the same, within a very small degree of variation. That is reliability—the consistency of an observation or a measure. If weighing yourself resulted in wild swings of 15 or more pounds, you would no longer trust your scale.

Now imagine getting on the same scale with the knowledge that your usual weight is around 150 pounds. The scale, however, reads 275 pounds. As before, you reweigh yourself several times, and several times it returns the weight, 275 pounds. This scale is reliable; you always get the same result. But it is not valid. That is, it is not good at measuring what it is intended to measure: your actual weight.

As in scales, in HFE we need research observations and measurements that are both reliable and valid. On a questionnaire, say, under the same circumstances, we would expect the same person to express the same responses. This would suggest that our questionnaire is reliable.

A useful metaphor for the relationship between reliability and validity is an archery target. Think of the center of the target (the "bullseye") as the concept or construct you are trying to measure. These concepts might be something like usability, interest in your product, task completion time, or error frequency. For each person you measure, you are taking a shot at the target. If you measure the concept perfectly for a person, you are hitting the center of the target. If you do not, you are missing the center. The more "off" you are for that person, the further you are from the center. Figure 2.2 shows three possible situations.

In the left hand target, you are hitting the target consistently, but you are missing the center of the target. In other words, you are consistently and systematically measuring the wrong value for all respondents. This measure is reliable, but not valid (it is consistent but wrong). The middle scenario shows a case where your hits are spread across the target and you are consistently missing the center. Your measure, in this case, is neither reliable nor valid. The right-hand figure shows the Robin Hood or William Tell scenario; you consistently hit the center of the target. Your measure is both reliable and valid.

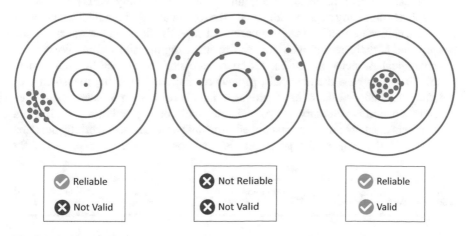

Fig. 2.2 Reliability and validity

The fields of measurement and psychometrics have identified and described various types of reliability and validity, which are too detailed to cover here, but it is important to gather evidence that the information you are collecting is both reliable and valid. That is, can you count on it?

2.4 Selecting Research Participants

Since it is impossible to study every member of a population, we need to focus on only a representative subset of people, called a *sample*, which is used to make judgments about the entire population. In journal publications (and elsewhere), a study's sample size is noted with an italicized *n*. For example, $n = 20$ means the study's sample consists of 20 participants.

The best way to ensure a representative sample is through random sampling, in which every member of the population has an equal chance of being selected for the study (Fig. 2.3). The more representative the sample, the better the results will generalize to the population. Unfortunately, random sampling is not always possible. Instead, HFE often uses samples of convenience, or groups of people who are easily accessible to the researcher. For example, many of the behavioral and medical sciences are criticized for using participants who are WEIRD (Henrich, Heine, & Norenzayan, 2010), that is, *W*estern, *E*ducated, and from *I*ndustrialized, *R*ich, and *D*emocratic countries. As a result, it is possible that many findings in various academic journals would not apply to people with other characteristics.

2.5 Ethical Standards

It is critical to protect research participants from harm. Usually, HFE practitioners ensure this by adhering to guidelines set by the American Psychological Association (APA), which state that researchers carry out investigations with respect for the people who participate and with concern for their dignity and welfare (American Psychological Association, 2017). Often researchers use an institutional review board (IRB) to review the materials, procedure, and confidentiality measures provided by the study. At the core of ethical standards for human research is the idea that participation is voluntary. No participant should be coerced into participating. Researchers must obtain informed consent from all participants, and briefly describe the goals of the project, the potential risks and/or benefits, the procedures for

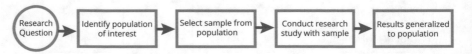

Fig. 2.3 Selecting research participants

maintaining confidentiality, and the incentives or payments offered. Once this information has been communicated, participants can agree to participate in the study.

The researcher should also make sure to do no irreversible harm to participants. Research using human participants should be private and confidential. Privacy refers to the participants' control over how their information is shared, and methods for ensuring privacy are usually indicated in the informed consent. Confidentiality refers to the participants' rights to not have their data revealed to others without their permission. Confidentiality is usually maintained by substituting codes for names and storing data in locked cabinets. APA guidelines are listed below.

1. Do no harm
2. Accurately describe risks to potential participants
3. Ensure that participation is voluntary
4. Minimize discomfort to participants
5. Maintain confidentiality
6. Do not unnecessarily invade privacy
7. Use deception only when absolutely necessary
8. Provide debriefing to all participants
9. Provide results and interpretations to participants
10. Treat participants with dignity and respect
11. Allow participants to withdraw at any time for any reason

2.6   Literature Review

Often, you can answer some of your research questions without collecting any new data at all by conducting a literature review. Many times, others have conducted research that is relevant to your current project, informing you about what is known, and what has been done. This information is usually available in journal articles, books, and conference proceedings, but it is also helpful to look at previous projects your organization has already completed. Hundreds of thousands of dollars can be invested in research investigations (e.g., contractor fees, opportunity cost, participant compensation, facility rentals). It is worth taking some time to check whether similar problems have been addressed through other research.

The literature review is intended to distill information from all sources, capturing the essence of previous research or projects as they might inform the current project. The review need not summarize everything from each source, but should synthesize the work by drawing connections among previous findings.

Internet resources such as search engines, digital journals, e-mail, and interlibrary loans expedite literature searches significantly, allowing researchers to access libraries from around the world. However, it is important to be discerning, ensuring that the literature selected for inclusion is not only relevant, but also from credible sources. Particular caution should be exercised if including website or blog resources, which may not be vetted or peer-reviewed for credibility.

2.7 Case Study

A case study is an in-depth analysis of a small number of events or people (Cacioppo & Freberg, 2013). Case studies are common in medicine, law, and business, and are sometimes used by human factors engineers when large numbers of participants are not available. For example, imagine the need to learn about people with a very rare disease (say an incidence of 1 in 200,000). It might be helpful to start by studying one individual at a time. One case will likely lead to hypotheses for future case studies or larger studies with more people. Because so few participants are involved (often only one), it affords the opportunity to gain detailed information about that one participant, even applying several research methods to more fully examine the person, small group, or event of interest.

2.8 Naturalistic Observation

Naturalistic observation is an in-depth study of potential users in their natural setting. A classic example of naturalistic observation in primatology was Jane Goodall's long-term study of chimpanzees in Tanzania beginning in 1960. Though our observations may not take us to Tanzania, there is a wealth of insight to be gained from studying medical device users in their natural habitats, so to speak. Observing surgical techs in the surgical suite or physician's assistants in the exam room provides insight into how they interact with the environment, including what materials they use, who they communicate with, what forms they fill out, what makes them nervous, which tasks are most frustrating, and so on. This is information not easily uncovered using some other research techniques (Fig. 2.4).

Even though naturalistic observation is unstructured, it is still critical to begin with research questions. These questions dictate the goals and approach for your work. In preparation, you should be able to answer the following questions before starting your observations.

- What questions do I intend to answer?
- Who (what work roles) would I like to observe?
- What activities would I like to observe?
- What communications would I like to observe?
- What devices or tools would I like to observe being used?
- What support materials (e.g., Instructions for Use) should I pay attention to?

  Along the way, you will likely learn about:

- Responsibilities—for example, who does what and when
- Sequences of activities
- What problems can occur
- What activities are difficult and why
- How people troubleshoot problems

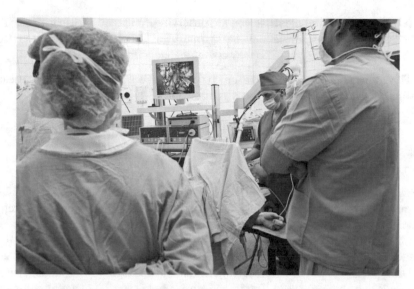

Fig. 2.4 Naturalistic observation in the operating room (Image by Farferros/shutterstock.com)

- How these problems are addressed
- How a task or activity is initiated
- How people know when they have completed the task or activity correctly

Observational research produces so much data that it is difficult to know what information to capture. Rick Robinson, Ilya Prokopoff, John Cain, and Julie Pokorny (Robinson, 2015) had the same challenge in their observational work. To organize research questions and information, they relied on a mnemonic using all English vowels: *A*ctivities, *E*nvironments, *I*nteractions, *O*bjects, and *U*sers.

- (A) Activities: Describe sequences of actions that enable people to achieve a goal. It is worth documenting not only the activity itself but also the sequence that comprises that activity.
- (E) Environments: Describe the environment in which activities take place. Make sure to describe crowding, organization of the space, traffic flow, lighting, noise heating, and interruptions.
- (I) Interactions: Describe the interactions among people. Who talks to whom? What information do they exchange? When do people help others with their tasks? What types of questions are asked? What other media are used to interact? Do people discuss things only in person or also by telephone, e-mail, Slack, or Zoom? How do these interactions differ depending on the medium? Be sure to capture, where possible, both verbal and non-verbal communication.
- (O) Objects: Describe the tools, devices, forms, notes, and other artifacts that are used. How do these objects relate to the user's goals, activities, interactions, and work in general?

- (U) Users: Describe the users you are observing. What are their characteristics, such as age, gender, educational background, goals, behaviors, and preferences? What are their roles? Who do they communicate with?

This framework can be applied to any type of observational or contextual research to simplify data collection, and focus your efforts.

2.9   Design Ethnography

Design ethnography adapts traditional ethnographic methods used in anthropology to study the social life of various cultures. It relies on observational research of people in their natural environments, but it also involves unstructured interviews and document analysis. Particular attention is paid to characterizing users' lives, behaviors, language, and artifacts.

Design ethnography is different from traditional ethnography in that it is shorter and tends to be more focused. Traditional ethnographic studies in anthropology can last for years, whereas a typical design ethnography session will last only a few hours. As in all research methods, ethnography begins with a research question. As Fetterman (2019) pointed out, "The ethnographer enters the field with an open mind, not an empty head" (p. 1). Design ethnography is different from naturalistic observation because it relies on gathering all kinds of data. It uses not just observation, but interviews, artifacts (tools, etc.), photographs, sketches, document analysis, investigation of rituals, and so on. In other words, design ethnography is not limited to any one type of data collection method. What really matters is gaining insight, usually in the context of use.

2.10   Interviewing

Interviews fall into one of three approaches: structured, semi-structured, and unstructured. Each approach has different intentions, advantages, and disadvantages. The following sections describe each one, as well as a few tips about when it may or may not be appropriate to use them.

Structured Interview

A structured interview involves asking participants a specific set of questions, in a specific order. Every participant experiences the exact same procedure. As a result of this systematic approach, structured interviews often produce results intended for quantitative analysis: the frequencies of times a participant "agreed" vs. "disagreed" on a topic, for example.

Structured interviews require considerable planning to identify the questions to be asked, and the order of asking them. Furthermore, researchers must use techniques to reduce participant fatigue—which can set in easily during longer structured interviews. These might be in the form of frequent breaks, or introducing other activities (e.g., surveys, questionnaires) to break up the monotony of a one-sided conversation.

Semi-Structured Interview

Semi-structured interviews are the most common approach taken in HFE. It represents the middle ground between a "structured interview" and an "unstructured" one, allowing researchers to follow a core set of questions, and fill in gaps as they are identified (Fig. 2.5).

The semi-structure gives the moderator freedom to go off-script to explore areas of interest based on the conversation, as long as this new direction remains in scope for the study objectives. This can be a challenging task for inexperienced study moderators. It is especially important that they define the scope of the interview beforehand.

Semi-structured interviews are useful for investigating the specific steps that users complete with a device. For example, imagine trying to understand a Sterile Processing Technician's intake procedure for a used medical device. The "structured" part of the interview might come from trying to create a linear sequence of tasks. The "unstructured" part comes into play when clarification is needed.

One important part of designing a semi-structured interview is to build in enough time to permit the moderator to go off-script. Inexperienced researchers tend to design their protocols and moderator guides with too many pre-baked questions. They do not realize that answering these questions is time consuming.

Unstructured Interviews

Unstructured interviews have no predefined questions or topic order. Often, the moderator begins with a simple, open-ended question, and the interview (hopefully) progresses naturally from there. This is best done by a study moderator who is an expert—or, at least very knowledgeable—on the topic at hand. A moderator with expertise knows where gaps, uncertainties, or inconsistencies exist in product opinions or workflows. They intuitively lead the conversation in these directions to ferret out the information they need. Additionally, an expert can often spot another expert (or non-expert) based on the questions they ask. Thus, if the study moderator asks a

Fig. 2.5 Usability testing interview at research facility with a moderator, patient, and their caregiver

bunch of "softball" questions, the participant is less likely to explain important details that they anticipate may be outside the moderator's purview.

Yet even when the moderator is an expert in the field, unstructured interviews are rare. There are simply too many opportunities for the interview to go astray, including situational factors such as the participant being tired after a long day of work. Stanton, Salmon, and Rafferty (2013) also point out that unstructured interviews have the potential for crucial information to be neglected or ignored.

Interview Questions

There are two basic types of interview questions: closed-ended and open-ended. A closed-ended question presents the participant(s) with a limited set of response options. This might be something as simple as a "yes" or "no" to a question such as: "Do you feel that other healthcare professionals would be able to use this product safely?" Additional response options can also be incorporated, such as survey questions to assess a participant's agreement or disagreement with a statement.

Closed-ended questions are underappreciated, probably because they create bottlenecks in a line of questioning. That is, the participant must first choose a "yes" or "no" response, then the interviewer can ask a follow-up question they really care about—why or how? In some cases, having a participant commit to a choice prepares them to justify their selection. However, it may be better to save time and just get to the question you intended to ask in the first place.

You can often tell that a question is closed-ended if it begins with one of the following:

- Can you…?
- Do you…?
- Will you…?
- Have you…?

These can always result in a Yes or No response. Of course, in natural conversation, most people know to elaborate more. For example, a question such as, "Can you show me how you did that again?" will almost always be returned by the participant with an action, rather than the simple response of, "yes, I can."

Closed-ended questions must be thought through carefully before use. They are so common in everyday speech that we often do not realize we are using them with participants in the first place. And, try as we might, a closed-ended question will slip its way into your interview whether you like it or not. This is to be expected given the interviewer's responsibility to lead the interview, remember the important details, take notes, and stay engaged with the participant. To limit these slip-ups, the interviewer should offload as many tasks to note takers and fellow observers as possible. They should also complete several "dry runs" with pilot participants and colleagues to identify areas where they may get stuck.

An open-ended question enables participants to explain things in their own terms, and to elaborate. That is, they are not bound by the choices or options presented by the study moderator or testing materials. Open-ended questions often begin with one of the following:

- How…?
- Why…?
- What…?

Keep in mind that open-ended questions do not actually have to be questions at all. In fact, just telling the participant what you want them to talk about is perfectly acceptable, and, in many cases, more natural during conversation. For example, a good go-to statement to prompt an open-ended response is one of the following:

- Tell me (more) about….
- Walk me through….
- Explain how….

There are several benefits of using open-ended questions. As Stanton et al. (2013) point out, they allow the participant to answer the question in whichever way they choose, often resulting in more pertinent information than you would receive through a closed-ended question. And, perhaps just as important, open-ended questions keep the conversation moving forward. By contrast, a closed-ended question really only has one follow-up question: "Why did you pick that choice?" Open-ended questions are also more conversational and can feel less intimidating. It feels like there is less of a spotlight on them, or less like they are being "tested."

Despite these benefits, one drawback to open-ended questions is that they can take longer to analyze. You cannot simply count up who chose what. The researchers may have to incorporate qualitative analysis techniques, such as affinity diagramming, to tease apart similar and dissimilar concepts.

Ensuring Interviews Are Productive

Below are a few tips for productive interviews.

- Make participants comfortable: Interviews are unfamiliar and frankly an odd situation for most participants. The participants will vary in their degree of extroversion, among other traits, so some may find it intimidating. Your ability to put them at ease will improve their responses and provide you with better information.
- Start broad and funnel to more specific questions: It is often helpful to think about the structure of an interview as a funnel with general questions at the top and asked at the beginning, then funneling down to more specific questions as the interview proceeds. This enables the participants to express what is top of mind first, then follow with more specific issues.
- Avoid leading questions: Leading questions encourage the answer you are interested in. We once observed a software developer ask a participant after a usability test, "that was not too unusable, was it?" That is an example of a leading question.
- Ask follow-up questions as the interview unfolds: The first answer from the participant is not usually the most insightful one. We often use laddering (Reynolds, Dethloff, & Westberg, 2001), which requires us to ask "Why" questions in succession after the first response. This enables us to understand the reasoning behind the first, superficial, answer.
- Do not be afraid of silence: If you need a moment to gather your thoughts, it is perfectly acceptable to tell a participant that you need a second to think about how to phrase your next question. Silence is also a natural cue that the person speaking last (e.g., the participant) should elaborate more. Participants themselves might need a moment to process their thoughts before speaking.

2.11 Focus Groups

A focus group is an interview with several participants at the same time, either in person or remotely. This method lends itself to covering a wide range of topics, ranging from the perceived usability of a system, common workflows followed by users, or to simply elicit feedback on a new design concept (Stanton et al., 2013). One benefit of the focus group is that it enables group interactions, allowing participant responses to build upon each other (Farnsworth & Boon, 2010). It also allows large amounts of qualitative data to be collected quickly.

Traditionally, focus group samples are composed of homogeneous groups of participants (i.e., all from the same background). This might include subject matter experts (SMEs), C-suite executives, or people with Type II diabetes who use continuous glucose monitors (CGMs). Topics usually follow a semi-structured interview format to allow time for open discussion.

Less commonly, focus groups are made up of heterogeneous groups (i.e., people from different backgrounds). Heterogeneous groups can be useful in the early requirements-gathering phases of designing a medical device, as specific end-user groups may not be defined yet. This approach can also be useful when multiple user groups work with the same medical device, but may not work with each other on a regular basis. For example, a respiratory therapist and a reprocessing technician could both work with the same flexible endoscope in an average day, but may not have direct contact with each other. As such, they may not understand the constraints and issues associated with each other's experience with the device. Sometimes having one user group hear the "pains" of another can generate opportunities for redesign.

Although this approach generates a lot of data quickly and easily, it is prone to a variety of social biases (e.g., groupthink behaviors) and may not be valid or reliable. Technologies such as audience response systems—also known as clickers to university students—or similar smartphone apps can be a work-around to this issue. These platforms allow focus group participants to cast their selections privately, thereby reducing the likelihood of bias. In general, however, focus groups have progressed toward the more qualitative end of the spectrum—especially in the field of medical HFE.

Klein, Tellefsen, and Herskovitz (2007) point out that a focus group can be completed in one of four ways, though only three of these combinations actually make sense from a logistical standpoint. These variations are outlined in Table. 2.1.

In-Person Focus Groups (Synchronous, Co-Located)

An in-person focus group is the traditional format that people imagine when they hear the term focus group. Usually, it involves a large room with a conference table surrounded by seats so participants can see each other while conversing (Fig. 2.6).

Table 2.1 Variations of focus group formats

Synchronous, co-located At the same time, in the same location	Asynchronous, co-located (not common) At different times, in the same location
Synchronous, distributed At the same time, in different locations	Asynchronous, distributed At different times, in different locations

Fig. 2.6 A focus group with ten participants and two study moderators

An in-person focus group should be led by at least one study moderator who is both trained and experienced. It is recommended that additional moderators should be considered for groups with more than eight people. The moderator keeps participants on track, and guides the progress at appropriate times. Not surprisingly, the study moderator has to be familiar with and kept apprised of the study objectives and research questions well in advance in order to do their job well and to elicit the most useful information out of the focus group.

In-person focus group sessions should ideally accommodate five to eight participants. The session length should be about 15–20 min per participant involved. For example, this would warrant that a five-participant focus group be approximately 90 min long (Stanton et al., 2013). The obvious caveat to this recommendation is that the topics of the study itself may demand more (or less) time. If the study requires participants to view a lengthy video, for example, the session length will likely change (Table 2.2).

Remote (Online) Focus Groups (Synchronous, Distributed)

Recently, focus groups have begun to make use of remote video conferencing platforms instead of conference rooms and lab environments. Unlike in-person focus groups, each participant in a remote focus group meets at a designated time online. This is especially useful for hard-to-reach participants (e.g., patients with rare disorders or specialized surgeons), as all they need to participate is a laptop, camera, and stable Internet connection. Additionally, remote focus groups can cost less (Reid & Reid, 2005; Summanen, Liikkanen, Laakso, & Leisti-Szymczak, 2013).

Table 2.2 Advantages and disadvantages of in-person focus groups

"Pros" of an in-person focus group	"Cons" of an in-person focus group
• Elicits responses from several end-users in a single study session • Sharing opinions and experiences can "connect the dots" in ways that individual experiences cannot provide • Useful way to elicit information when team's current knowledge on the subject may not be sufficient to create a survey or questionnaire • Flexibility to shift topics quickly on the fly • Avoids wasting time in one-on-one interviews talking about the same issues with individual participants	• Group dynamic may intimidate some participants from speaking • Some participants may withhold feedback due to fears of criticism or scrutiny • You may get a "dominating" voice in the group that stifles collaboration or biases others • Quality of session depends heavily on the capabilities of the study moderator • Data gathered does not lend itself well to common forms of statistical analysis • Some data collection methods are subject to bias • Audio transcriptions may mix up speakers in the room (i.e., which participant actually said that?)

Remote focus groups can be slightly more challenging to moderate, since participants do not have access to the same non-verbal cues, such as facial expressions, hand movements, and verbal pauses that people naturally pick up on in-person. Furthermore, it can be challenging to ensure that all participants engage equally in the conversation. Some participants may be afraid to interrupt others or may choose to remain idle in the background until called upon. For these reasons, remote focus groups should be conducted with fewer participants than in-person groups. Despite these potential shortcomings, however, Abrams, Wang, Song, and Galindo-Gonzalez (2015) found online focus groups using webcams produced similar results (i.e., data richness, word count) to in-person focus groups following the same discussion topics (Table 2.3).

Bulletin Board (Online) Focus Groups (Asynchronous, Distributed)

A third form of focus group—the Bulletin Board Focus Group (BBFG)—also uses the Internet but forgoes audio and video components altogether. Instead, a BBFG (Fig. 2.7) is completed asynchronously over multiple days on a platform similar to an Internet forum-style website. Unlike a typical website forum, however, a BBFG is by invitation only, and is closed to public viewing and contributions.

Table 2.3 Advantages and disadvantages of remote focus groups

"Pros" of a remote focus group	"Cons" of a remote focus group
• Elicits responses from several end-users in a single study session • Sharing opinions and experiences can "connect the dots" in ways that individual experiences cannot provide • Useful way to elicit information when team's current knowledge on the subject may not be sufficient to create a survey or questionnaire • Flexibility to shift topics quickly on the fly • Avoids wasting time in one-on-one interviews talking about the same issues with individual participants • Easier for hard-to-reach participants to commit to the study • Easier to set up than in-person focus groups • (Usually) costs less than in-person focus groups	• Group dynamic may intimidate some participants from speaking • Some participants may withhold feedback due to fears of criticism or scrutiny • You may get a "dominating" voice in the group that stifles collaboration or biases others • Quality of session depends heavily on the capabilities of the study moderator • Data gathered do not lend itself well to common forms of statistical analysis • Some data collection methods are suspect to bias • Audio transcriptions may mix up speakers in the room (i.e., which participant actually said that?) • Technology factors (i.e., Internet loss, participant not familiar with platform) can eat up session time

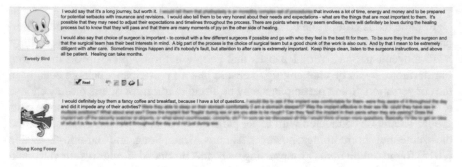

Fig. 2.7 Bulletin Board (online) Focus Group using Hot Potato™ (Courtesy of Bradley Peacock at Peacock Nine, with permission)

A BBFG study begins by the moderator laying the groundwork and expectations for the study. This often involves a brief tutorial of the BBFG platform (i.e., how does it work?). Then, the moderator presents the first of several topics to participants, who can respond in writing, brief videos, or images. Depending on the platform settings, participant identities can be replaced with a pseudonym to promote anonymity. Likewise, the research team may opt to prevent participants from seeing others' comments until after they have responded to the moderator's initial inquiry. This can reduce social loafing and other biases throughout the study. It can also help create "lower stakes," or judgment-free environment to assuage participant concerns about criticism from peers.

Table 2.4 Advantages and disadvantages of Bulletin Board Focus Groups

"Pros" of Bulletin Board Focus Groups	"Cons" of Bulletin Board Focus Groups
• Elicits responses from several end-users in a single study session • Sharing opinions and experiences can "connect the dots" in ways that individual experiences cannot provide • Useful way to elicit information when team's current knowledge on the subject may not be sufficient to create a survey or questionnaire • Avoids wasting time in one-on-one interviews talking about the same issues with individual participants • Participants do not have to think on the fly; they can think about their responses before replying to the group • Easier for hard-to-reach participants to commit to the study • Participants can pick up and stop during the day at their leisure • Ability to run more participants than other types of focus groups • (Usually) costs less than in-person focus groups • Anonymity helps reduce participant concerns over criticism or scrutiny • Ability to control how and when participants can see each other's information • Web-based platforms lend themselves to presenting media (e.g., photos, videos)	• No audio/video of participants (unless study tasks are designed with that intent ahead of time) • The "richness" of hearing and seeing someone speak is lost • You may get a "dominating" voice in the group that stifles collaboration or biases others • Quality of session depends heavily on the capabilities of the study moderator • Data gathered do not lend itself well to common forms of statistical analysis • Some data collection methods are suspect to bias • Difficult to redirect group when they get off-topic • Inability to shift topics quickly on the fly • Difficult to follow-up or clarify on specific topics or comments

One benefit of a BBFG over in-person or remote focus groups is that several dozen people can participate in a single study. This can be a quick way to capture lots of opinions and information in one study effort. Bear in mind, however, that more information does not necessarily mean your study elicited higher-quality information. Similar to in-person and remote focus groups, having a well-trained study moderator at the helm can make the difference between low and high participant engagement (Table 2.4).

2.12   Diary Studies

A diary study asks research participants to record their daily experiences, activities, and events in writing or video. Diaries are useful for collecting information over a period of time when you cannot be there to observe directly. Nowadays, most diary studies are conducted online. Participants may be asked to document each time they engage in a particular behavior, encounter a product or situation, or have specific

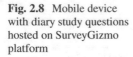

Fig. 2.8 Mobile device
with diary study questions
hosted on SurveyGizmo
platform

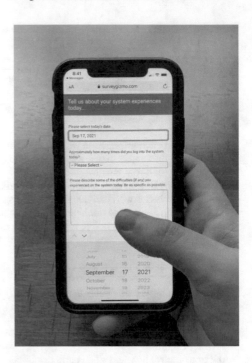

types of interactions. Other studies may require regular entries at particular times of
day, or a log of items in summary at day's end (Martin & Hanington, 2012) (Fig. 2.8).

A research method similar to a diary study—referred to as *experience sampling*
(Hektner, Schmidt, & Csikszentmihalyi, 2006)—involves collecting data about
behaviors, thoughts, or feelings from people at various times of the day. At random
or fixed time intervals, participants receive texts querying them about what they are
doing, how they are feeling, what is on their mind, and so on. Observations are often
entered into an application resembling a diary or journal, and entries can include
text, images, or recordings. This is different from a diary study, in which the participant is free to choose when they would like to submit their diary entry, which often
makes diary studies more enjoyable and easier for participants to fit into their daily
routines. However, it also comes with the risk of participants (occasionally) forgetting to complete their assignments.

Keep in mind that since these types of studies often run over several days, weeks,
or months at a time, you may lose participants along the way—a concept referred to
as *attrition*. While attrition is a common occurrence in any type of longitudinal
study, in our experience you will lose more participants when there is little "skin in
the game" for them to stay invested beyond monetary compensation. Remote diary
studies lack the face-to-face contact, which reminds participants that there are actual
people invested in this study with them. For these reasons, it is usually a good practice to over-recruit your diary study sample by 10–20% participants to help offset
attrition. Otherwise, you run the risk of not meeting your study sample size
objectives.

2.13   Critical Incident Technique

Originally used in aviation safety, the Critical Incident Technique (CIT; Flanagan, 1954) is really a set of research techniques. A commonly used approach is to ask seasoned users to describe a time period or event of significance—either good or bad. For example, you might ask the participant to tell you about a situation in which things almost went terribly wrong. What happened? Why did it happen? What did you do? How did you feel? How did you solve the problem? What do you think could have prevented it? Seasoned users' responses to these questions are particularly helpful in the design of medical devices.

Similarly, you could ask about positive critical incidents. For example, "tell me about a time that you were really happy when using the device." What happened? Why did it happen? How did you feel? and so on. Collecting a large volume of these stories (50–100) enables you to analyze each one according to cause and outcome, as well as the user's responses, emotions, and motivations.

2.14   Participatory Design

Participatory design involves designers and users working together to develop solutions, ensuring that the product is useful, usable, and desirable. Participatory design appears to have started in the 1970s in Norway where computer scientists worked with ironworker and metalworker unions to integrate technology (Kuhn & Winograd, 1996). Participatory design has since expanded in scope and methods, gaining in design research. It encompasses several methods all dedicated to active consultation with users in the design process, ideally through face-to-face contact in activity-based co-design engagements.

There are a variety of participatory design methods used to elicit different types of information from research participants. For example in a recent study we had physicians select emotionally evocative images from a large database to describe their current and ideal healthcare experiences. In these exercises, the images serve as a vocabulary for participants to express how they feel. The images chosen are important, but we find the real value can be found in the reason participants provide for choosing the images they did. It often leads to rather rich discussions.

In Fig. 2.9, a participant sketches his ideal user interface for a medical device. Sketching involves using simple tools like large sheets of paper, markers, and Post-it Notes for participants to describe their ideal designs on paper. Again, the value of this exercise is often obtained from the interview discussing the artifact after it has been created.

Figure 2.10 shows the result of a Velcro modeling session. During these participatory design sessions, we provide a variety of ambiguous shapes (usually foam core), and objects covered in felt and Velcro. Participants are instructed to create a device that would help them achieve their ideal experience. As you can see, the

Fig. 2.9 Sketching the participant's ideal device

Fig. 2.10 Velcro modeling

details can seem almost comical, but the reasoning behind them are informative. Follow-up interviewing enables us to probe:

- Which features were included, and which were not.
- The reasoning behind the placement of items.
- The relationships and grouping among items.

- The size of each item. For example, often participants will make the most important or most frequently used items larger than others.

2.15   Contextual Inquiry

Contextual inquiry combines observational and interview methods with a set of
analysis methods to learn about users, their tasks, environments, and challenges. Its
strength derives from its combination of observational and verbal response methods, its focus on context of use to understand how people work, and techniques for
analyzing and visualizing the resulting data. As with observational methods, one
goal of contextual inquiry is to observe users in their natural environments.

Contextual inquiry results can be used to identify needs, define requirements,
improve processes, and learn what is important to users. These sessions can enable
the researcher to document:

- Sequence of tasks
- Artifacts and tools people use to accomplish work
- Communication flows
- The impact and influence of the physical environment on the work
- Impact and influence of the culture on the work

Contextual inquiry achieves its goals by adhering to four principles:

- Focus: This refers to identifying the purpose of the research, the questions you
  are attempting to answer, and the information you seek to collect. This focus may
  change over time, but you always want to remember what your focus is at any
  one point. Any time you are surprised, or detect a contradiction, provides an
  opportunity to refocus inquiry.
- Context: This involves observing and interacting with users in their natural environment as they conduct their day-to-day activities. For surgeons, this environment will be the surgical suite, for home users this is likely to be the home. The
  fact that observers see the task as it occurs naturally means they can see the
  actual experience rather just hear about the user's memory (a summary) of the
  experience.
- Partnership: In our opinion, partnership is one of the greatest strengths of contextual inquiry. This approach often employs a master/apprentice model, in which
  the user behaves as a master teacher and the researcher serves as an apprentice.
  This encourages the user to demonstrate what they are doing, think out loud, use
  accessible language, and expect naive questions from the researcher. It puts both
  of them on the same team.
- Interpretation: After data are collected, researchers spend significant time and
  effort to interpret the information collected, and potentially to verify their interpretations with the participants.

Beyer and Holtzblatt (1997) use two mnemonics—FACES and BASIC—to remember the types of information to attend to during contextual inquiry interviews.

- (F) Flow—refers to work transactions.
- (A) Artifacts—these are objects people use or produce during their work. Analyzing the artifact, including its location, whether it is public or private, whether it is well worn or pristine, and so on, at a later time can help provide insights into the user, goals, tasks, challenges, and culture.
- (C) Context—this can refer to pressures or constraints on the user.
- (E) Environment—physical layout of work space and work structures. This can reveal priorities, task sequences, and challenges related to things like storage, communication, and so on.
- (S) Sequence—this provides information about tasks, steps, task triggers (what tells us the task is ready to be conducted), the order in which they are conducted, goals task, and flow. This can help point out activities that might need support.
- (B) Breakdowns—situations in which things go wrong.
- (A) Anecdotes—stories about situations that have happened before.
- (S) Scenarios—situations that occur during observation.
- (I) Insights—information gained.
- (C) Custom tools—these are artifacts that users create themselves to improve the efficiency or effectiveness of their work.

2.16   Analyzing Qualitative Data

Overview

Obviously, user research can generate a great deal of data quickly. This raises the question of how to make sense of all this information. Beyer and Holtzblatt (1997) suggest that the research and design teams gather within 48 h of the contextual inquiry session for a 2-h interpretation session. During this session, the team members use meeting notes, artifacts, and other data to replay their visit, and transfer the information onto notes and work models. The goal is to share the visit information and to develop a shared understanding of the data. They start by assigning the following roles:

- Storyteller—who reviews the notes out loud and represents what they learned from the user.
- Note taker.
- Facilitator—who keeps the process on task. This person uses their team management skills to get data down and ensure that team members feel heard.
- Modeler—who captures and generates data models that illustrate what was learned.

During the interpretation sessions, it is important to avoid abstracting or summarizing the data until later. It is also important to avoid critiquing or arguing over the data at this point. Instead, teams should default to the interviewer for the final interpretation or mark it as a data question and follow-up with users later.

After initial interpretation sessions, it is time to consolidate the data into an Affinity Model. Affinity diagramming identifies patterns or clusters of data collected during the contextual inquiry. Researchers start by writing each observation, insight, quote, or idea on Post-it Notes (one observation per Post-it). Notes might include the following types, and be assigned the labels in Table 2.5.

Each Post-it Note is based on something the researcher directly observed or heard from participants. This might be an important quote, a part of the interface that gave someone trouble, or a couple words that address what happened (e.g., could not find the power button). Post-its are initially placed randomly on a wall or table (i.e., no affinity to one another). As more interviews or sessions are conducted over time, researchers will add more Post-its to their collection. After an agreed upon point, the researchers take stock of the assets they have created and begin to look for relationships or patterns among them.

Stronger and more reliable insights from affinity diagramming are generated by drawing information from multiple studies or observations, rather than just one. For example, Martin and Hanington (2012) recommend that affinity diagrams created from ethnographic studies should include interviews or observations from at least four to six different locations. Each interview or observation will usually yield a large amount (e.g., 50–100) of observations, depending on the length of the interview and the content addressed in it. As a result, a full round of interviews and observations will often provide researchers with several hundred items for their affinity diagram.

Effective affinity diagram content can be challenging to develop for inexperienced researchers. Often, novices will put too much information on a Post-it, or will

Table 2.5 Codes used for Post-it Notes in contextual inquiry

Note label	Meaning
#	Unique note number
U#	Unique user number
What	What is happening
Why	Why something happened. This depicts intent and importance
P	User profile note
WN	Work note
BD	Work breakdown
DI	Design idea
I	Major insight
H	Data hole
C	Context note
A	Artifact note

use language that is non-descriptive and hard to understand after the study is finished. Experienced researchers also know to incorporate additional factors into each Post-it, such as a participant identity (ID) number, the user group, and the study from which the information originated. These details can be vital after the fact to trace back the original data.

Keep in mind that affinity diagramming—like other types of analytic exercises and researcher-generated tools—is subject to bias. One way to reduce bias is to avoid sorting important criteria by Post-it color. This seems counterintuitive; after all, why not sort all the "nurses" in one color, and all the "physicians" in another? While group-color distinctions make it easier to see group patterns from afar, they can also lead researchers to read into these differences too much. It is easy to detect group differences when there really are none. This also encourages teams to think of problems as user group-specific, rather than determining whether the problem might apply to all—or, at least other—user groups (Fig. 2.11).

Task Analysis

Task analysis (Kirwan & Ainsworth, 1992) studies what a user (or team) must do (actions and cognitive processes) to achieve their goals using a system. It provides a step-by-step description of each important task conducted with a device. Its purpose is to understand and represent human and device activities, human–device interactions, and performance in a specific task or scenario. There have been over

Fig. 2.11 Affinity diagramming using photos as reference (This exercise was conducted in March 2020 during the COVID-19 pandemic)

100 task analysis techniques described in the literature (Diaper & Stanton, 2004), which makes it a bit hard to determine which one is best for your project. Because of this variety, we will just cover one, probably the most common, here—hierarchical task analysis (HTA; Stanton, 2006).

HTA analyzes the goals and tasks of a system, treating those goals and tasks as a hierarchy. Specifically, every task is a means of achieving some goal. HTA starts by describing the purpose of the system at the highest level. It then lists the highest-level tasks (and their concomitant goals) required to achieve that purpose. Once it has done that, it further decomposes those tasks into subtasks and subgoals. This means that in order to satisfy the goal in the hierarchy its immediate subgoals have to be satisfied, and so on (Stanton, 2006).

Annett (2004) described the process as continual reiteration and refinement. First, you describe a rough outline of the hierarchy. Then you decompose those tasks and goals into a subgoal hierarchy. Then you do the same thing again for that hierarchy, and so on. Theoretically, this could go on forever; in fact, knowing when to stop the analysis is "one of the most difficult features of task analysis" (Annett, 2004). Eventually, you reach a subtask level that is granular enough to analyze. At this point, breaking down the subtask any further begins to feel almost ridiculous. We wish there were a more scientific way to describe this, but that is really how it works. In fact, Annet (2004) urges task analysts not to pursue re-description unless it is absolutely necessary. That is, once you reach a satisfying and useful description, you will reach a point of diminishing returns. Stop there (Fig. 2.12).

The flowchart resulting from HTA provides an exhaustive description of the activity, which can serve as a precursor to other HFE analysis methods. Each terminal subtask (the leaves of the tree) in that flowchart can then be analyzed according to:

- Criticality of each subtask
- Ease or difficulty of completing the subtask
- Stimuli that initiate step
- Actions required

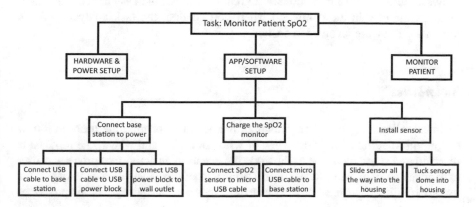

Fig. 2.12 Flowchart resulting from a hierarchical task analysis

- Criteria for successful performance
- Interrelationships between people, objects, and actions
- Decisions that need to be made
- Information needed to make those decisions
- Feedback requirements
- Potential problems
- Potential use-errors
- Potential interruptions
- What mistakes might the user make?
- What must the user do when they make a mistake?
- What strategies do the users employ to improve their performance?
- Whether the next subtask is immediately obvious?
- Whether the series of steps is efficient?
- What must the user remember between subtasks?

The advantages of task analysis are numerous:

- It enables you to create a structure that helps make sense of all the observation and interview data
- It makes the data useful for design or training
- It helps identify the goals and functions of the system
- It highlights critical steps in the users' tasks
- It identifies problems with the ways that users conduct their tasks
- It helps us to learn more about the user's environment
- It shows how a new system might fit in with old ones
- It shows how a new system will fit with the workflow
- It inspires new design ideas

The disadvantage is that it is time consuming, and can be a bit unreliable. Because it is a qualitative analysis method, different analysts may develop different flow-charts and analyses. Anecdotally, researchers often report that the value of task analysis resides in actually conducting the activity rather than the flowchart and analysis table that results. You learn a great deal about the system (perhaps more than anyone else in the company) when you conduct the task analysis. This is important if you are involved in design.

Swimlanes

Swimlane diagrams (Fig. 2.13) illustrate the activities of multiple users in a flow diagram, providing a bird's-eye view of who does what and when. The top lane is reserved as a storyboard lane to illustrate the events in an activity or process. A user experience lane (or set of lanes) uses a flowchart of boxes and arrows to depict the users' activities. A business process lane describes how business goals and processes

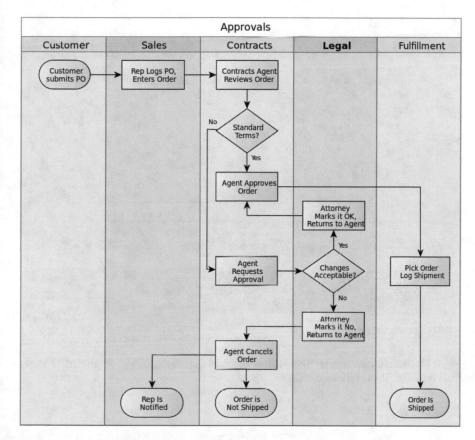

Fig. 2.13 Swimlane diagram (Image by Paul Kerr is licensed under Creative Commons, CC BY 1.0, and in Universal Public Domain)

fit in with the activities, and finally a tools and systems lane documents the other devices and systems materials that would be used along the way.

Journey Maps

Journey maps (Fig. 2.14) illustrate the experiences people have when interacting with a product or service. It includes the actions, decisions, feelings, perceptions, and frame of mind—including the positive, negative, and neutral moments (Martin & Hanington, 2012).

The map should represent an experience, including moments of indecision, confusion, frustration, as well as delight. Multiple maps will need to be created for multiple roles, as each role will have different tasks and goals, and will experience different breakdowns and successes on their journey.

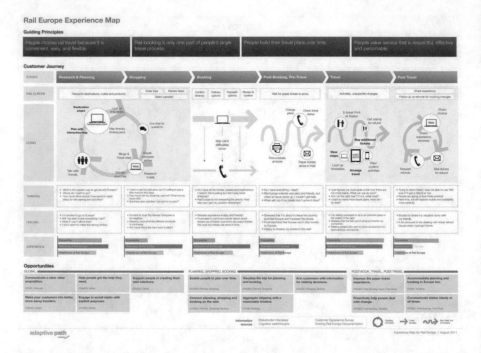

Fig. 2.14 Journey map (Image "File:SD053- Fig. 5.14 (8462249080).jpg" by Rosenfeld Media from Brooklyn, USA is licensed under CC BY 2.0)

Scenarios

You can use some of the qualitative data you have collected to form scenarios. A scenario (Carroll, 2000) is a story describing the future use of a product from the user's point of view. This is done to make design ideas explicit and envision the ways the device will be used. It is also helpful to depict exactly what happens in suboptimal situations, such as high stress, high noise, and when time constrained. Scenarios should follow a traditional story arc, beginning with a trigger, preconditions, discussing activities, and providing resolution.

User Profile

Often the qualitative data you collect will tell you about various types of users who use your device. A user profile lists the most important characteristics of your users, so that they can be accommodated in the device design. For example, consider the user profile of an electrophysiologist using a programmer device for a pacemaker/defibrillator. That profile might describe the user as a well-educated medical professional, who is a good reader, medically literate, adept at reading graphs, and between the ages of 25 and 60. The profile might go further to indicate that they are time constrained, dislike reading instructions for use, are highly motivated to learn, and technologically savvy. Note that the same device may have several user groups, and thus would require several user profiles. For example, a technician who interrogates the device may have different characteristics from the electrophysiologist who ultimately uses the data.

Many variables contribute to how users or different user groups perform with a device, and it is up to you as the researcher or designer to determine which variables are most important for performance and satisfaction with your device. Some sample variables are listed below.

- Job type
- Reading ability
- Attitude toward technology
- Motivation to learn
- Typing skill
- Height
- Task experience
- Training method
- Application experience
- Native language
- Computer literacy
- Frequency of use
- Turnover rate
- Education level
- Major in college
- Knowledge and experience
- Tasks and needs
- Mandatory or discretionary use
- Task or need importance
- Task structure
- Social interactions
- Primary training
- Attitude and motivation
- Patience
- Stress level
- Expectations
- Vision, hearing, cognitive processing, manual dexterity
- Handedness
- Disabilities

Prototyping

A prototype is a model of a product that enables you to demonstrate what the product looks like, how it is organized, or how it works. Early prototypes usually have low fidelity or low resolution, just demonstrating key aspects of the design. These early prototypes usually do not work; that is, they do not actually perform a medical function. On the other hand, they might appear to work. For example, a prototype of a software user interface might switch from one screen to another when a particular button is pressed. This may enable a research participant to interact with the prototype like it was the real product. This can be helpful during feedback sessions or usability testing. Over time, and with additional development, the design team may develop a high-fidelity prototype, which closely resembles the end product and may even include some basic functionality.

Resources

- Holtzblatt, K., & Beyer, H. (2016). Contextual design. In *Design for life* (2nd ed.). San Rafael, CA: Morgan & Claypool.
- Laurel, B. (2003). *Design research: Methods and perspectives*. Cambridge: MIT Press.
- Martin, B., & Hanington, B. (2012). *Universal methods of design: 100 ways to research complex problems, develop innovative ideas, and design effective solutions*. Beverly, MA: Rockport Publishers.

References

Abrams, K. M., Wang, Z., Song, Y. J., & Galindo-Gonzalez, S. (2015). Data richness trade-offs between face-to-face, online audiovisual, and online text-only focus groups. *Social Science Computer Review, 33*(1), 80–96.

American Psychological Association. (2017). Ethical principles of psychologists and code of conduct (2002, amended effective June 1, 2010, and January 1, 2017). Retrieved August 24, 2020, from http://www.apa.org/ethics/code.

Annett, J. (2004). Hierarchical task analysis (HTA). In *Handbook of human factors and ergonomics methods* (pp. 355–363). Boca Raton, FL: CRC Press.

Beyer, H., & Holtzblatt, K. (1997). *Contextual design: Defining customer-centered systems*. Amsterdam: Elsevier.

Cacioppo, J. T., & Freberg, L. A. (2013). *Discovering psychology: The science of mind*. Boston, MA: Wadsworth Cengage Learning.

Carroll, J. M. (2000). *Making use: Scenario-based design of human-computer interactions*. Cambridge: MIT Press.

Farnsworth, J., & Boon, B. (2010). Analysing group dynamics within the focus group. *Qualitative Research, 10*(5), 605–624.

Fetterman, D. M. (2019). *Ethnography: Step-by-step* (Vol. 17). Washington, DC: SAGE Publications, Incorporated.

Flanagan, J. C. (1954). The critical incident technique. *Psychological Bulletin, 51*(4), 327.

Gould, J. D., & Lewis, C. (1985). Designing for usability: Key principles and what designers think. *Communications of the ACM, 28*(3), 300–311.

Hektner, J. M., Schmidt, J. A., & Csikszentmihalyi, M. (Eds.). (2006). *Experience sampling method: Measuring the quality of everyday life.* Thousand Oaks, CA: Sage Publications.

Henrich, J., Heine, S. J., & Norenzayan, A. (2010). Most people are not WEIRD. *Nature, 466*(7302), 29–29.

Kirwan, B., & Ainsworth, L. K. (Eds.). (1992). *A guide to task analysis: The task analysis working group.* Boca Raton, FL: CRC press.

Klein, E. E., Tellefsen, T., & Herskovitz, P. J. (2007). The use of group support systems in focus groups: Information technology meets qualitative research. *Computers in Human Behavior, 23*(5), 2113–2132.

Kuhn, S., & Winograd, T. (1996). Participatory Design. In T. Winograd (Ed.), *Bringing design to software.* Boston: Addison-Wesley.

Martin, B., & Hanington, B. (2012). *Universal methods of design: 100 ways to research complex problems, develop innovative ideas, and design effective solutions.* Beverly, MA: Rockport Publishers.

Reid, D. J., & Reid, F. J. (2005). Online focus groups: An in-depth comparison of computer-mediated and conventional focus group discussions. *International Journal of Market Research, 47*(2), 131–162.

Reynolds, T. J., Dethloff, C., & Westberg, S. J. (2001). Advancements in laddering. In T. J. Reynolds & J. C. Olson (Eds.), *Understanding consumer decision making: The means-end approach to marketing and advertising strategy* (pp. 108–134). UK: Psychology Press.

Robinson, R. E. (2015). Building a useful research tool: An origin story of AEIOU. In *EPIC: Advancing the value of ethnography in industry.* Retrieved July 29, 2020, from https://www.epicpeople.org/building-a-useful-research-tool.

Stanton, N., Salmon, P. M., & Rafferty, L. A. (2013). *Human factors methods: A practical guide for engineering and design.* UK: Ashgate Publishing, Ltd..

Stanton, N. A. (2006). Hierarchical task analysis: Developments, applications, and extensions. *Applied Ergonomics, 37*(1), 55–79.

Summanen, I., Liikkanen, L. A., Laakso, M., & Leisti-Szymczak, A. (2013). Online focus groups: A new tool for concept testing with remote users. From the Proc. of Co-Create 2013. Espoo, Finland. 16–19 June (pp. 685–696).

Chapter 3
Quantitative Human Factors Research

Chapter 2 introduces qualitative research methods, which are helpful for developing an initial understanding of users, their tasks, needs, and desires. Qualitative research often helps generate hypotheses about user behavior, but is not great for testing those hypotheses. For testing hypotheses, you need more formal quantitative methods. Quantitative methods are more specific and constrained, and so tend to provide you with more reliable data. That is, they can yield the same answer if the research is conducted over and over again, giving us more confidence in the results. This section describes several quantitative methods. Some, like questionnaires, are based on verbal report, whereas others like formal experiments are based on human-system performance.

3.1 Questionnaires

Questionnaires enable us to ask large numbers of people about their attitudes and behaviors (Cacioppo & Freberg, 2013), providing substantial information quickly and inexpensively. The way a question is constructed plays a key role in the type of response and analysis. For example, open-ended questions provide opportunity for richer responses, whereas closed-ended questions are easier to analyze numerically and communicate.

Because questionnaires rely on self-report results, they can be influenced by peoples' natural desire to appear socially appropriate (Corbett, 1991). Like any self-report instrument, questionnaires may not accurately reflect true thoughts, feelings, perceptions, or even behaviors. What people say and what they do can be different. This means that the design, administration, and analysis of the questionnaire should be done with caution. Further, you should also employ complementary research methods such as observations. For example, it is common to survey a large group of users using questionnaires, and then pair this with in-depth observations, contextual

© Springer Nature Switzerland AG 2021
R. J. Branaghan et al., *Humanizing Healthcare – Human Factors for Medical Device Design*, https://doi.org/10.1007/978-3-030-64433-8_3

inquiries, or participatory design sessions. Questionnaires include various question types, or scales, each with its strengths and weaknesses.

Likert Scale

A Likert scale (Spector, 1992) is the most widely used question type on question-naires. Likert scales ask participants to express their level of agreement or disagree-ment with a statement, such as the example shown in Fig. 3.1. The response options in a Likert scale can be odd or even numbered, and almost always use "strongly disagree" and "strongly agree" as the outermost options. Scales with an even num-ber of response items (e.g., six options) are referred to as a *forced choice* scale, as they "force" the participant to choose to side towards agreement or disagreement. In other words, they cannot choose a middle ground between the two sides. This can help researchers see which way an opinion leans, but can come at the cost of inac-curately representing participant thoughts or beliefs. After all, sometimes people truly are split between agreement and disagreement.

More commonly, however, Likert scales use an odd-number of response options (e.g., 5-items, 7-items). This creates a mid-point that lies between the sides of "agreement" and "disagreement." The two most common labels for the midpoint are "neutral" and "neither agree nor disagree," though others have been used by researchers in the past for specific applications (see Baka, Figgou, & Triga, 2012; Klopfer & Madden, 1980; Masters, 1974).

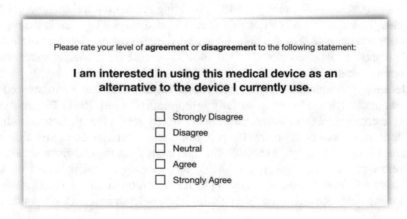

Fig. 3.1 A Likert item with a 5-point scale (including a "neutral" option)

Semantic Differential

Semantic differential questions (Osgood, Suci, & Tannenbaum, 1957) ask participants to rate their attitudes on a scale of antonym word pairings. Usually it employs a seven-point scale so that participants can choose a neutral response. Other times, however, you can force the participant to choose one side or the other by using a six-point scale. An example of a few semantic differential questions is shown in Fig. 3.2. A graph of the output from a semantic differential is shown in Fig. 3.3.

The value of a semantic differential scale is that researchers can quickly gleen patterns. On the administrative side, semantic differential scales are straightforward and fast for participants to complete.

Ranking

Ranking asks participants to place items in order, according to some dimension. For example, you might ask participants to rank 10 product features in order of importance to them. This can force them to make important tradeoffs and fine distinctions, among the candidates. This type of question can be cognitively difficult, and can force respondents to really consider their choices.

Ranking activities are often a good precursor to a follow-up interview that discusses the participants' choices. However, because it requires so much effort, ranking should be used rarely. Furthermore, even if an item is ranked as the most

There are six antonym word pairings listed below. Please select one box between each word pairing that best matches how you feel about the continuous glucose monitor setup process:

Task: Continuous Glucose Monitor Setup

Negative	□ □ □ □ □ □ □	Negative
Unstructured	□ □ □ □ □ □ □	Unstructured
Boring	□ □ □ □ □ □ □	Boring
Difficult	□ □ □ □ □ □ □	Easy
Unpleasant	□ □ □ □ □ □ □	Unpleasant
Stressful	□ □ □ □ □ □ □	Stressful

Fig. 3.2 Sample set of semantic differential items

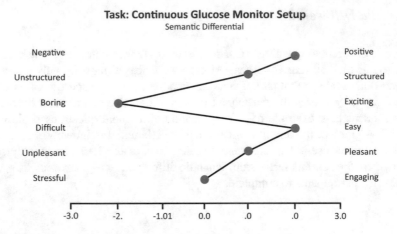

Fig. 3.3 Visualization of semantic differential results

important, we still do not know exactly how important it is; we simply know it is more important than all of the other items.

Constant Sum

Constant sum is a useful variation of the ranking method, in which participants assign a certain number of points to each item in a list. For example, imagine participants are provided with 100 points. Their job is to distribute those 100 points according to how desirable each feature is to them. This is more satisfying than just plain ranking. An example of an online version of this is shown in Fig. 3.4.

3.2   Biometric Research

As our bodies operate throughout the day, all kinds of physiological operations occur. Guided by attention, our eyes dart from one place to another, fixating on some stimuli briefly, only to dart around some more. Brain waves vary according to our mental concentration. Our pupils dilate or constrict to admit more or less light into our eyes. We blink our eyes more or less depending on how vigilant or sleepy we are, or how much we are concentrating. Our heart speeds up or slows down, we produce more or less perspiration, and we make facial expressions, hardly noticeable to us, in response to various situations and feelings.

These operations are unconscious, outside our purposive control, and vary according to important psychological and physiological constructs such as interest, attention, concentration, cognitive workload, fatigue, and surprise. These are

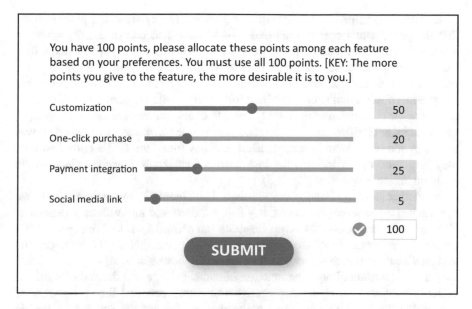

Fig. 3.4 A variation in the constant sum method using poker chips

precisely the kinds of physiological variables human factors engineering (HFE) is interested in measuring. Further, unlike interview or questionnaire responses, these variables are almost impossible for someone to fake or hide from the sensitive biometric sensors that researchers use.

Biometric research uses these biometric sensors to measure otherwise hidden features related to emotion, attention, cognition, and physiological arousal (Farnsworth, 2017). These include brain signals (electroencephalography, EEG) facial expressions (facial expression analysis), eye movements, pupil changes (pupillometry) heart rate changes, and perspiration changes (galvanic skin response, GSR).

Eye Tracking

Eye tracking (Rayner, 1998) records exactly where and for how long people look at a stimulus. So, if you want to know what people are paying attention to, eye tracking is ideal (get it, eye-deal?). If you wonder if people noticed a certain feature in the product, eye tracking can help you determine that as well. On a recent project, we usability tested instructions for use (IFU), and the participants frequently failed to complete the task, though it was not clear why, even after thorough interviews. There were two possibilities: One, the participant could not find the portion of the IFU that provided the proper instructions, or two, they found the right section, but were confused by the instructions.

Eye tracking helped answer this question easily. The eye tracking path revealed that the participant's eyes did not dart around from one part of the document to another, but instead focused on a few sentences in the first paragraph. Indeed, they followed the sentence several times, only to return to the beginning of the sentence each time. This tells us that the problem was likely one of comprehension. We flagged this observation in our notes and followed up with the participant during the debrief at the conclusion of their study tasks to check our assessment. Sure enough, the participant clarified that the sentence was worded awkwardly (passively), was too long, and used unfamiliar nomenclature. They found the IFU section just fine, they just could not understand the instructions themselves. Once we clarified the problem, fixing it was easy (Fig. 3.5).

In another study (O'Brian, Orn, Sheehan, Hildebrand, & Branaghan, 2020), we eye tracked hemophilia patients as they self-administered intravenous infusions of clotting factor (a process they complete about three times a week). We expected that patients' eyes would fixate frequently and for long durations on the infusion site, and data showed that this was true. Unexpectedly however, analysis revealed that they also closely monitored the junction between syringe and the catheter for the butterfly needle. Follow-up retrospective interviews revealed that patients were monitoring the last bit of medicine going through the syringe and the air bubble (usually rather small) that followed it. Patients were making sure that they did not infuse air into their veins, but they were also making sure to get every last drop of clotting factor. Clotting factor is one of the most expensive drugs in the pharmacopeia. Each mL of the patients' 10 mL injection costs about $300. This means it costs about $9000 per week to prevent them from bleeding under normal circumstances. Although we will not go into detail, this may hold some promise for redesigning the process.

In Fig. 3.6, the yellow circle represents where the eyes are fixated—the larger the circle, the longer the fixation. There is also a small number in the middle of each yellow circle. These represent the running count of fixations that have occurred thus

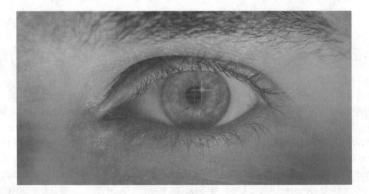

Fig. 3.5 The human eye is a complex thing—a surprising amount can be learned about what a person is experiencing based on subtle movements in their gaze and pupil dilation (Image courtesy of iMotions)

Fig. 3.6 Eye tracking with hemophilia patients as they self-infuse clotting factor

far. A brief, five-minute eye tracking study may produce hundreds of fixations. Each one might last about 250 ms up to several seconds long.

Facial Expression Analysis (FEA)

The ability to read people's emotional expressions is an essential and inherent skill that has been refined through evolution. Facial expressions are so inextricably linked to emotion (Ekman, 1992), that even infants as young as just 7 months can reliably match emotional cues to a person's facial expressions and tone of voice (Grossmann, 2010). However, tracking those emotions on a moment by moment basis while people are performing tasks is untenable. Further, often detecting the exact moment of onset for the facial expression is difficult due to the expressions' fleeting nature; they are there briefly and gone just as quickly. For these reasons, psychologists developed a systematic way to code the facial cues that work in tandem to express specific emotions—a quantitative research method known as Facial Expression Analysis (FEA).

Until just a few decades ago, almost all FEA efforts were completed manually by small teams of trained researchers. Each researcher would review and edit video footage of a participant's face as they completed specific tasks or answered queries. Individual stills of this video footage would be evaluated by researchers who coded facial movements, shapes, and landmarks using methods such as the Facial Action Coding System (FACS; Friesen & Ekman, 1978). Coding models like FACS broke down facial expressions into bite-sized components called *Action Units* (AUs). In total, researchers would need to reference anywhere from 30–44 AUs, plus an additional set of "action descriptors" (De la Torre & Cohn, 2011). For example, "AU1" indicated an "inner brow raiser," whereas "AU44" mapped to "squint." Specific combinations of these AUs would add up to indicate an emotion such as joy or fear.

A single participant's video footage could take hours to code. What's more, however, there would often be problems with *inter-rater reliability* when multiple researchers coded the same participant. In other words, two (or more) researchers would reach different conclusions, despite using the same video footage for reference. This problem could be (partially) mitigated through extensive training across the research team to promote greater consistency. But these efforts came at the expense of vast amounts of time and money.

More recently, FEA has been championed by computer scientists who have developed robust algorithms to detect these AUs automatically in mere milliseconds (Lewinski, den Uly, & Butler, 2014). No more messy coding efforts. No more manual review. No more hours of time parked in an office chair with a pad and pencil. Computer-evaluated FEA uses the same underlying rules and constructs developed in coding models like FACS. And, it still looks at participants' subtle facial changes on a frame-by-frame basis (Fig. 3.7).

Keep in mind, however, that FEA is not some "silver bullet" that will help you make sense of all your users' emotional responses and experiences with a medical device. Indeed, like all research methods, it has its fair share of strengths and weaknesses. Table 3.1 provides an overview of the things we've observed or experienced with FEA over recent years.

Fig. 3.7 Facial expression analysis study (Image courtesy of iMotions)

Table 3.1 Strengths and limitations of FEA

FEA strengths	FEA limitations
• Relatively inexpensive • Low barrier to entry (for data collection) • Automatically coded • Real-time coding • Consistent coding • Uses any off-the-shelf video camera with 720p or higher resolution capabilities • Useful for tasks or events that you are confident will elicit a real (facial) reaction • Video footage is compelling for project managers and stakeholders • Easy, intriguing, and innovative topic that can get non-researchers interested in your findings	• Does not work as effectively for tasks that do not elicit an observable facial (emotional) response. Participants must physically "emote" for the process to work well • Some automatic analysis platforms may not explicitly tell you which emotion the participant is experiencing—you have to interpret this for yourself • Participants must keep their face "in frame" at all times • Participants have to be instructed (and reminded) to not cover their face with their hands (e.g., resting their chin on their hands)

Galvanic Skin Response (GSR)

Your skin is rich with thousands of tiny sweat glands occupying every inch of your body. These sweat glands—referred to as *eccrine sweat glands*—are different from the larger sweat glands under your armpits (and other places). Areas along your feet, fingertips, forehead, wrists, and ankles contain the greatest concentration of eccrine sweat glands. For example, some areas of our skin will have as many as 530 eccrine sweat glands per square centimeter (Taylor & Machado-Moreira, 2013).

One way that eccrine sweat glands differ from other sweat glands is that they produce trace amounts of sweat in response to your arousal; which simply means any type of sympathetic nervous system response to a stimulus. In this sense, *lots* of things cause arousal, such as the joy of seeing a friend's face, or the fear of thinking you left your oven on. Eccrine sweat produced as a result of cognitive demand is called *mental sweating* (Ohmi, Tanigawa, Yamada, Ueda, & Haruna, 2009).

Your arousal is constantly fluctuating in small ways, even when you are not aware of it. Indeed, your eccrine sweat glands produce small changes in sweat production several times per second in concert with your arousal level—the greater the arousal response, the more sweat. The key to GSR is that your sweat promotes electrical conductivity, since it contains electrolytes like chloride and sodium. This means that as our arousal increases on a second-by-second basis, the surface of our skin becomes slightly more conductive.

Most GSR measuring devices consist of two electrical leads placed in close proximity to each other (i.e., about 2 in. or less). A very small electrical current—less than 0.5 V—is transmitted from one lead, and based on the skin's conductance at that moment, a lot or a little of the electrical current reaches the other lead. The GSR unit samples this conductance several dozens or hundreds of times per second. The data output from this measurement is a line graph that rises and falls in response to arousal. Importantly though, there is a minor delay between when your sympathetic nervous system "responds" to a stimulus, and when the eccrine sweat glands are sufficiently activated. In most cases, this delay is about 1–3 s after detection of the stimulus.

One study on lead placement by Van Dooren and Janssen (2012) compared GSR data sampling quality from over a dozen areas on the body. Each participant was instructed to watch a series of brief videos intended to elicit different emotional responses (e.g., happiness, fear, disgust). That study reported that GSR sampling from the participants' feet and fingertips provided the best quality data overall. By contrast, samples collected from the abdomen, back, and upper arm provided the least reliable data.

While these findings imply that GSR is best collected from the fingers or feet, it's important to remember that not all studies in the medical domain will lend themselves to this setup. Minor movements in the hands and feet can throw off GSR data. As a result, participants cannot do things like walk around or interact with handheld devices while sampling from these areas of the body. Likewise, things like gloves (PPE) can interfere with GSR lead placement. In these situations, researchers should plan ahead of time to sample GSR data from less ideal locations (e.g., wrist).

Electroencephalography (EEG)

Electroencephalography (EEG) places multiple electrodes on the scalp to measure voltage fluctuations as neurons in the brain communicate. This provides insight into what the brain is doing when an individual encounters a stimulus (Farnsworth, 2017). Various EEG patterns can determine if someone is engaged, distracted, or how hard their brain is working (Fig. 3.8).

Electrocardiography (ECG)

Electrocardiography (ECG) provides important information about how we are responding physiologically. Generally, our hearts beat faster as tasks become more cognitively difficult. As a result, heart rate could serve as a good measure of cognitive difficulty. Unfortunately, of course, our heart rate also increases with physical exertion and emotional stress. So, instead of relying solely on heart rate, researchers have begun to employ measures of heart rate variability. That is, heart rate variability usually declines as workload increases (Matthews & Reinerman-Jones, 2017).

Fig. 3.8 EEG study (Image courtesy of iMotions)

3.3 Correlational Research

Correlational research measures the direction and strength of a relationship between two variables, like a person's height and weight. We compare the values of one variable to those of the other and compute a correlation coefficient. The analysis can yield a positive, negative or zero correlation. In a positive correlation, high levels of one variable are associated with high levels of the other variable. An example of this is height and weight: Typically, taller people weigh more than shorter people, even though the relationship is not perfect. Alternatively, two variables can be negatively correlated. For example, the number of hours of watching television can be negatively correlated with grade point average in school. The third possible outcome is a zero correlation, in which the two variables do not have any systematic relationship with each other. When variables have no correlation, knowing the value of one variable tells you nothing about the value of the other variable (Fig. 3.9).

The strength and direction of a relationship is expressed as a correlation coefficient, number between +1.00 and −1.00. If the number is zero or close to zero, the association between two measures is weak or nonexistent. A correlation of +1.00 is a perfect positive relationship, whereas a correlation of −1.00 is a perfect negative correlation. The strength of the correlation is not influenced by its sign. That is, a correlation of −0.5 is just as strong as a correlation of +0.5.

Importantly, correlation is different from causation. Our colleague, Professor Scotty Craig often uses the example of the relationship between ice cream sales and murder. It turns out that, as the sales of ice cream increase, so does the homicide rate. Most of us would be surprised and disappointed to conclude that ice cream causes murder; and thankfully this is not the case. Instead ice cream sales and

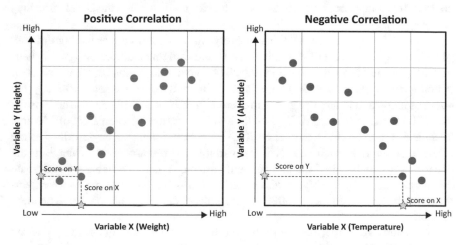

Fig. 3.9 Positive and negative correlations

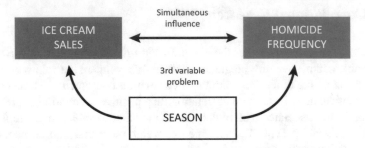

Fig. 3.10 The third variable problem

homicide both increase during the summer. Of course, summer is when people are outside, when schools are on break, and when heat can make people more confrontational. It is also a great time to grab an ice cream cone. This is called the third variable problem. Yes, there is a relationship between ice cream sales and homicide, but it is probably caused by a third variable that you were not originally paying attention to; namely, season. We need to be cautious interpreting correlational data because of this third variable problem (Fig. 3.10).

3.4 Experiments

The weakness of correlational research is that we cannot determine causality. This is where experiments come in. Experiments enable us to study variables of interest, while controlling the effects of variables we are not interested in and may in fact confuse the situation. This control enables us to discover causal relationships between variables.

To simplify matters you can think of every simple experiment as having the title, "The Effect of X on Y." For example, if you were studying a device design for rheumatoid arthritis (Schwarzenbach et al., 2014), your title might be, "The Effect of Autoinjector Type on Patient Compliance." Or imagine you wanted to study the effect of medical technical jargon (Zimmermann & Jucks, 2018) on instructions for use (IFU) comprehension. Well, actually that is your title—"The Effect of Medical Technical Jargon on IFU Comprehension." Each experiment can be described as the effect of X on Y. In each case X is the variable the researcher can change in the experiment. This is called the independent variable. Y is the variable we anticipate might be changed by modifying X. This is the dependent variable. So, every experiment can be described as the effect of the independent variable on the dependent variable.

You begin each experiment with a hypothesis; an educated guess ideally based on systematic observations, a review of previous research, or a scientific theory. Because we are interested in causal relationships, we need to make sure that the

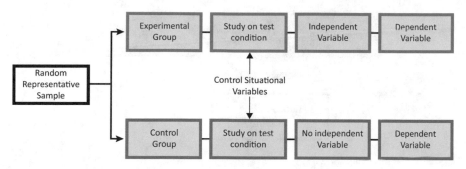

Fig. 3.11 Design of a two-condition controlled experiment

only difference among conditions is the level of the independent variable. Everything else needs to be exactly the same (Fig. 3.11).

In most experiments, we want to know how simply going through the procedures of being in an experiment influences our dependent variable. Perhaps traveling to the laboratory and knowing you will be observed changes our behavior. To evaluate these effects, participants are randomly assigned to a control group—a group that experiences all experimental procedures except the independent variable. We make the experience of the control group as near as possible to the experimental groups, which experiences the independent variable.

Of course, each participant is different from the other. To prevent these individual differences from obscuring or distorting the effects of our independent variable, participants are randomly assigned to experimental or control groups. In random assignment, each participant has an equal chance of being assigned to any group in the experiment. This makes it so any differences between the behavior of one group and that of another is not likely caused by individual differences among the participants. These individual differences tend to cancel each other out. Individual differences among participants are one type of confounding variable (Fig. 3.12); that is a variable that is irrelevant to the hypothesis being tested but that can still affect our conclusions.

Random assignment to groups controls for individual differences, but there can still be other problematic confounds. For example, imagine running all of the experimental group participants early in the morning, and all of your control group participants late in the morning. If there were differences among them, according to their performance on the dependent variable, you would not know if that was caused by the independent variable of interest or by the time of day. Perhaps people do not perform as well before a certain amount of coffee in the morning. It is critical to run your experiments under the most constant circumstances to rule out confounding variables.

Another potential confound is experimenter bias—unintentional effects of researchers on experimental results. To prevent experimenter bias, you could use a

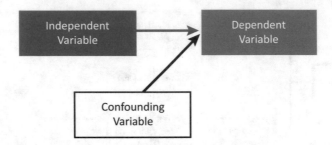

Fig. 3.12 Confounding variable

Table 3.2 Two-condition experiment

	Shallow menu structure (two levels)	Deep menu structure (four levels)
Performance measure (time to find and select target item)		

double-blind design, one in which both the research participants and those giving the treatments are unaware of (blind) who gets which condition of the independent variable. Only researchers with no participant contact have this information, and they do not reveal it until the experiment is over.

The Two-Condition Experimental Design

Imagine, we were conducting an experiment on the depth versus breadth of a computerized device's menu system. Imagine further, that we anticipated that participants would be able to find and select an important option faster if the menu system was shallow (a hierarchy two levels) versus deep, (a hierarchy with four levels). This experiment has only two conditions (shallow or deep), as depicted in Table 3.2.

Multiple Condition Design

Now imagine you wanted more detail than this. Perhaps, for example, you think a middle (intermediate) level between shallow and deep would be best. You would simply add a third condition to your experiment as shown in Table 3.3.

Table 3.3 Multiple condition experimental design

	Shallow menu structure (two levels)	Intermediate menu structure (three levels)	Deep menu structure (four levels)
Performance measure (time to find and select target item)			

Table 3.4 Factorial design

	Shallow menu structure (two levels)	Intermediate menu structure (three levels)	Deep menu structure (four levels)
32 menu items			
64 menu items			
96 menu items			

Factorial Design

So far, we have talked about the effects of only one independent variable on a particular dependent variable. However, often you want to explore more than one independent variable at a time. This enables you to explore simultaneous relationships among variables. A factorial design enables you to study every combination of two or more independent variables. Moreover, it enables you not only to determine how these independent variables affect the dependent variable by themselves, but it enables you to determine how these independent variables interact. This is more economical than running these as separate experiments, because it enables us to study several things at once. For example, in our study of breadth and depth of menu hierarchies, maybe the results depend on the number of total items in the menu hierarchy. So, you might add another variable called number of items (32, 64, or 96 items). This is represented in Table 3.4.

Between Subjects and Within Subjects Designs

A between subjects design uses different groups of participants for each level of the independent variable. So, if you used a between subjects design for the first experiment we discussed (the one with only two conditions), each group of participants would be placed in one condition or the other.

Conversely, a within subjects design has the same subjects participate in each of the experimental conditions (i.e., all levels of the independent variable). In essence, this enables you to compare each subject with him or herself across the various conditions. The within subjects design has the benefits of requiring fewer participants and of being more sensitive, making it easier to detect statistical differences among conditions. On the other hand, it requires you to counterbalance the order in which participants are presented with each condition. For example, in our simple two factor experiment, Participant 1 might complete the shallow menu task first followed by the deep menu task, whereas Participant 2 would complete the deep menu task first, followed by the shallow.

Mixed Designs are factorial designs in which one or more of the variables are within subjects and one or more are between subjects.

3.5   Analyzing Quantitative Data

Descriptive statistics summarize and highlight patterns in data sets. These contribute a rough snapshot of the data. Typically, they're used to summarize the results and illustrate trends or patterns in the data.

Central Tendency

Descriptive statistics usually provide information about two characteristics of the data: central tendency and dispersion. Central tendency is a description of how the typical participant performed on some measure. Three measures of central tendency are: mean, median, and mode.

The mean represents the numeric average of a data set. To calculate the mean, simply add all of the scores and divide by the number of scores. One problem with the mean is that it can be skewed by extreme values. Imagine, for example, that you timed participants as they completed what you assumed to be a very simple task. The times to complete the task are shown in Table 3.5.

Adding all of the times and dividing by the number of participants (20) provide the mean time of 9.6 s. However, examining the scores you notice that participant 9 had a difficult go of it. Whereas everyone else completed the task in around 6 s, participant 9 required about 68 s. Maybe using their score to calculate central tendency makes the typical participant actually seem slower than they really are. In this case, we should use the median to describe central tendency (Fig. 3.13).

The median represents a halfway mark in the data set, with half of the scores above and half below, and is less affected by extreme scores than the mean. In many cases, the mean and the median are very close together, but in some cases, like the one above where there is an outlier, they can be different, and the median is a better representation of central tendency. The median—the score in which half of the

Table 3.5 Sample completion times for a common task

Participant	Time (s)
1	7.2
2	5.6
3	8.4
4	6.1
5	6.1
6	8.4
7	5.7
8	8.1
9	67.7
10	5.4
11	6.1
12	7.7
13	5.6
14	6.4
15	6.1
16	5.6
17	7.3
18	5.8
19	6.6
20	6.1

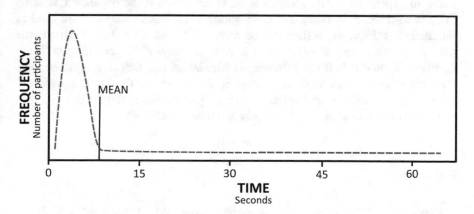

Fig. 3.13 An illustration of how extreme scores affect the mean

scores are above and half are below—in this case is 6.1 s. The participants are actually faster than the mean led us to think.

The mode of a data set refers to the score that occurs most frequently. For example, you may notice that the time of 6.1 s occurred five times in the data. This most frequently occurring score is the mode.

Fig. 3.14 Equation for calculating standard deviation for a sample

$$s = \sqrt{\frac{1}{N-1} \sum_{i-1}^{N} (x_i \bar{x})^2}$$

Dispersion

Measures of dispersion indicate how much the individual scores vary from the most typical score. That is, are all of the scores really similar to the mean or are they all over the place? A simple measure of dispersion is the range. This simply lists the lowest and highest scores, along with the mean. So, in the example above, the range is 5.4–67.7.

The most commonly used measure of dispersion is the standard deviation, which is the average deviation (whether above or below) of individual scores from the mean. Larger standard deviations represent larger dispersions. A simple equation that captures the concept rather intuitively is shown in Fig. 3.14. In essence, it takes each score and subtracts the mean. Since half of those results will be negative, it squares each of them to turn them positive (recall that the square of a negative number is a positive number). It sums all of those and then divides them by $n - 1$. Then finally, it takes the square root of all of those.

In most HFE research, some measure of central tendency and dispersion is provided. In experiments, data analysis aims to determine if the dependent variable changed as a result of the independent variable. Inferential statistics are used to determine how likely any differences between conditions are real or are simply the result of chance. People show random variation in behavior and performance. These statistics help to decide if the difference is big enough that we can rule out chance. The inferential statistics required are beyond the purview of this book, but we will mention that for two-group designs we typically use a t-test, and for designs with more than two groups we use an analysis of variance (ANOVA).

Resources

- Field, A. (2017). *Discovering statistics using IBM SPSS statistics: North American edition*. Thousand Oaks, CA: Sage.
- Imotions offers several short booklets introducing several biometric measures. These are worth taking a look at. Go to https://imotions.com/guides/
- Myers, J. L., Well, A., & Lorch, R. F. (2010). *Research design and statistical analysis*. Abingdon: Routledge.
- Patten, M. L. (2016). *Questionnaire research: A practical guide*. Abingdon: Routledge.
- Pyrczak, F., & Oh, D. M. (2018). *Making sense of statistics: A conceptual overview*. Abingdon: Routledge.

References

Baka, A., Figgou, L., & Triga, V. (2012). 'Neither agree, nor disagree': A critical analysis of the middle answer category in voting advice applications. *International Journal of Electronic Governance, 5*(3–4), 244–263.

Cacioppo, J. T., & Freberg, L. A. (2013). *Discovering psychology: The science of mind.* Boston, MA: Wadsworth Cengage Learning.

Corbett, M. (1991). *American public opinion.* London: Longman.

De la Torre, F., & Cohn, J. F. (2011). Facial expression analysis. In *Visual analysis of humans* (pp. 377–409). London: Springer.

Ekman, P. (1992). Are there basic emotions? *Psychological Review, 99*(3), 550–553.

Farnsworth, B. (2017) *What is biometric research?* Retrieved July 25, 2020, from https://imotions. com/blog/what-is-biometric-research.

Friesen, E., & Ekman, P. (1978). Facial action coding system: A technique for the measurement of facial movement. *Palo Alto, 3.*

Grossmann, T. (2010). The development of emotion perception in face and voice during infancy. *Restorative Neurology and Neuroscience, 28*(2), 219–236.

Klopfer, F. J., & Madden, T. M. (1980). The middlemost choice on attitude items: Ambivalence, neutrality or uncertainty? *Personality and Social Psychology Bulletin, 6*, 97–101.

Lewinski, P., den Uyl, T. M., & Butler, C. (2014). Automated facial coding: Validation of basic emotions and FACS AUs in FaceReader. *Journal of Neuroscience, Psychology, and Economics, 7*(4), 227.

Masters, J. R. (1974). The relationship between number of response categories and reliability of Likert-type questionnaires. *Journal of Educational Measurement, 11*, 49–53.

Matthews, G., & Reinerman-Jones, L. (2017). *Workload assessment: How to diagnose workload issues and enhance performance.* Santa Monica, CA: Human Factors and Ergonomics Society.

O'Brian, J. S., Orn, A. G, Sheehan, N. A., Hildebrand, E. A. & Branaghan, R. J. (2020). Through the eyes of hemophilia patients: Eye tracking hemophilia patients' intravenous self infusion. In *Human Factors and Ergonomics Society 2020 Health Care Symposium.*

Ohmi, M., Tanigawa, M., Yamada, A., Ueda, Y., & Haruna, M. (2009). Dynamic analysis of internal and external mental sweating by optical coherence tomography. *Journal of Biomedical Optics, 14*(1), 014026.

Osgood, C. E., Suci, G. J., & Tannenbaum, P. H. (1957). *The measurement of meaning* (No. 47). IL: University of Illinois Press.

Rayner, K. (1998). Eye movements in reading and information processing: 20 years of research. *Psychological Bulletin, 124*, 372–422.

Schwarzenbach, F., Trong, M. D., Grange, L., Laurent, P. E., Abry, H., Cotten, J., et al. (2014). Results of a human factors experiment of the usability and patient acceptance of a new autoinjector in patients with rheumatoid arthritis. *Patient Preference and Adherence, 8*, 199.

Spector, P. E. (1992). *Summated rating scale construction.* Thousand Oaks, CA: Sage.

Taylor, N. A., & Machado-Moreira, C. A. (2013). Regional variations in transepidermal water loss, eccrine sweat gland density, sweat secretion rates and electrolyte composition in resting and exercising humans. *Extreme Physiology & Medicine, 2*(1), 4.

Van Dooren, M., & Janssen, J. H. (2012). Emotional sweating across the body: Comparing 16 different skin conductance measurement locations. *Physiology & Behavior, 106*(2), 298–304.

Zimmermann, M., & Jucks, R. (2018). How experts' use of medical technical jargon in different types of online health forums affects perceived information credibility: Randomized experiment with laypersons. *Journal of Medical Internet Research, 20*(1), e30.

Chapter 4
Usability Evaluation

4.1 Introduction

Usability evaluation methods are neither strictly qualitative or strictly quantitative. They keep one foot planted squarely in each camp, and then hop from one side to the other depending on the study objectives, and specific method used. As a result, we did what any rational person does when they find an orphan sock like this topic in their clean laundry basket: you set it in its own special little pile.

Fortunately, usability evaluation methods are widely discussed in the human factors literature and warrants a deep discussion of how to plan and conduct this type of research successfully. It's the one group of methods in human factors—both in and outside the medical domain—that is a constant go-to-for practitioners regardless of experience level. Indeed, it's such a pervasive part of human factors that multiple, stand-alone books have been published on the topic. We provide a list of these in the resource section.

This chapter provides an overview of usability evaluation methods, along with several recommendations for what to do and not do in each one. It walks you through what are referred to as the "discount usability" methods, all the way up through validation usability testing for US Food and Drug Administration (FDA) (or similar) review. While we aim to provide information that is useful to novice and seasoned HFE practitioners alike, we anticipate that the former will benefit the most from this discussion. Additional resources listed at the end of the chapter should be your next stop to fill in any remaining questions or concerns you might have.

© Springer Nature Switzerland AG 2021
R. J. Branaghan et al., *Humanizing Healthcare – Human Factors for Medical Device Design*, https://doi.org/10.1007/978-3-030-64433-8_4

4.2 Usability Inspection

In a usability inspection, evaluators review a user interface to identify problems. These techniques, sometimes referred to as discount usability methods, are not actually research methods per se, but are useful precursors to data collection, providing help in identifying usability problems. During inspections, evaluators identify problems with the user interface. However, they do not usually complete tasks while doing so. It is more similar to a code review or a design review. We discuss three of these inspection methods below.

Heuristic Evaluation

Heuristic evaluation is a usability method in which one or more reviewers inspect a user interface (software, IFU, hardware, etc.) against a list of best practices or heuristics (rules of thumb) in user experience and user interface design. The evaluators identify places in the user interface that violate the heuristics. Ideally, four to six evaluators conduct independent evaluations, then the lead researcher convenes a meeting of the evaluators to discuss the issues, resolve disagreements, reach consensus, and write a short report. Because evaluators have the list of heuristics handy as they complete their evaluations, they do not need to be HFE experts. On the other hand, HFE experts may be more sensitive to violations than other evaluators.

Heuristic evaluation became popular during the 1990s, after Nielsen and Molich (1990) published their original set of usability heuristics and Shneiderman (1998) published his golden rules of interface design. Later, Zhang, Johnson, Patel, Paige, and Kubose (2003) modified these two contributions to produce the 14 usability heuristics for medical devices listed below.

1. *Consistency and standards.* Users should not have to wonder whether different words, situations, or actions mean the same thing. Standards and conventions in product design should be followed.

   (a) Sequences of actions.
   (b) Color.
   (c) Layout and position.
   (d) Font, capitalization.
   (e) Terminology and language.
   (f) Standards.

2. *Visibility of system state.* Users should be informed about what is going on with the system through appropriate feedback and display of information.

   (a) What is the current state of the system?
   (b) What can be done in the current state?
   (c) Where can users go?

    (d) What change is made after an action?

3. *Match between system and world.* The image of the system perceived by users should match the model the users have about the system.

    (a) User model matches the system image.
    (b) Actions provided by the system should match actions performed by users.
    (c) Objects on the system should match objects of the task.

4. *Minimalist.* Any extraneous information is a distraction and a slow-down.

    (a) Less is more.
    (b) Simple is not equivalent to abstract and general.
    (c) Simple is efficient.
    (d) Progressive levels of detail.

5. *Minimize memory load.* Users should not be required to memorize a lot of information to carry out tasks. Memory load reduces users' capacity to carry out the main tasks.

    (a) Recognition vs. recall (e.g., menu vs. commands).
    (b) Externalize information through visualization.
    (c) Perceptual procedures.
    (d) Hierarchical structure.
    (e) Default values.
    (f) Concrete examples (DD/MM/YY, e.g., 10/20/99).
    (g) Generic rules and actions (e.g., drag objects).

6. *Informative feedback.* Users should be given prompt and informative feedback about their actions.

    (a) Information that can be directly perceived, interpreted, and evaluated.
    (b) Levels of feedback (novice and expert).
    (c) Concrete and specific, not abstract and general.
    (d) Response time.

        •   0.1 s for instantaneously reacting;
        •   1.0 s for uninterrupted flow of thought;
        •   10 s for the limit of attention.

7. *Flexibility and efficiency.* Users always learn and users are always different. Give users the flexibility of creating customization and shortcuts to accelerate their performance.

    (a) Shortcuts for experienced users.
    (b) Shortcuts or macros for frequently used operations.
    (c) Skill acquisition through chunking.
    (d) Examples include abbreviations, function keys, hot keys, command keys, macros, aliases, templates, type-ahead, bookmarks, hot links, history, default values, etc.

8. *Good error messages*. The messages should be informative enough such that users can understand the nature of errors, learn from errors, and recover from errors.

   (a) Phrased in clear language, avoid obscure codes. Example of obscure code: "system crashed, error code 147."
   (b) Precise, not vague or general. Example of general comment: "Cannot open.
   (c) Constructive.
   (d) Polite. Examples of impolite messages: "illegal user action," "job aborted," "system crashed," "fatal error," etc.

9. *Prevent errors*. It is always better to design interfaces that prevent errors from happening in the first place.

   (a) Interfaces that make errors impossible.
   (b) Avoid modes (e.g., vi, text wrap). Or use informative feedback, e.g., different sounds.
   (c) Execution error vs. evaluation error.
   (d) Various types of slips and mistakes.

10. *Clear closure*. Every task has a beginning and an end. Users should be clearly notified about the completion of a task.

   (a) Clear beginning, middle, and end.
   (b) Complete seven-stages of actions.
   (c) Clear feedback to indicate goals are achieved and current stacks of goals can be released. Examples of good closures include many dialogues.

11. *Reversible actions*. Users should be allowed to recover from errors. Reversible actions also encourage exploratory learning.

   (a) At different levels: a single action, a subtask, or a complete task.
   (b) Multiple steps.
   (c) Encourage exploratory learning.
   (d) Prevent serious errors.

12. *Use users' language*. The language should be always presented in a form understandable by the intended users.

   (a) Use standard meanings of words.
   (b) Specialized language for specialized groups.
   (c) User defined aliases.
   (d) Users' perspective. Example: "we have bought four tickets for you" (bad) vs. "you bought four tickets" (good).

13. *Users in control*. Do not give users that impression that they are controlled by the systems.

(a) Users are initiators of actors, not responders to actions.
(b) Avoid surprising actions, unexpected outcomes, tedious sequences of actions, etc.

14. *Help and documentation*. Always provide help when needed.

(a) Context-sensitive help.
(b) Four types of help.

- Task-oriented;
- Alphabetically ordered;
- Semantically organized;
- Search.

(c) Help embedded in contents.

This simple evaluation method can identify numerous usability problems quickly. For example, Hildebrand et al. (2010) asked five independent evaluators to use these heuristics while assessing the usability of a flexible endoscope. As illustrated in Fig. 4.1, the evaluators identified 324 unique usability problems. The evaluation identified that memory, feedback and visibility were the most common problems, which enabled future development efforts to focus on improving those aspects of the design.

Heuristic evaluation reports usually identify the problems found, the heuristics violated, and potential solutions to the problem. Screenshots or photographs of the problem often help communicate the issue.

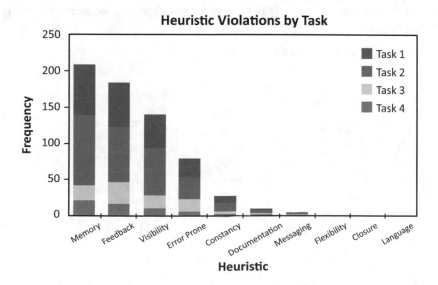

Fig. 4.1 Heuristic violations for endoscope reprocessing (Hildebrand et al., 2010)

Cognitive Walkthrough

Cognitive walkthroughs evaluate whether an interface is understandable and easy to learn, especially for a first-time or one-time user (Polson, Lewis, Rieman, & Wharton, 1992). To begin, the researcher develops representative tasks written from the user's perspective and described in a set of steps. Then, one or more evaluators work through tasks while asking themselves a set of questions (Wharton et al., 1994) such as:

- Will users want to produce whatever effect the action has?
- Will users see the control (button, menu, label, etc.) for the action?
- Once users find the control, will they recognize that it will produce the effect that they want?
- After the action is taken, will users understand the feedback they get, so they can confidently continue on to the next action?

The success of the interface is judged according to how well it helps or hinders users to achieve their goals. The researcher then critiques each step and evaluates it according to the aforementioned questions. The cognitive walkthrough was originally designed to evaluate walk-up-and-use systems like postal kiosks, automated teller machines (ATMs), and interactive exhibits in museums. However, it has also been employed successfully with more complex systems and works well for medical devices.

In a recent public workshop, we taught a group of HFE practitioners how to conduct a slightly modified type of cognitive walkthrough. The only modification was that we tried to simplify the questions the evaluator asked at each step. Participants were shown a hypothetical product like the one in Fig. 4.2. Specific design problems were purposely built into the prototype, and the questions evaluators asked were:

1. Would the user know what to do?

Fig. 4.2 Sample wireframe of heart pump screen, with purposely added usability problems

2. Would the user know how to do it?
3. Would the user recognize when it was done correctly (or incorrectly)?
4. Did the system behave as the user would expect?

Evaluators had a field day identifying problems (they were HFE engineers, after all), and several opined that this would save them a great deal of time; identifying these problems in an inspection method means they could be fixed before using real research participants.

4.3   Usability Testing

What Is Usability Testing?

In our previous discussion of human-centered design (HCD), we mentioned that, according to Gould and Lewis (1985), HCD has three tenets:

1. Early and constant focus on users and their tasks.
2. Using data to guide design decisions.
3. Iteration.

Often, the second item on that list comes from usability testing. In a usability test, representative users perform realistic tasks with a medical device, system, or process. The content of what's being tested (i.e., medical device, system, process) can vary in terms of its fidelity. For example, it's perfectly acceptable to conduct a usability test on something as simple as a crudely drawn paper prototype (i.e., low fidelity) all the way up to something as complex as a production-equivalent medical device (i.e., high fidelity).

Usability tests are almost always conducted in individual sessions. In other words, only one participant participates in a test at a time with one study facilitator (i.e., moderator). Meanwhile, one or more note takers, camera operators, and passive observers may view the study session from behind a one-way mirror or remotely if there is a broadcasted video stream (see Fig. 4.3).

One common exception to this "one participant per session" structure is when device responsibilities are shared between the patient and a caregiver. This dynamic is often found in an at-home setting among certain pediatric patient populations, older adults, or those with certain types of disabilities or impairments. For example, a 10-year-old pediatric patient with diabetes may be responsible for dialing up a dose and injecting themselves with insulin, but their parent or guardian may help interpret the dosing parameters and guidelines from the physician. In this situation, a usability study may benefit from having both the pediatric and caregiver participate jointly in the same usability study session.

Observers who are unfamiliar with user research may need to be (politely) reminded about how their behaviors or questions could influence the participant. Even when observers are hidden behind a one-way mirror, participants may still be

Fig. 4.3 Study observers and note takers in the observation room

able to hear them talking, laughing, or even the occasional cough or sneeze. You should never trust that the one-way mirror will block out all of the sound. These little cues remind participants that they are being "watched," which can add a great deal of stress on top of the pressure some participants feel from attempting tasks with an unfamiliar device. Figure 4.4 lists out several observation room guidelines for quick reference.

Relatedly, this up-front conversation with observers is a good opportunity to establish whether the study moderator is comfortable with observers asking questions at all. This recommendation mainly applies to usability studies where observers must be in the same room as the participants for logistical reasons. In almost all cases, observers should be instructed to hold their questions until the moderator concludes all of their study session objectives. Your authors have experienced a few situations where an overzealous project manager or designer will interrupt a participant mid-task when they see them do something "wrong". This is an understandable behavior for someone who is personally invested in the design or outcome of a medical device, but it's also fodder for bad research.

If you find this up-front conversation to be too intimidating with certain people (e.g., your boss, a client), there are a few ways to address it effectively. One way is to address the group as a whole before your first study session begins. This would include all the people on the research team, plus any observers. The advantage of this approach is that nobody feels singled out like the *cause* of the issue. Plus, it's a good reminder for experienced researchers too. If you still encounter unwanted behavior from observers, don't let the issue fester—address it early on. This might mean restating your initial request, or even suggesting that they move to another room if need be, asking an observer to leave your study if they cannot follow the rules. Though, the issue rarely escalates to this latter point. The goal should be to

All phones should be set to vibrate or silent while you are observing live participant sessions.

Use headphones if you need to listen to audio from your computer or other media devices.

Conversations should be minimized when possible. However, if you do need to chat with others in the room, please be mindful of how loud you speak as not to disturb participants.

Sometimes participants do or say odd things. Do your best to avoid audibly reacting to their behavior, so that they don't feel like they are doing something the wrong way.

Try to minimize how frequently you enter and exit the room. The light and noise from the door can be distracting for participants.

The one-way mirror works best when the observation room is kept as dark as possible. Please do not turn on the light, unless it's necessary.

The light from your laptop and other electronic devices can show through the one-way mirror if the brightness is set too high. Aim to keep it it low, but still useful for you.

Near the doorway, you will find a knob to control level of the lab room's volume. Adjust this accordingly — some participants are low talkers.

Fig. 4.4 Observation room guidelines

always remain calm and professional—keep the attitude that you want everyone to work together to facilitate a good study.

When you have observers in your observation room (or in the same room as your participant), it's recommended that you address ground rules. This will minimize the risk of the participant hearing laughter or catching an odd look that in turn may cause them to lose confidence or leave the room in frustration or embarrassment.

Researchers observe and keep track of a few different types of quantitative and qualitative measures during a usability test. On the quantitative side of things, they might keep a tally of task successes, close calls, difficulties, and use-errors. Qualitatively, they'll explore factors like the root cause(s) of deleterious task outcomes, as well as the participant's perceived ease of use of the device or system itself.

Usability Study Tips and Pitfalls

Before we get too far into our discussion of usability testing, here are a few general tips to help you conduct an effective usability study. Note that these tips also broadly apply to many other types of research methods discussed elsewhere in this book too. Try to keep these in the back of your mind as you read through the rest of this chapter.

Seven Tips for a Successful Usability Test

1. Make the participant feel comfortable, wanted, and respected—they are your *expert.*
2. Identify the study scope and problem statement a priori.
3. Develop an easy-to-follow task table and moderator's guide.
4. Ensure you have enough time for each session.
5. Remain calm, objective, and nonreactionary.
6. Build a rapport and define responsibilities with your notetaker (and observers).
7. Develop an easy and quick way to manage tasks that need debriefing.

Top Seven Mistakes in Usability Test

1. Just giving a demo of the device or system.
2. Developers and engineers testing their own products.
3. Selecting only the "best" participants.
4. Failing to pilot test your usability study protocol.
5. Failing to identify task completion criteria.
6. Trying to moderate and take notes at the same time.
7. Failing to debrief on observed difficulties, use-errors, etc.

Categories of Usability Tests

HFE practitioners in healthcare usually place usability testing in two broad categories: formative and validation. Formative usability testing is conducted early and iteratively throughout product design. In general, formatives are aimed at identifying usability problems so that they can be fixed in successive iterations of the user interface. In the initial design phases, formative usability tests can be conducted with paper prototypes, low fidelity digital prototypes (e.g., via Adobe XD), or higher fidelity prototypes as they become available.

Importantly, there are no hard and fast rules for how a formative usability test *must be* conducted. It's up to the research team to determine what tasks and features are evaluated, how many participants are included, where the test is conducted, and what and how metrics are scored. For example, researchers might design a formative usability test to only investigate the perceived satisfaction of a medical device among eight participants. In a different study, they might look at safety, effectiveness, and satisfaction altogether with a sample of six participants.

Validation studies (also referred to as a "summative") are more formal and focus on safety and effectiveness. For example, for FDA submission purposes, a validation usability test must adhere to the following criteria unless there are good reasons that this isn't possible (U.S. Food and Drug Administration, 2016):

- Evaluates all safety "critical" tasks defined in the device's Use Risk Analysis (URA).
- At least 15 representative end-users from *each* user group.
- Representative use-environment (e.g., materials, lighting, sound, distractions).
- Production-equivalent device and labeling (e.g., IFU, training materials) (Fig. 4.5).

Fig. 4.5 A researcher conducting a usability study with a participant, using a medical mannequin to help identify and discuss anatomical landmarks

Beyond the lens of this "formative" vs. "validation" distinction, there are other types of usability test categories worth considering. In fact, the two types described a moment ago only represent one of three categories of usability tests (a category referred to as "assessment"). Rubin and Chisnell (2008) propose that the three types of usability tests include the following:

Exploratory—Users conduct tasks with low fidelity simulations of the user interface. These simulations might be as simple as paper prototypes or foam core models. Exploratory tests often use think aloud techniques to identify where users might have difficulty or misunderstanding of how the system works.

Comparison—In a comparison test, two or more design alternatives or even competing products are used to conduct realistic tasks. This enables the researcher to understand which design alternative or product is better overall in terms of usability, as well as what things each alternative does well, and what things it does poorly.

Assessment—Assessment tests are designed to identify usability problems early and continuously throughout the design cycle. This is what we typically think of when we think about formative and validation usability testing. Assessment tests can identify the parts of an interface that confuse people (and why), so the design team can develop mitigations against them in the future.

We have noticed that in practice, exploratory and comparison tests are underutilized, especially given the value they can provide. The exploratory test often provides you with information needed for later assessment tests. And, comparison tests can provide information about where you stand in relation to your competition and how to either catch up or capitalize. It is a shame these are not used more often.

Components of a Usability Test

A usability test is made up of scenarios, tasks, and subtasks, each one progressively smaller in scope and (often) how long they will take a participant to complete. The focus is on providing insights into the mindset and working methods of end-users as it relates to a specific device, system, or process (Nielsen & Landauer, 1993). The following subsections provide an overview of scenarios, tasks, and subtasks, as well as a few tips and tricks to help make them effective in a usability study.

Scenarios

Usability test scenarios provide context for the tasks that the participants will perform, and are written to provide all the information necessary to complete them. It's important to take the time up front to carefully plan out the scenarios so that all moderators are comfortable with what they will say.

Scenarios should be as realistic as possible, given the resources available. Every now and again you'll hear a participant say at the end of a long (and painful) usability test that the reason they didn't follow the study scenario was because, "they didn't think it mattered because it's a simulation," or "that would just never happen at my facility". The testing scenarios can have a lot to do with those comments. Effective test scenarios require knowledge about what users deal with on a day-to-day basis. If you aren't sure about certain details, consider consulting with your colleagues or a Subject Matter Expert (SME) while planning your study.

A good scenario says what it needs to say and nothing more. Here are a few things that make for an effective scenario—some of which are borrowed from (Rubin & Chisnell, 2008):

- Use simple language.
- Use brief sentence(s).
- Use statements/commands (not questions).
- Identify language that is consistently interpreted.
- Give a reason for performance.
- Use actual data and names.
- Use realistic events or circumstances the user normally experiences.
- Provide a clear goal or end point to the scenario.
- Avoid influencing or "coaching" the participant.
- Avoid jargon or internal company references.
- Avoid using the same wording or cues found on the user interface.
- Begin at a natural breaking point in a workflow.
- Cover (slightly) more ground than the area of interest in the study.

This last point is worth discussing slightly further. By "covering more ground," we are referring to having the participant complete a few more tasks before the topic your team wants to study as it relates to usability. To help illustrate this point, let's imagine that you are working with a medical device team that is developing a new type of syringe injector. The drug going into this syringe has been recently modified to be less viscous than before. Your team has discovered that this thicker fluid requires that the user must push harder on the syringe plunger during an injection.

Rather than simply filling the syringe for the participant and asking them to inject it into a mannequin or injection pad, consider having them complete the steps leading up to this point themselves. For example, this might involve assembling components, and drawing medication into the syringe from a vial. These initial steps can help the participant get a feel for the force levels involved, which in turn, can help avoid the "startle" of discovering the injector resistance on the target task itself. Plus, it's more similar to what a user would do in the real world.

It is also important to think about the order in which participants are presented with scenarios. There are three ways to do this:

- *Natural order*—Scenarios (and their tasks) tend to unfold into each other naturally. For example, in reprocessing a flexible endoscope, a precleaning phase will precede the cleaning solution and high-level disinfectant phases. If conducting a second scenario depends on obtaining the results of the first, then you should follow that natural order.

- *Importance*—Sometimes you run out of time in a usability session, and you want to make sure that the most important scenarios (the one you were most interested in or concerned about) actually gets done. It is best to make sure this is one of the first few tasks. Consider rearranging scenarios so there is ample time to cover the most important ones.
- *Randomization or counterbalancing*—Randomization and counterbalancing the scenario order can help limit experimenter bias and learning effects. This is a good idea if you think interacting with the system on one task will make it easier to conduct the following tasks.

Tasks

Participants complete tasks in pursuit of completing the overarching scenario. In other words, several tasks can comprise a scenario. Here's a quick example to demonstrate the distinction between a scenario and a task:

Scenario: "You have recently completed a blood draw on your patient. Use the system to produce a 20 mL sample of platelet rich plasma (PRP) using the blood sample collected from your patient."
Task: [participant realizes they need to…] Assemble the transfer needle and syringe.

In this example, the participant must complete this task in order to complete the scenario. Any other method of completing this scenario (e.g., using off-label products) would not count as a success—though, it still may technically result in producing the desired output.

Tasks must have clear success criteria to allow personnel to determine whether tasks were performed correctly or incorrectly. Pragmatically speaking, you might not be able to test all of the tasks that users can accomplish with your device all in one usability test. You will need to evaluate these tasks over time in different tests. So, how do you choose which tasks to test first? Consider testing tasks that:

- Are associated with critical harm (i.e., safe and effective use).
- Are essential to the operation of the device.
- Are frequent.
- Are conducted under time pressure.
- Are new or recently modified.
- Seem to require a great deal of explanation (i.e., during training, past study debriefs).
- Seem to produce errors that are difficult to recover from.
- You intuitively think might cause difficulty for users.
- Tasks that designers suggest or are wary of.

With enough iterations of formative usability tests, you should be able to cover all of your tasks at least once. As you approach the last one or two formative usability tests prior to your validation study, consider running those as the "real deal"; meaning, evaluate *all* of your (critical) tasks from start to finish, just like you would in a validation study. You might also consider including tasks that are on the cusp of criticality as well, on the off chance that you need to revise your risk assessments later on in a product's development as you learn more about your users and their use-environments.

Subtasks

Subtasks are the building blocks of tasks; one task may have many or just a few subtasks. Usually, subtasks are not "scored" like a task would be. Instead, a subtask helps isolate any observed or reported issues to specific points in a task. Here's a quick example using the same information listed above for consistency purposes:

Task: Assemble the transfer needle and syringe.
Subtask #1: Remove the transfer syringe from its packaging.
Subtask #2: Remove the needle guard from the syringe.
Subtask #3: Identify the correct needle spacer based on the injection parameters.
Subtask #4: Place the needle spacer on the transfer syringe's needle.

The exact granularity between tasks and subtasks can be confusing. Sometimes you will have a strong "gut" reaction to what is a task and what is a subtask. Consider working with a (small) team to figure this out.

What Is Measured in a Usability Test?

Here's a quick overview of the most common measures scored and/or collected during a usability test.

- *Success/fail*—This binary process for noting performance outcomes is pretty straightforward in most cases.
- *Success, difficulty, close call, and use-errors*—A multi-part performance scoring system like this one affords better granularity of details. It also happens to match up well with the FDA's expectations for validation usability test reports for submission purposes. The FDA guidance for human actors research provides general guidelines on what each of these categories mean. "Success" and "Use-error" are easy to understand, though, practitioners often debate the differences between a "close-call" and a "difficulty". It's up to the researcher's judgment to place observations into appropriate categories. In many cases, this will depend on what is discussed during the debriefing with the participant while identifying potential root causes to issues.
- *Time on Task (ToT)*—The exact amount of time that a participant takes to complete a task. For ToT to be effective, researchers must have a careful and consistent way of monitoring time, meaning that each task must have clear starting and stopping points.

- *Post-task and post-test questionnaires*—Surveys and questionnaires that are administered to participants following completion of a task or the entire study, respectively. One widely used option is the System Usability Scale (SUS), which is a usability measurement tool consisting of 10 Likert scale items. These particular items have been carefully selected and refined over the past several decades, and have been translated into multiple languages successfully (see Gao, 2019; Sharfina & Santoso, 2016). Once sold internally to stakeholders as a "quick and dirty" measurement of (perceived) usability (Brooke, 1996, 2013), the SUS has become a staple measurement tool across many domains and industries. Several other off-the-shelf questionnaires exist for assessing the usability of a device as well. Consider reviewing the resources listed at the end of this chapter for other recommendations. Of course, you can always create your own questions if these options don't suffice.

How Many Participants Do You Need for a (Formative) Usability Test?

The question of how many participants to include in a research study is a topic of widespread discussion in HFE (and beyond). As this question pertains to usability testing, several researchers have set out to find a solution over the past decades. Here are a few of the main take-aways from this research:

- Some findings indicate that one participant can uncover about 1/3 of the total usability issues (see Nielsen & Landauer, 1993). This, however, is a broad recommendation that stems from studies conducted outside the context of medical devices. For example, many of these pioneering studies focused on word processing, timesheets, and general computer software issues. Your results will vary as a function of things like your device's complexity and how many of these usability "bugs" your team has already caught and fixed.
- Participants often "uncover" the same usability issues as each other. Expect to discuss the same problem over and over again with different participants in the same study. If you are lucky, a participant will encounter a unique issue no other participant has seen. But this is less common than you would hope in a lot of cases.
- Participants tend to find the most egregious usability issues right away (Virzi, 1992). For example, white text on a light gray background is an easy issue to spot.
- Although some usability issues occur in pairs (i.e., cascading failures), most models estimating usability study sample sizes do not take this into account. Instead, they assume that usability issues are mutually exclusive.
- Several authors suggest about 80% of usability issues can be detected with a sample size of four to five participants (see Nielsen & Landauer, 1993; Virzi, 1992). Importantly, however, other researchers argue that these findings are too

generous, and that four to five participants may only uncover 25–35% of the total issues in a system farther along in development (see Faulkner, 2003; Hudson, 2001; Macefield, 2009; Perfetti, 2001; Spool & Schroeder, 2001).

Estimating the number of participants needed is a function of your device or system's unique probability of finding an error in the first place (see Fig. 4.6). For example, if device A has been researched for months and has many of the kinks worked out already, the probability of a participant finding a new kink is fairly low (though, *plausible*). On the other hand, if device B is in its first iteration and has never been tested before, there are likely lots of kinks to be found. You might think of the blue or orange line in Fig. 4.6 as a "production-equivalent" device, whereas the red or green line are more representative of a medical device in its first few iterations.

Nielsen and Landauer (1993) propose that estimating the number of participants needed to find usability issues can be explained by a Poisson model. They recommend the following formula for calculating sample size for usability studies:

$$\text{Usability issues found}(i) = N\left(1 - (1 - \lambda)^i\right)$$

- N is the total number of usability problems in the device or system.
- λ is the probability of finding the average usability problem when running a single, average subject (Nielsen & Landauer, 1993).
- i is the number of participants (i.e., "usability issues found *as a function of* eight participants").

Based on Fig. 4.6, we can estimate that it would take about seven to eight participants in a usability study to identify approximately 80% of the issues when each individual participant uncovers an average of 20% of issues. It would take well over 30 participants to identify the same percentage of total issues when each participant's

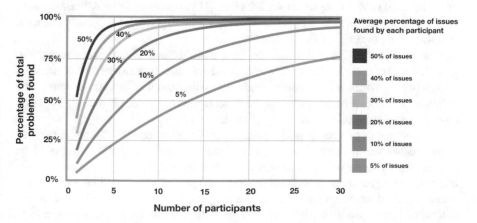

Fig. 4.6 Estimating sample size needs—percentage of total usability issues found vs. number of participants required. (Data are based on the formula described in Nielsen & Landauer, 1993)

hit rate for issues is only 5%. Think of each line on this graph as a representation of each iteration in a medical device's development cycle. The product is getting progressively more "usable," and as a result, it gets increasingly difficult for a participant in a usability study to detect issues. Of course, you will not know the magnitude or total count of all the usability issues a priori. However, estimates can be made with increasing accuracy as you move further into the product development cycle and complete several usability studies.

The guiding principle here is this: expect that a product (or feature) that is further along in development will require more participants in an upcoming formative usability study than it did in the last one. Start with just a few participants in your early studies (i.e., four to five participants), then gradually increase your sample size as your device (or feature) is iterated upon and improved (Wiklund, Michael, Kendler, & Strochlic, 2011). The assumption here is that since things are becoming "better" over time, the issues yet to be discovered are increasingly remote and difficult to find. You simply need more people to uncover them.

Training Prior to Usability Testing

Device training is often a necessary prerequisite to using a medical device. In some cases, training can be delivered in self-paced videos that users view prior to initial use. Other times, a more formalized version of training is provided by a qualified medical worker or a representative from the manufacturer. All of this depends on how, or if, the manufacturer intends to communicate and manage training with its users.

Although formative usability testing does not require training per se, it's usually a good idea to include it if the manufacturer provides it (or intends to provide it) in the real world. After all, the FDA views training materials as one of the many parts of the medical device or system user interface (U.S. Food and Drug Administration, 2016). It's a good idea to test it out at least a few times prior to your validation usability study (see Fig. 4.7).

One important point about training in a usability test is that you should also include a decay period at the conclusion of the training. A decay period is simply a gap in time between when training ends and when the usability test begins. The length of the decay period is up to the researchers to decide; however, the FDA advises a minimum 60-minute decay period.

The purpose of this decay is to replicate the "information loss" that occurs when users are trained in the real world. For some devices and systems, it would be unusual for a user to receive training, then immediately start using the device. For example, a surgeon new to Robotic Assisted Surgery (RAS) would never receive a 1-day "crash course" on how to use the surgical system, then operate on a real patient with that system the next day. Realistically, it would take dozens or hundreds of hours working with this system before they would even think about taking on a patient case. And, even then, the standard practice would be to have several

Fig. 4.7 System training on a smartphone app (moderator was provided with a training checklist to review with the participant)

preliminary cases proctored by a surgeon with substantial experience on that system (Van Der Sluis, Ruurda, van der Horst, Goense, & van Hillegersberg, 2018; Zorn et al., 2009). Of course, the opposite pattern with respect to training is found as well. For example, a person with diabetes may receive training from a nurse educator on a new blood glucose testing unit, and begin using it just hours later on their own.

Deciding on a satisfactory decay period depends on factors such as the ones below. The goal for a valid and realistic usability study is to balance realism with these logistical constraints:

- What is a realistic timeframe between training and initial device use for your device/system in the real world?
- Can training and evaluation be split up over multiple days or weeks?
- What would be the resource (i.e., time, cost, opportunity) burden by having a longer decay period?
- Are necessary training resources (e.g., simulators) logistically moveable?
- What is the trainer(s) availability?
- What is the researcher(s) availability?

Estimating Time Needs for a Usability Test

There is a famous saying that is often misattributed to Abraham Lincoln. The saying goes something like this: "Give me 6 hours to cut down a tree, and I'll spend the first 4 hours sharpening my axe." (quick aside: in what world does it take 4 h to sharpen an axe?!). Although this saying never actually came from the United States 16th president—it likely came from Jaccard (1956)—its sentiment is still a good one: the more time you spend preparing for a task, the easier that task will be. Indeed, the same can be said about preparing for a usability test. Here are a few tips and considerations for estimating how much time you will need for a usability test session.

Internal Pilot Testing

Once you have your final materials ready to go, consider running a full usability study test session with a colleague to get a sense of how long things will take. Plan to run through all of your materials, including the introduction script and notes about the study, all the way through any debrief questions you have at the conclusion. And, treat it like the real deal (even though it might seem odd). Although you are not collecting or analyzing data from this session, it will help you figure out where slow spots exist in the study, as well as ways to rectify them.

This is also a great chance to test out all of your study-related equipment in action. For example, this might entail running the usability lab's Pan Tilt Zoom (PTZ) cameras, streaming audio and video, and practicing communication strategies with your note taker in the observation room. You'd be surprised at how often these little details get overlooked until minutes before an actual participant usability study session begins (Fig. 4.8).

Fig. 4.8 A PTZ camera on a slide mechanism from Intuitive Surgical's human factors usability lab allows for easy adjustments to the camera's placement in a surgical lab environment (Image courtesy of Intuitive Surgical, Inc. with permission)

Plan for Extra Time During Remote Usability Testing

Remote usability testing is increasingly popular in HFE, as it eliminates many of the costs and headaches associated with travel. Often, however, remote research will take longer to set up and get your participant oriented. Delays can happen due to things like the participant forgetting to install the host computer program, not having requisite materials downloaded, or even having to complete a mandatory system update and restart mid-study.

Less tech-savvy participants struggle most with remote usability tests. If you know ahead of time that your user population may have difficulties with computer or smartphone use, consider developing a walkthrough guide or video for setup ahead of time and emailing it to them. Alternatively, set aside extra time in the beginning and end of the study session to help them install and remove software, respectively.

"It's an Easy to Use System (I Swear)"

We've worked with many companies who have developed "production-equivalent" medical devices before our involvement in usability testing. Sometimes, these companies are so confident in their product's superiority that they lose sight of whether the device is even usable. Not surprisingly, these companies have high expectations for their usability tests as well. Lo and behold, however, the results are occasionally less than ideal. In some cases, they are flat out bad.

When budgeting for time for a usability test, don't take what others say at face value. They may be biased. They might not know what usability testing entails. They may forget about training time, or the need for a decay period. They may expect for you to stop participants mid-task and give them the correct answer at the slightest indication of confusion.

Instead, treat each iteration of a device, system, or process as an independent thing. Ensure that you have enough time to get your hands on the device and related materials to help you estimate what issues a real user might encounter. Generally speaking, more issues means you'll need more time to debrief the participant to uncover *why* they did what they did. If you rely on someone else's judgment for these estimates, you may schedule sessions too short and not get through everything you need.

Complicated Scenario Setups

Some usability studies require careful arrangement and orchestration of test materials before or during a participant session. For example, you may need to empty fluid containers, fill sinks, change system parameters on a GUI, or implement "work arounds" to trigger alarms or quasi-hazardous events. These activities take time away from the study session, especially if the participant cannot work on a separate task while the study moderator sets up. Be sure to set aside extra time for this choreography.

True Pilot Test

Run at least one (real) pilot participant. A pilot participant is an actual end-user of the medical device or system you plan to test. Unlike a colleague standing in as an internal pilot, this participant can give you the most accurate sense of how other users will perform. Treat their session like the real thing—introduction, training, scoring, breaks, follow-up questions, etc. Depending on whether you want this usability test to be similar to a validation study, you may choose to be as "hands off" as possible. Understand that this choice can lead to some participants to go down some serious rabbit holes. Your session length needs to account for this (real) possibility.

Consider running your pilot participants at least a few days before the real participants. This gap will give you a little bit of time to modify the study schedule in the event that more (or less) time is needed. Afterwards, you and your team can contact your scheduled participants with an update. Though, be mindful and respectful of participants that are traveling long distances for your usability study. For example, a last-minute change request for someone flying across the country in a few days would not be wise.

Plan for the Worst, Hope for the Best

It can be difficult to estimate time requirements for especially long, large-scale usability studies. Every task adds one more possibility of an issue being observed, adding to the laundry list of tasks that the participant and moderator may need to debrief at the end of the session. As a result, your best and worst performing participants will have incredibly different usability study session lengths.

As a guiding principle, plan all of your usability study sessions as if they will be your worst. This will help you provide enough time on the off chance this situation comes to fruition. As painful as these error-riddled sessions can be from a moderating standpoint, they are also the most helpful and informative sessions. In all likelihood, they reveal new insights about use-errors or risks. You don't want to short-change this learning process by running out of time.

The Iceberg Paradox

As discussed earlier, there are two types of usability tests with respect to medical devices: formative and validation. Interestingly, however, the latter receives disproportionate attention, resources, and fear among device manufacturers. In some ways, this focus on validation testing does have merit. After all, it's similar to the final exam at the conclusion of an intensive college course. Everything else before it—all the quizzes, homework, and tests—matter little if you bomb the final exam. The same can be said about a validation usability test for submission to the FDA. If things go belly up during your validation study, that could mean the difference between your device receiving FDA clearance or not.

Formative usability testing, on the other hand, can and often will go "poorly". But that's okay. These tests are there to literally "form" the design specifications of a device. More specifically, these studies help ferret out issues (observed or anticipated) and supply your team with the information it needs to mitigate against them in the future. Even if you observe formative usability test outcomes that are less than stellar, these results do not bring your proverbial grade down in the eyes of the FDA. On the other hand, if you observe a pattern of issues and do nothing to mitigate them, then that's another problem entirely.

The iceberg paradox of device testing stems from this relationship. In the majority of cases, manufacturers will conduct one validation usability test, but potentially dozens of formative usability tests beforehand. In a sense, manufacturers fear the one test they know is ultimately lying in wait in the murky waters ahead. They can see it; they know the dangers it presents to their device's success. Yet many do not apply the same standards and scrutiny to the multitude of formative tests leading up to this icy precipice. Furthermore, they tend to ignore or dismiss key findings and trends from formative usability studies that indicate danger ahead (i.e., running into large chunks of ice upon approach).

Instead, formative usability tests are often viewed too casually, with little regard for how they're conducted, which tasks are evaluated, or the experience level of the people conducting the studies. In some cases, suboptimal findings are dismissed altogether as irrelevant or inconsequential, rather than causing the team to band together and develop solutions. It's almost as if the formative usability study is a checkbox some manufacturers feel obligated to tick as it makes their HF/E submission to the FDA appear more robust and thorough (Fig. 4.9).

Similar to an iceberg, many medical device teams and manufacturers focus only on what they can see; forgetting that the bulk of the work lies beneath the surface. In the authors' experience, the most successful design teams embrace early and iterative research efforts. They invite uncertainty and defeat, as it means they have a new and unexpected opportunity to turn a good device into a great one. They are also the ones that view the researchers not as the "people telling us what we're doing wrong," but rather as the people who have helped us collect information we need. This shift in perspective facilitates success. Once this shift occurs, teams begin to see their validation usability study less as an obstacle, and more as a victory lap; a testament to the hard work, ingenuity, and grit that's gone into the medical device.

Fig. 4.9 Full view of an iceberg (Image by Romolo Tavani/shutterstock.com)

Counterproductive Outlooks About Formative Usability Testing

Of course, the rose-colored perspective described above is less common than you would hope. Unfortunately, HFE must continue to fight for its position in the medical device industry. Here are a few examples of (short-sighted) attitudes that tend to come up from inexperienced HFE practitioners, project managers, and other executive stakeholders during the final few weeks or months preceding validation testing. Each example is followed by a response explaining why and where the thinking is flawed.

"All of our other formative usability studies up to this point have only had 5 participants in them. Why should we spend more time and money now to run more participants? Our validation usability study is already going to cost us an arm and a leg as it is!"

First off: we get it—usability studies take time and money. If you're part of a start-up, an extra $10K can mean life or death for your company, especially towards the end of development. But five participants may simply not be enough to detect all the usability issues rooted in a device, system, or process. Indeed, many studies have demonstrated this point over the past several decades (see Faulkner, 2003; Hudson, 2001; Macefield, 2009; Nielsen & Landauer, 1993; Perfetti, 2001; Spool & Schroeder, 2001).

Difficulties and use-errors are like targets of varying size. Your participants are like archers shooting arrows. The catch is: *you* don't know where those targets are "hidden," and all of your archers are blindfolded. The most glaring difficulties and use-errors will receive a lot of hits from a lot of participants. But it may take a lot of archers, time, and a lucky stray arrow to hit the smallest targets. You don't want those rogue targets to be hit during your validation study when you are running three times as many people as your earlier formatives. And, you certainly don't want this issue to come to bear after your product goes to market when even more users get their hands on it.

"We have an internal company product fair coming up in Toledo next week. Let's send some HFE people out there and run a quick formative usability study with our sales team before the validation."

There is a classic saying in human-centered design: you are not the user. And, neither is your sales colleague at the product fair in Toledo. As the designer of a medical product, there is a good possibility that you know more about your device than your user will (for a while at least). Occasionally, you will know more about all the devices that your users interact with on a regular basis than they do. People on your team—even those not directly involved in product development—will know and do things differently than your actual users. Sometimes they do things better. Sometimes a lot worse. It's not worth chancing the results of this quasi-formative usability study on people who do not live and breathe the same constraints, background, education, and capabilities as your users or user groups.

"Our validation study is coming up soon. The development team just completed a system wide software design change that affects the shape, location, and label of several buttons in the GUI. For the sake of time, let's just spot check those changes through a usability test on a couple of the main tasks users would complete. The team feels pretty confident it's not going to have an impact on testing anyways."

There are a couple of easy to miss issues with the perspective described above. The first one is the fact that this person wants to "spot check" a system wide change. There is no telling if these design changes impact some tasks more than others. This could be in the form of not sufficiently mitigating an issue in a particular task, or possibly introducing new issues altogether.

The second problem is an attitudinal one. If you are part of the design team on a medical device, it can be hard to tell where your device has shortcomings. It's personal. You've worked on it for months or years. Many of your colleagues are in the same position as well. You have to be willing to accept the possibility that you (and your team) may be overconfident about the positive impact that the latest design idea will afford. The take-home message here is that despite your colleagues' confidence, you need to remain objective and weigh all relevant options.

"Your HFE team can run a final formative usability study if you want, but we need the full amount of time between now and the validation usability study to complete the design. Any findings you have will just go into the 2.0 version following FDA approval and commercial launch."

This one can be especially painful and personal to hear. It feels as if the person is saying, "you go play in your own sandbox, while us cool kids play in this one". It also feels like this person doesn't really understand the purpose or value of usability testing. (Which, may or may not be true). For what it's worth, this is a perennial issue in medical device design. Your team runs short on time and scrambles to meet the deadline.

The best strategy is a proactive one: talk with the project manager or team as a whole and figure out when specific design milestones will be ready for formative testing. Make sure they build in time for a design freeze or parallel paths during your study. Work with the team as you approach these dates and create your test protocol based on what's been concretized. Then, communicate when the team can expect your report. This discussion should clarify the scope and type of results your report will include. The fluffier and less defined you are here, the less confidence and value they'll see in setting aside this time for testing.

Lastly, be sure that your project manager remains on top of deadlines that put your formative usability tests at risk. You cannot do your job, if you don't have things to evaluate. The long-term objective—like any member on a design team—is to demonstrate your value and contribution to the team's objectives. Some of the best designers and developers we've worked with are ones who bring up the topic of usability testing themselves. No pulling teeth. No twisting arms. They do the arguing for you.

Of course, not every team will have followed the proactive strategy described above. You should also have a reactive strategy in your back pocket as well. A good approach is to start by taking stock of the questions listed below. These will help you and your team decide the feasibility and value of conducting a last-minute formative study before the design is finalized before the validation study. The key is striking a balance between all three, and remaining flexible.

- What were the outcome patterns across past formative usability studies?
- Where are the greatest risks according to your use risk analysis (URA)?
- What design or device aspects are still flexible?

An additional, long-term strategy is to have these naysayers view a usability study first-hand. Generally speaking, an in-person session is more effective than a remote one, because you know for sure that you have their undivided attention. Make their passive observation a more active role by encouraging feedback and questions from them in between study sessions. Having them weigh in with their thoughts is a great way to build rapport and confidence among those who may not have seen value in usability testing. Anecdotally, your authors can tell you that we've heard the same thing from dozens of (initially skeptical) stakeholders, project managers, designers, biomedical engineers, and executives:

"I learned more about the issues with our device from watching this one participant than months of team meetings and multiple reports. I get it now."

Resources

- Kortum, P. (2016). *Usability assessment: How to measure the usability of products, services, and systems*. USA: Human Factors and Ergonomics Society.
- Nielsen, J. (1994, April). Usability inspection methods. In *Conference companion on human factors in computing systems* (pp. 413–414).

- Rubin, J., & Chisnell, D. (2008). *Handbook of usability testing: How to plan, design and conduct effective tests*. Indianapolis, IN: Wiley.
- Wiklund, M. E., & Wilcox, S. B. (2005). *Designing usability into medical products*. Boca Raton, FL: CRC Press.
- Wiklund, M., Kendler, J., & Strochlic, A. Y. (2015). Usability testing of medical devices. Boca Raton, FL: CRC Press.

References

Brooke, J. (1996). SUS: A "quick and dirty usability". *Usability Evaluation in Industry, 189*.

Brooke, J. (2013). SUS: A retrospective. *Journal of Usability Studies, 8*(2), 29–40.

Faulkner, L. (2003). Beyond the five-user assumption: Benefits of increased sample sizes in usability testing. *Behavior Research Methods, Instruments, & Computers, 35*(3), 379–383.

Gao, M. (2019). Multi-cultural usability assessment with system usability scale. Doctoral dissertation, Rice University.

Gould, J. D., & Lewis, C. (1985). Designing for usability: Key principles and what designers think. *Communications of the ACM, 28*(3), 300–311.

Hildebrand, E. A., Branaghan, R. J., Wu, Q., Jolly, J., Garland, T. B., Taggart, M., et al. (2010, September). Exploring human factors in endoscope reprocessing. In *Proceedings of the human factors and ergonomics society annual meeting* (Vol. 54, No. 12, pp. 894–898). Sage, CA/Los Angeles, CA: SAGE Publications.

Hudson, W. (2001, May/June). *How many users does it take to change a website?* SIGCHI Bulletin.

Jaccard, C. R. (1956). Objectives and philosophy of public affairs education (No. 761-2016-51596, pp. 12–18).

Kortum, P. (2016). *Usability assessment: How to measure the usability of products, services, and systems*. USA: Human Factors and Ergonomics Society.

Macefield, R. (2009). How to specify the participant group size for usability studies: A practitioner's guide. *Journal of Usability Studies, 5*(1), 34–45.

Nielsen, J. (1994). Heuristic evaluation, w: J. Nielsen and R. L. Mack (Eds.), Usability inspection methods.

Nielsen, J., & Landauer, T. K. (1993). A mathematical model of the finding of usability problems. In *Proceedings of the INTERACT'93 and CHI'93 conference on Human factors in computing systems* (pp. 206–213).

Nielsen, J., & Molich, R. (1990). Heuristic evaluation of user interfaces. In: *Proceedings of ACM CHI'90* (pp. 249–256).

Perfetti, C. (2001, June). Eight is not enough. Retrieved August 23, 2020 from https://articles.uie.com/eight_is_not_enough/.

Polson, P. G., Lewis, C., Rieman, J., & Wharton, C. (1992). Cognitive walkthroughs: A method for theory-based evaluation of user interfaces. *International Journal of Man-Machine Studies, 36*(5), 741–773.

Rubin, J., & Chisnell, D. (2008). *Handbook of usability testing: How to plan, design, and conduct effective tests*. Indianapolis, IN: Wiley.

Sharfina, Z., & Santoso, H. B. (2016, October). An Indonesian adaptation of the system usability scale (SUS). In *2016 International Conference on Advanced Computer Science and Information Systems (ICACSIS)* (pp. 145–148). Washington, DC: IEEE.

Shneiderman, B. (1998). *Designing the user interface* (3rd ed.). Boston: Addison-Wesley.

Spool, J., & Schroeder, W. (2001). *Testing web sites: Five users is nowhere near enough in CHI 20 Extended Abstracts* (pp. 285–286). New York: ACM Press.

U.S. Food and Drug Administration. (2016). *Applying human factors and usability engineering to medical devices: Guidance for industry and Food and Drug Administration staff*. Washington, DC: U.S. Department of Health and Human Services Food and Drug Administration, Center for Devices and Radiological Health, Office of Device Evaluation.

Van Der Sluis, P. C., Ruurda, J. P., van der Horst, S., Goense, L., & van Hillegersberg, R. (2018). Learning curve for robot-assisted minimally invasive thoracoscopic esophagectomy: Results from 312 cases. *The Annals of Thoracic Surgery, 106*(1), 264–271.

Virzi, R. A. (1992). Refining the test phase of usability evaluation: How many subjects is enough? *Human Factors, 34*(4), 457–468.

Wiklund, P. E., Michael, E., Kendler, J., & Strochlic, A. Y. (2011). *Usability testing of medical devices*. Boca Raton, FL: CRC Press.

Wharton, C., Rieman, J., Lewis, C., & Polson, P. (1994). The cognitive walkthrough method: A practitioner's guide. In Usability inspection methods (pp. 105–140).

Zhang, J., Johnson, T. R., Patel, V. L., Paige, D. L., & Kubose, T. (2003). Using usability heuristics to evaluate patient safety of medical devices. *Journal of Biomedical Informatics, 36*(1–2), 23–30.

Zorn, K. C., Gautam, G., Shalhav, A. L., Clayman, R. V., Ahlering, T. E., Albala, D. M., et al. (2009). Training, credentialing, proctoring and medicolegal risks of robotic urological surgery: Recommendations of the society of urologic robotic surgeons. *The Journal of Urology, 182*(3), 1126–1132.

Chapter 5
Visual Perception

5.1 Information Processing

Every moment, day or night, awake or asleep, we are battered with sensory stimulation from the world. Eyes are stimulated by photons of electromagnetic energy, ears by vibrations in the air, the nose and tongue by chemical molecules, and the skin by pressure and temperature changes. This process, called *sensation*, delivers information from the world to our sense organs. We then combine these sensations, with previously stored knowledge to make sense (*perceive*) the stimuli. Next, we can think about, manipulate, and make decisions about the information, until finally we choose a response and actually execute the response. Figure 5.1 provides a model of how we do these things. Obviously, this is done by neurons, electrical impulses, and chemical activity, rather than boxes and arrows, but the model shows the processes so we can discuss the capabilities and limitations of each mechanism.

The icons on the left represent stimuli (e.g., light, sound) impinging on our sensory organs (e.g., eyes, ear, skin). These stimuli are stored for a brief time in a sensory register, a high capacity, but brief duration, holding cell. Since most stimuli we encounter are irrelevant to us at any one time, they do not capture our attention and are not selected for further processing. Instead, they decay quickly from the sensory register. This is what happens when you hear an alarm so many times that you get used to it, and no longer even notice it. You have habituated to it; your brain has determined it is not even worth dedicating the attention to notice it any longer.

Some stimuli, however, are important to us, capture our attention, and continue to perceive—the process of understanding or assigning meaning to, the stimuli we sense. Information that is attended to (or focused on) in perception can proceed to further processing in working memory (WM). WM houses all of the information we are currently thinking about, including stimuli we have just perceived, as well as currently activated contents of long-term memory. WM provides the ability to think about the information, interpret it, predict what might happen next, make judgments,

© Springer Nature Switzerland AG 2021
R. J. Branaghan et al., *Humanizing Healthcare – Human Factors for Medical Device Design*, https://doi.org/10.1007/978-3-030-64433-8_5

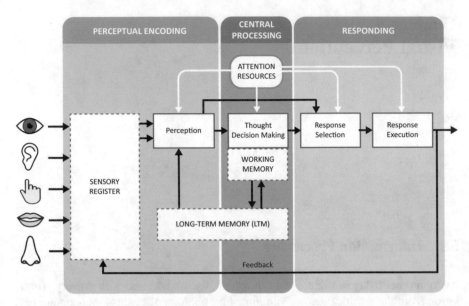

Fig. 5.1 Information processing model. (Adapted from Lee, Wickens, Liu, & Boyle (2017) with permission)

and make decisions. Actually, WM sounds a lot like what most people describe as thinking. We say that working memory is what you usually think about when you think about thinking.

Some information in WM will be rehearsed enough to be transferred to LTM, a very long-term and durable store of memory. LTM houses knowledge of facts, skills, events, and so on, which is used as needed for activities conducted in WM and for mechanisms involved in perception.

All this machinery exists so that we can react to our environment. In the response selection phase, people choose how to respond to the stimulus, and that response is executed, thus reacting to the original stimulus. Finally, that action causes a change in the environment, and that change is fed back to our sensory organs.

Critically, several processes and stores in Fig. 5.1 depend on a limited pool of attention resources, which must be allocated to the right processes at the right time. For example, attention keeps important information in the sensory register for further processing via perception. In this sense, attention is a selection mechanism to choose what information is important enough for further processing. This is referred to as selective attention.

In other situations, attention must be divided among information processing stages, as a limited amount of mental energy used to fuel the stages of information processing. This is often referred to as divided attention. As you can see, attention is selective in the early stages of cognition, and divided in the latter stages.

Every stimulus we encounter, every noticeable thing in our environment, everything we are surprised by, think about, commit to memory, fear, make a decision about, or take action on, goes through this process. Each mechanism has strengths and weaknesses. When we design for people, we design for these. Require people to do more than these processes allow, and you decrease performance and satisfaction while increasing errors. But get it right; guide the user's attention to the right place at the right time, group information accordingly, provide sensible nomenclature, use color intelligently, and users will safely and happily improve their effectiveness, and efficiency.

5.2   Bottom-Up and Top-Down Processes

Sensation starts when a stimulus (some type of energy) impinges on a sensory receptor cell. Different types of receptor cells respond to different types of physical energy. For instance, photoreceptors on the retina of the eye respond to light (electromagnetic) energy, whereas the hair cells in the cochlea of the ear respond to vibrations in the air. These receptors then transduce (convert) that physical energy into neural signals that the brain can use.

Once transduced, chains of neurons (neural pathways) ferry these signals to the brain, successively propagating the stimulus according to characteristics such as shape, color, and intensity. Next, the pathways usher the neural signals to the thalamus that serves as a switchboard for the brain, sending signals to the appropriate parts of the cerebral cortex. The cortical neurons then relay signals to brain areas responsible for motor coordination, memory, emotion, and other functions.

Though they are often discussed together, *sensation* and *perception* are not the same. Whereas sensation senses or detects a stimulus, perception recognizes, interprets, and understands the stimulus. Consider this example: Sometimes I fail to notice when I have left the patio lights on during the daytime. At night, though, I have no problem detecting these exact same lights. In each case, the number of photons impinging on my retina are the same. The sensation during the day and the night are identical, but the perception is entirely different. Even though I see the same stimulus, I "see" two completely different things.

Here is another example, and maybe this has happened to you. You hear a song for the first time, and immediately decide you do not like it. Then, after hearing it several more times, it grows on you, becoming a summertime favorite. The sensation itself did not change; you still sense the same vibrations of air in each case. It is your perception that has changed, based on contextual factors such as familiarity.

This highlights a critical aspect of perception: Perceiving something—recognizing, interpreting it, deriving meaning from it—involves not only the stimuli impacting our sensory cells (bottom-up processing) but also previous knowledge, experience, and expectations provided by our cerebral cortex (top-down processing). So, when you examine the results of digital diagnostic imaging, your bottom-up processing indicates that there are light, dark, sharp, and fuzzy components, your top down processes notice that a particular spot looks suspiciously like a tumor (Fig. 5.2).

As another example, consider the following sentence from Johnson (2014).

Fig. 5.2 Bottom-up and top-down processing

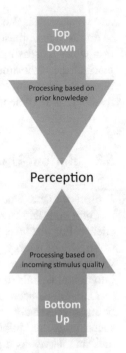

All you hvae to do to mkae a snetnece raedalbe is to mkae srue taht the fisrt and lsat letrtes of ecah wrod saty the smae. Wtih prcatcie, tihs porcses becoems mcuh fsater and esaeir.

This is a great example of top down processing. The stimuli are incorrect, making bottom-up processing difficult. Yet the brain makes sense of it nonetheless through top-down processing. Indeed, design itself is a matter of facilitating both bottom-up processes (providing stimuli of adequate size, contrast, and color) and top down processes (matching expectations, using sensible item placement, grouping, and nomenclature). Designers who accomplish these two things design well. Now that we understand how sensation and perception work in general, we now discuss how these principles are recognized in visual perception.

5.3   Light Energy and the Eye

Vision, the act of processing light emitted from light sources or reflected from objects, is our central mechanism for understanding the world. Half of our cerebral cortex processes visual information, whereas only 3% is used for hearing and 11% for touch and pain (Wurtz & Kandel, 2000).

Light energy is simply radiation traveling in waves of small particles called photons. It may surprise you, but humans can see only a small fraction of all light energy. We can see wavelengths between 360 and 760 nm (billionth of a meter) but we cannot see any other wavelengths such as X-rays, ultraviolet, infrared, and microwaves. This highlights an important fact about perception: We do not perceive the world as it is. Instead, our nervous system evolved to sense a small portion of reality and to ignore the rest. Even Isaac Newton himself wrote that light waves themselves are colorless (Levitin, 2014); it is only the machinery of our eyes and brain that interpret certain waves as certain colors. Obviously, this small portion is what has been necessary for use to survive and procreate (Fig. 5.3).

The wavelength of these photons is the distance from one peak to another, and determines the wave's hue (color). Longer wavelengths are interpreted by our visual system as red, whereas shorter wavelengths yield violet or blue (Goldstein & Brockmole, 2016) (Fig. 5.4).

A handy mnemonic for naming the colors (in decreasing wavelength order) is ROYGBIV for red, orange, yellow, green, blue, indigo, and violet. In reality, most colors we experience are not pure wavelengths, but a combination of wavelengths, which we perceive as a composite. White light, for example, consists of approximately equal amounts of all wavelengths, it is the ultimate composite.

The amplitude (height) of the wave is determined by the number of photons present and is interpreted by our visual system as brightness, with large amplitudes appearing bright and small amplitudes appearing dull (Fig. 5.5).

Seeing begins when waves of light enter the eye (Fig. 5.6) and pass through the *cornea*, a protective layer on the outside of the eye. Next, the light waves reach the *pupil*, an opening controlled by the muscles of the iris (the colored part of the eye). Depending on lighting conditions, the pupil dilates to allow more light in or constricts to keep more light out. When the pupil is completely dilated, it allows about 16 times more light in than when it is fully contracted (Young & Biersdorf, 1954). Well, at least this is the case when we are young. This starts to change drastically by the ripe old age of 30.

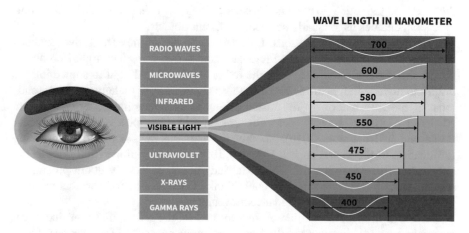

Fig. 5.3 The electromagnetic spectrum. (Image by MicroOne/shutterstock.com)

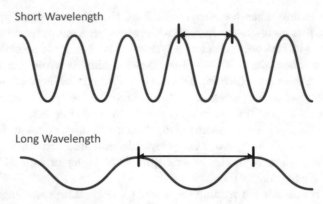

Fig. 5.4 Light wavelength and color

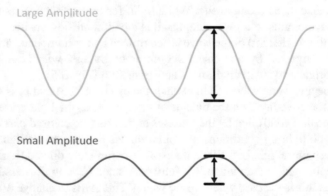

Fig. 5.5 Light amplitude and brightness

From there, light passes through the *lens*. The lens uses ciliary muscles to focus the image on the retina, located at the back of the eye. This process, called accommodation, enables us to clearly see near or distant objects.

After passing through the lens, light travels to the retina. The retina contains specialized photoreceptors, called rods and cones, that transduce light into neural signals that the brain can use. Once the retina has translated light to something the brain can understand, neurons transmit these signals to the thalamus, which sends most of the information to the visual cortex of the occipital lobe. Here, more advanced processes such as detecting colors, edges, shapes, and motion occur (Goldstein & Brockmole, 2016). Notably, the thalamus does not send all incoming information to the visual cortex. Roughly 10% of visual information goes to the amygdala, which makes quick emotional judgments, especially about potentially harmful stimuli (Ware, 2008). You could think of this as a neural shortcut to allow quick, reflexive, and danger-avoidance movements.

As we mentioned, the photoreceptors located on the retina, that transduce light energy into neural signals, are called rods and cones because of their shapes. These two types of photoreceptors have different, but complementary, characteristics, which we describe in the next section.

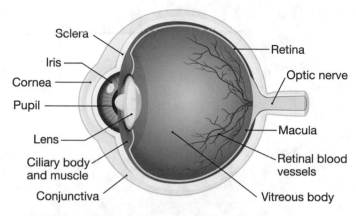

Fig. 5.6 The eye. (Image by solar22/shutterstock.com)

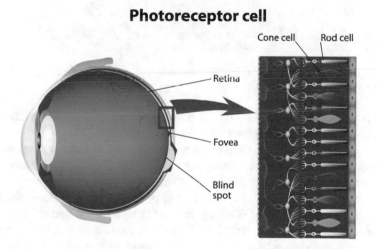

Fig. 5.7 The retina, fovea, rods, and cones. (Image by Designua/shutterstock.com)

5.4   Rods, Cones, and Color Perception

The photoreceptors on our retinas come in two types. Rods are plentiful and occupy mostly the periphery of the retina, whereas cones are rarer and occupy the central part of the retina, called the fovea. Cones work mostly in the day and in conditions of high ambient light (photopic conditions). Naturally, they require ample light to operate.

As you can see in Fig. 5.7, there are three types of cones, each sensitive to different wavelengths. Short-wave cones detect blue most effectively, whereas middle-wave cones detect green, and long-wave cones detect red. Figure 5.8 illustrates that colors toward the middle of the visual spectrum, green, for example, require less intensity (lower amplitudes, less energy) for us to detect than colors at the ends of the spectrum, such as blue and red. In other words, it is easier for us to see greens

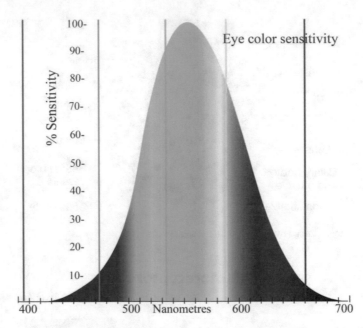

Fig. 5.8 The visible spectrum. ("File:Eyesensitivity.svg" by Skatebiker, vector by Adam Rędzikowski) is licensed under CC BY-SA 3.0)

than it is to see reds and blues. This is why construction crews and cyclists wear lime green vests when on the road; they are easier to see!

Sometimes it is difficult to distinguish between colors, especially pale ones. This is even worse when we need to distinguish between small patches of colors that are far away from each other (requiring eye movement).

For example, take a look at Fig. 5.9. It may be hard to tell that states that are far away from each other, such as Washington and Maine have the same level of pollen (Low-Medium). New Mexico and North Dakota are also far from each other, but are one level different (ND = Low, NM = Low-Medium), so it is difficult to detect. On the other hand, Colorado is next to New Mexico, making it easier to distinguish the same one level difference (CO = low, NM = Low-Medium).

5.5   Color Deficiency

Not all people see the same. For example, 8% of males and 0.5% of females are color deficient (many people mistakenly call this color blind). This does not mean these people do not see color at all, but that their visual cortex does not enable them to distinguish certain pairs of colors. The most common form of color deficiency is red-green color deficiency, which makes it difficult to distinguish dark red from black, blue from purple, and light green from white. People with red-green color

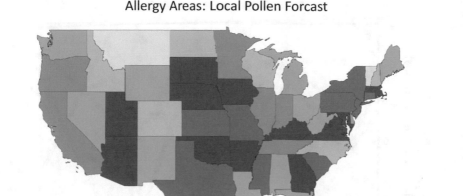

Fig. 5.9 Pollen count by state

deficiency would have difficulty detecting the differences in the colors of the lines in Fig. 5.10 (Johnson, 2014).

5.6 Contrast

Because our hunter-gatherer ancestors needed to detect changes quickly, we are masters at detecting contrast and edges rather than absolute colors. The gray bar in the middle of Fig. 5.11 is the same shade across the whole graphic. But the visual cortex, with its sensitivity to contrast, tells us it changes. This visual illusion, called the simultaneous contrast illusion, tells us that the brightness of the bar is less important to our visual cortex than the difference between the bar and its surroundings. This contrast helps us navigate the world, identifying the beginnings of one object and the ends of another.

Contrast is important to vision because it enables us to recognize shapes by discriminating between the figures (the object we are trying to detect) and the background of a scene. Contrast is defined as the difference between the light and dark luminance as shown below:

$$\text{Contrast} = (L - D)/(L + D)$$

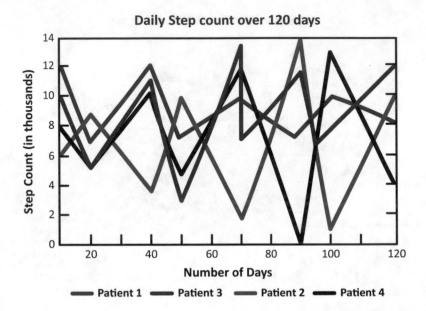

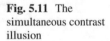

Fig. 5.10 Graph that demonstrates users with red-green color deficiency would have difficulty detecting the differences in these colors

Fig. 5.11 The simultaneous contrast illusion

where L is the luminance of the light area and D is the luminance of the dark area. To enable people to easily discriminate between light and dark objects, the contrast should be between 3:1 and 7:1 (International Standards Organization, 2008).

5.7   Image Size and Visual Angle

The size of an object's image on our retina is measured by its visual angle, which is created by imagining two lines extending from the eye. One line extends to the top or left of the image, and the other extends to the bottom or right. The visual angle, the angle subtended by those two lines, is usually measured in degrees, minutes and seconds. One minute of arc is 1/60th of one degree of arc. Similarly, one second of arc is 1/60th of one minute of arc. Large angles describe large or close objects, whereas small angles describe small or faraway objects.

Figure 5.12 shows how to calculate visual angle, where H is height of the object and D is the distance to the object. For this to work, you must use the same unit of measurement for both H and D. Thankfully, the World Wide Web provides numerous visual angle calculators using exactly the equation provided below. Just plug in your distance from the object and height of the object to be rewarded with the visual angle.

According to Association for the Advancement of Medical Instrumentation (2018), the minimum perceptible visual angle is about 1 s of a degree for a thin wire against a bright sky. The preferred angle for reading English text is 20–22 minutes of arc, whereas marginally acceptable angles range from 16 to 18 minutes of arc, with 12 minutes considered the threshold of readability.

5.8   Visual Accommodation

If you want to see something in great detail, you must focus the image on the retina. To accomplish this, the lens actually bends and flexes to create a sharp image. Just like a camera, images focused too far in front or behind the retina look blurry.

This process of the lens changing shape is called accommodation (see Fig. 5.13). Unfortunately, the lens can only flex so much. For example, when focusing on an object about three meters away, the lens becomes flat, and cannot become any flatter. This is called the far point. Conversely, when the object is within about 20 cm, the lens cannot become any more spherical. This is called the near point. Looking at objects closer than this causes discomfort and blurriness. What's worse, alternating

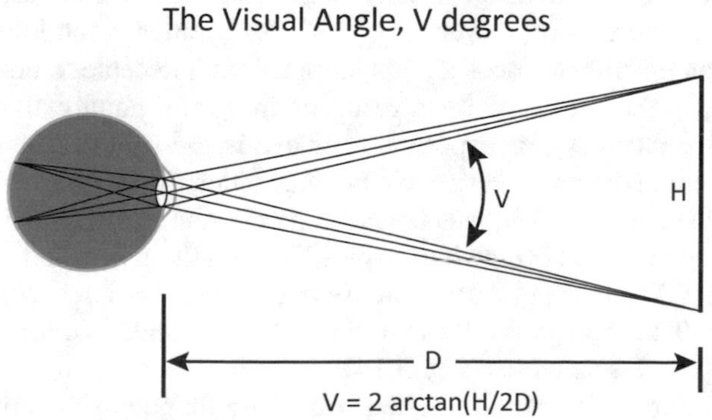

The Visual Angle, V degrees

$$V = 2 \arctan(H/2D)$$

Fig. 5.12 Calculating visual angle

The near response of the eye

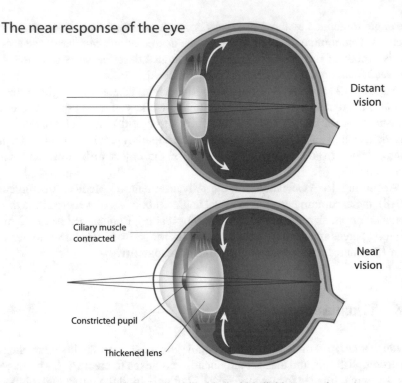

Distant vision

Ciliary muscle contracted

Near vision

Constricted pupil

Thickened lens

Fig. 5.13 Visual accommodation. (Image by Alila Medical Media/shutterstock.com)

between close and far distances takes time (about 1 s in young healthy people) and substantial effort, resulting in eyestrain.

5.9   Vision Problems

The inability to focus on an image usually stems from the shape of the eye (Llorente, Barbero, Cano, Dorronsoro, & Marcos, 2004). An eye that is too long or too short can result in an inability to focus an image on the photoreceptors. For nearsightedness, or myopia, the eye is too long, resulting in a focal point that is short of the retina. For farsightedness, or hyperopia, the eye is too short, resulting in a focal point that is behind the retina. As people become older, the speed and extent of their accommodation decrease. The lens becomes harder and less responsive to the ciliary muscles, and people become hyperopic. This condition is called presbyopia, or old-sightedness. The near point can increase from as close as 10 cm for 20-year-olds to as far as 100 cm by age 60. Presbyopia is usually treated with reading glasses (bifocals) or progressive lenses (Fig. 5.14).

Eye discomfort or eye strain is often caused by fatigue of the ciliary muscles, especially among people who engage in close work or spend a lot of time at a com-

VISION DISORDERS

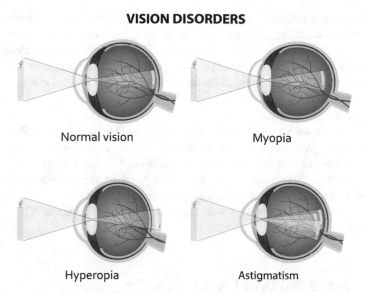

Normal vision   Myopia

Hyperopia   Astigmatism

Fig. 5.14 Common accommodative vision problems. (Image by Neokryuger/shutterstock.com)

puter monitor. Displays close to the viewer require more accommodation and fixating on these displays for a long time can cause the eye muscles to become exhausted.

Another problem causing visual fatigue is that text on a computer monitor is different from printed text. Whereas printed text has sharp edges, the text on a computer monitor is sharp in the middle but blurry on the edges. This means that a person who spends a long-time reading computer text must continuously work to keep the text on the screen in focus by accommodating two different types of images in rapid succession, again exhausting our eyes.

A person can also have problems focusing because he or she has astigmatism, which is caused by problems in the shape of the eye that cause light to bend asymmetrically as it passes through the cornea. This means that contours in certain orientations will be clearly focused whereas those in other orientations will not. No matter how much the eye accommodates, some parts of the image will always be blurred.

5.10 Aging and Vision

Currently, about 16.5% of the US population is over 65 years old, and this percentage is expected to rise to 21% in the next 10 years (United States Census Bureau, 2020). Moreover, about 36% of the population is over 50 years, which also happens to be when vision begins to decline.

Many changes happen to our eyes as we age. The lens tends to yellow, allowing less light into the eye, and making the discrimination of the color blue difficult. The cornea can be scarred with imperfections, which results in decreased acuity and an increase in the scattering of light. This can cause glare around light sources, especially at night. Another common problem is cataracts, which are hard, cloudy areas in the lens that usually occur with age. As many as 75% of people over 65 years have cataracts, although in most cases the cataracts are not serious enough to interfere with the person's activities. Sometimes surgery is necessary to correct major corneal and lens problems.

In addition, as we age our pupils shrink, allowing less light to enter the eye. And the pupil's response to dim light decreases. Not only does our pupil diameter decrease, but the ability for our pupils to dilate (and let more light into our eyes) decreases drastically. By the time we are 80, our pupil diameter at night is just about the same as in the day, but these changes begin to occur as early as 30 years old. Finally, our ability to see contrast worsens as we age. This is especially problematic in low light conditions.

5.11   Central and Peripheral Vision

The rod and cone photoreceptors on the retina enable us to see, but different parts of the retina see differently. For example, the fovea, a small indentation in the center of the retina (Fig. 5.15) provides central vision, which is especially sharp. It is this central vision, with the best spatial and color resolution, that enables us to identify objects, focus clearly, see fine detail, and read text.

To pull off this feat, the 1.5 mm wide fovea contains almost exclusively cones specialized for acute visual perception. In fact, the fovea contains 140,000 cone cells per square millimeter, whereas the retinal areas outside the fovea contain only about 10,000 cones per millimeter (Johnson, 2014). The fovea has more cones than all of the retinal periphery, even though the periphery is much larger.

As a result, the resolution of detail in the fovea is much higher than the periphery of the retina. The reading acuity chart in Fig. 5.16 after Anstis (1974) illustrates the relative sizes of letters people can identify at the center, and at the edges, of the visual field.

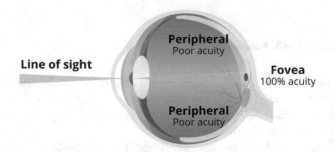

Fig. 5.15 Central (foveal) and peripheral vision in the retina

Fig. 5.16 Relative sizes of letters people identify at the center and edges of the visual field. (Image courtesy of Professor Stuart Anstis. http:// anstislab.ucsd.edu/ illusions/ peripheral-acuity/)

Further, cone cells in the fovea connect 1:1 to neural cells that process and transmit visual information. This is different from other parts of the retina, where many photoreceptors connect to each neural cell. Consequently, data from the visual periphery is compressed and subject to data loss, whereas data from the fovea is transmitted to the brain uncompressed (Johnson, 2014). Further supporting the fovea's special status is the fact that the brain devotes most of its visual processing to signals coming from the fovea. The fovea represents only about 1% of the area of the retina, but the visual cortex devotes about 50% of its area to the fovea (Grill-Spector & Malach, 2004).

As we move from the fovea toward the periphery, our visual acuity drops precipitously. Because cones and neural cells are so scarce in the periphery, and have less representation in the visual cortex, the periphery is poor at identifying details, color, and shape. Instead, peripheral vision seems best at recognizing familiar objects without needing to focus on them. The periphery provides the gist of the image, whereas the fovea provides the detail.

Failing to realize how this works frustrates designers and developers. Why, they wonder, do people fail to notice a button, label, or text box in the corner of the screen? The answer is that their fovea is focused elsewhere, and their peripheral vision fails to notice the item in the corner. Figure 5.17 illustrates this point. If a user is focused on entering their username or account number, their peripheral vision will not detect the "Incorrect ID" message in the upper left.

Fig. 5.17 Example login portal that demonstrates that the user may miss "Incorrect ID" due to fovea focus

5.12   How Visual Perception Works

We see with our eyes, but we also see with our brain. It feels like we see the world as a complete picture, in clear detail, but we know that this is an impossible illusion. For example, the image projected onto the retina is two dimensional like a sheet of paper, yet we perceive it as three dimensional. Also, each eye has a blind spot that should interrupt our vision, yet we do not even notice! Moreover, the images that fall on our retina are upside down and reversed, yet we perceive them as right side up and in the correct orientation. Finally, blood vessels and neuronal cells occupy much of the area directly in front of our photoreceptors, yet we do not notice those either. What we perceive is different from what we sense—the visual energy that hits our eyes.

Our brains are metabolically expensive, accounting for just 2% of our body weight and more than 20% of our energy demands (Raichle et al., 2001). Maintaining a representation of the entire world in our brains would be enormously wasteful, so evolution opted for a more frugal solution: cognitive economy (Levitin, 2014). This entails seeing only what we attend to, and attending to only what we need for the task at hand (Ware, 2008). The brain fills in the blanks based on knowledge.

Our eyes are constantly moving, so we use "just in time" (JIT) visual queries to make sense of our surroundings (Ware, 2008). This means that, at any one moment, we sense only a tiny fraction of our surroundings. This is usually enough to understand our environment; we sample our visual environment so rapidly that we think we have all of it at once in our consciousness experience (O'Regan, 1992). Since eye movements require about a tenth of a second, they feel instantaneous and provide the impression that we are always aware of everything. Our memory then serves as glue to bind these sensations into a perceptual whole. The whole provides meaning, something we recognize, that enables us to achieve our goals or complete our current task.

5.13 Attention's Role in Visual Perception

Attention directs our sensory activities based on conspicuity and context. Conspicuity refers to physical characteristics of stimuli that make them pop out as noticeable. Context refers to spatial characteristics like proximity, but also to psychological factors such as our expectancies, current tasks and current goals. We notice things that we expect to see, and we notice things that help us get our work done. Let's start by discussing conspicuity.

Fig. 5.18 Waldo. (Copyright © 1987–2020 Martin Handford from the Where's Waldo? Books by Martin Handford. Reproduced by permission of the publisher, Candlewick's Press, Somerville, MA on behalf of Walker Books, London)

5.14 Conspicuity

Some visual elements are conspicuous and tend to pop out at us. We see them with so little effort, it may be impossible *not* to notice them. As an example, consider the "Where's Waldo" puzzles, in which you examine a picture looking for Waldo (Fig. 5.18).[1]

First, look at a picture from "Where's Waldo" (Fig. 5.19). In this picture, Waldo swims in a sea of similarity: similar shapes, similar sizes, and similar colors. It is really challenging to pick him out. You have to inspect each and every object just to find him, in a slow and effortful process called "serial search" (Thornton & Gilden, 2007).

Now try to find Waldo in Fig. 5.20. This is easier because things that are different from their surroundings are easy to notice. They attract our attention immediately and automatically. The message is clear: If you want to direct peoples' attention to something, make it different from its surroundings. Making something different makes it conspicuous.

---

[1] This example was inspired by Jeff Johnson's excellent book, "Designing with the Mind in Mind: Simple Guide to Understanding User Interface Design Guidelines (Johnson, 2014).

Fig. 5.19 Where's Waldo? (Copyright © 1987–2020 Martin Handford from the Where's Waldo? Books by Martin Handford. Reproduced by permission of the publisher, Candlewick's Press, Somerville, MA on behalf of Walker Books, London)

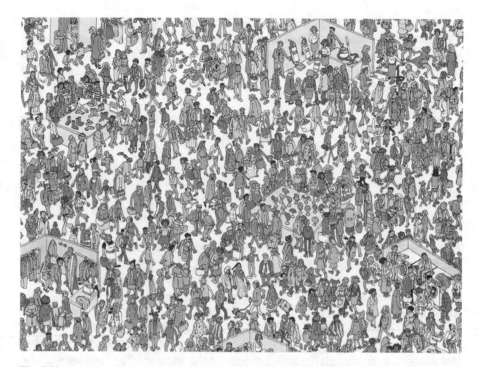

Fig. 5.20 An easier Where's Waldo. (Copyright © 1987–2020 Martin Handford from the Where's Waldo? Books by Martin Handford. Reproduced by permission of the publisher, Candlewick's Press, Somerville, MA on behalf of Walker Books, London)

CBC w/ Diff & PLT

Component	Your Value	Standard Range	Units
WBC	14.6	4.8 - 10.8	K/uL
RBC	4.9	4.70 - 6.10	M/uL
HEMOGLOBIN	10.0	12.6 - 17.4	g/dL
HEMATOCRIT	43.5	37.0 - 51.0	%
MCV	96.9	80.0 - 94.0	fl
MCH	30.5	27.0 - 31.0	pg
MCHC	33.5	33.0 - 37.0	g/dL
RDW	12.7	11.6 - 14.8	%
PLATELET COUNT	100	130 - 400	K/uL
SEGMENTED NEUTROPHILS #	3.4	2.40 - 7.60	K/uL
LYMPHOCYTE #	5.2	1.00 - 4.30	K/uL
MONOCYTES #	0.40	0.00 - 1.10	K/uL
EOSINOPHILS #	0.10	0.00 - 0.60	K/uL
BASOPHILS #	0.00	0.00 - 0.20	K/uL
SEGMENTED NEUTROPHILS %	37.0	50.0 - 70.0	%
LYMPHOCYTE %	52.0	20.0 - 40.0	%
MONOCYTES %	8.5	0.0 - 15.0	%
EOSINOPHILS %	1.9	0.0 - 6.0	%
BASOPHILS %	0.6	0.0 - 2.0	%

CBC w/ Diff & PLT

Component	Your Value	Standard Range	Units
WBC	14.6	4.8 - 10.8	K/uL
RBC	4.9	4.70 - 6.10	M/uL
HEMOGLOBIN	10.0	12.6 - 17.4	g/dL
HEMATOCRIT	43.5	37.0 - 51.0	%
MCV	96.9	80.0 - 94.0	fl
MCH	30.5	27.0 - 31.0	pg
MCHC	33.5	33.0 - 37.0	g/dL
RDW	12.7	11.6 - 14.8	%
PLATELET COUNT	100	130 - 400	K/uL
SEGMENTED NEUTROPHILS #	3.4	2.40 - 7.60	K/uL
LYMPHOCYTE #	5.2	1.00 - 4.30	K/uL
MONOCYTES #	0.40	0.00 - 1.10	K/uL
EOSINOPHILS #	0.10	0.00 - 0.60	K/uL
BASOPHILS #	0.00	0.00 - 0.20	K/uL
SEGMENTED NEUTROPHILS %	37.0	50.0 - 70.0	%
LYMPHOCYTE %	52.0	20.0 - 40.0	%
MONOCYTES %	8.5	0.0 - 15.0	%
EOSINOPHILS %	1.9	0.0 - 6.0	%
BASOPHILS %	0.6	0.0 - 2.0	%

Fig. 5.21 Two versions of results from a complete blood count (CBC) test

Figure 5.21 shows this principle applied to highlighting medical test results. The first sheet requires serial search of all numbers to identify items of concern, whereas the second highlights them for you, making them faster and easier to understand. In this case, the designer has guided the users' attention by making items that are "out of range" conspicuous; and they are conspicuous because they are different from their surroundings.

In Fig. 5.21, the image on the right clearly highlights values that are out of range; attracting attention is done simply by making those values look different from their surroundings.

5.15   Context

Experience influences our ability to perceive. Because our brain guides our vision, our perception is strongly influenced by context; for example, in the text below, the middle letter in the first word is an H, whereas the middle letter in the second is an A. Of course, the H and the A are precisely the same shape. The context of the surrounding letters influences the recognition of the middle letter (Fig. 5.22).

A similar context effect is illustrated in the Müller-Lyer illusion below. Which of the horizontal lines in the figure below looks longer? Viewers guess that the first looks longer, even though the lines are exactly the same length. The surroundings (i.e., context) affect our perception of line length (Fig. 5.23).

People tend to be goal oriented, especially when working with products such as medical devices. They are not doing it for fun, they are trying to achieve some goal. Our attention and perception is almost entirely focused on the goals we are trying to achieve.

We also tend to perceive what we expect to perceive. Because perception is an active process, we anticipate or make guesses about what we will see next. We have prior ideas of where an item might be located. One example of this is a traffic light. One of the authors (whose initials are Russ Branaghan) was on a business trip driving in San Francisco when he ran a red light. As is his luck, there was a police officer right there to see the whole thing. The problem was that, in some parts of San Francisco, the streetlights are mounted on posts on the street corner, rather than above the intersection as God intended. The officer seemed to understand the problem, having heard the explanation before, but he was not too understanding; Russ still received a $275 ticket (Fig. 5.24).

Fig. 5.22 Context effects in reading

Fig. 5.23 The Muller-Lyer illusion is an optical illusion in which two lines of the same length appear to be of different lengths. (Zhitkov Boris/shutterstock.com)

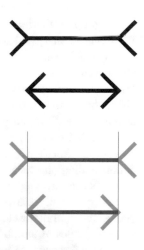

Fig. 5.24 The type of unexpected traffic light placement that earned Russ a traffic ticket. (Image by Thomas Haas/unsplash.com)

5.16   Gestalt Psychology

In the early 1900s, the prevailing paradigm in psychology required breaking down conscious experience into smaller and smaller ingredients, much like the approach to studying compounds in chemistry. Psychologists assumed that perceptions were simply constructed from rudimentary parts. They supposed that if they understood the parts, they could understand the whole.

A group of German and Austrian psychologists, including Max Wertheimer, Wolfgang Kohler, and Kurt Kofka were dissatisfied with that approach, arguing instead that breaking "whole" perceptions into components leads to a loss of important psychological information. They hypothesized that considering the world as organized, structured wholes would shed more light on perception (Koffka, 1935). Using the German term for "form" or "whole," this group would come to be known as the school of Gestalt psychology.

The mind groups patterns according to laws of perceptual organization, which reflect our experience and are used unconsciously. Gestalt theories are based on understanding objects as an entire structure, rather than the sum of its parts. Despite Gestalt's humble beginnings, its principles, such as proximity, similarity, figure-ground, continuity, closure, and connection have become a popular and well-known perspective within the study of sensation and perception. Gestalt principles describe heuristic "rules of thumb," which provide fast guesses about the identity and meaning of visual scenes. Although they are usually correct, they are not guaranteed to be correct. In fact, in visual illusions, they are sometimes misleading. We discuss these perspectives next.

Figure-Ground

One Gestalt principle states that we usually identify one primary object that stands out in a scene and assume the remainder of the scene is background. The foreground object appears to have more shape and substance, and our minds tend to pull the figure towards the front, while pushing the remainder to the back. This is illustrated in Fig. 5.25, which shows either a vase or two people facing each other depending on your figure—ground interpretation.

Fig. 5.25 Three figure-ground, face and vase relationships. (Image by Peter Hermes Furian/shutterstock.com)

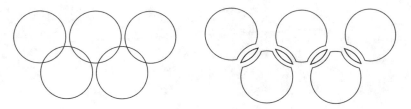

Fig. 5.26 The Olympic symbol on the right is perceived as five overlapping circles rather than nine shapes of two different kinds as on the right. Our perceptual system opts to interpret scenes in the simplest way possible. (Law of Pragnanz by Clint_2013 is licensed under CC BY 2.0)

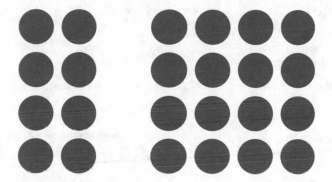

Fig. 5.27 An example of the principle of proximity

Law of Pragnanz

In an effort to simplify, understand, and seek structure in the visual world, we rely on the law of Pragnanz (also known as the law of good Gestalt or the law of good form). According to Pragnanz, we see patterns so that their structure is as simple as possible. In this way, scenes seem orderly, symmetrical, and simple, reducing complexity and unfamiliarity so we can experience the world in its simplest design.

For example, consider Fig. 5.26. In the figure on the left, we see five rings whereas in the figure on the right, we see nine separate objects separated by white lines.

Proximity

The principle of proximity states that objects arranged close together tend to be grouped together or related in peoples' minds. As an example, consider Fig. 5.27. Rather than one group of 24 dots, simply placing white space after the first two columns leads us to perceive two groups: one group of eight dots and one group of 16 dots.

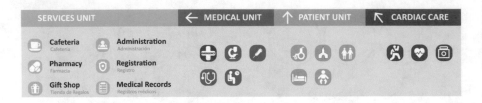

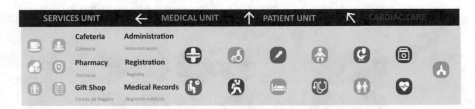

Fig. 5.28 Good use of proximity (*top*), and poor use of proximity (*bottom*)

Fig. 5.29 An example of
the principle of continuity.
(Photo by Javier
Calvete/shutterstock.com)

Another example can be found in the hospital signage shown in Fig. 5.28. The top provides a good use of proximity, whereas the bottom does not, resulting in confusion.

Continuity

Sometimes called the "principle of good continuation," continuity suggests that we assume that points forming smooth lines when connected probably belong together and construct the same object. We assume that lines follow the smoothest path as in Fig. 5.29. While the image is nothing more than horizontal black trapezoids and

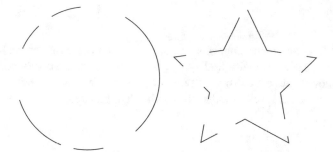

Fig. 5.30 An example of the Gestalt principle of closure

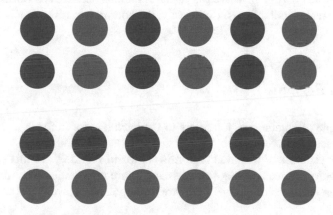

Fig. 5.31 Example of the Gestalt law of similarity

diagonal orange parallelograms, our brains create a two-sided ribbon encircling an invisible sphere.

Closure

The principle of closure suggests that we tend to see whole, closed objects, rather than collections of fragments. This occurs when we perceive a complete, unbroken image even when there are gaps in the lines forming the image. We "fill in blanks" to see a single object (Fig. 5.30).

Symmetry

The principle of symmetry suggests that our minds perceive objects as being symmetrical, and that we actually find symmetrical objects aesthetically pleasing. For example, shown below are different types of brackets. We tend to think of these as pairs of symmetrical brackets rather than individual brackets.

( ) { } [ ] ( )

Similarity

Similar objects tend to be grouped in our minds. For example, in Fig. 5.31 the top two rows of dots may be perceived as vertical columns because color is used in each column to create similar items. Conversely, the bottom two rows may be perceived as horizontal rows.

Common Region

The principle of common region refers to the tendency for items in a region to be seen as a group (Fig. 5.32).

Another example of common region is shown in Fig. 5.33. In this suture cart, common region is created using color, as each suture type comes in a different color box. Within each color group there are multiple sizes available of the same suture type.

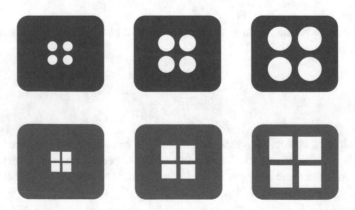

Fig. 5.32 An example of the Gestalt Principle of Common Region where items surrounded by a common border appear to be grouped together

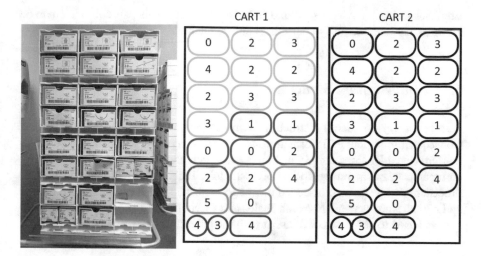

Fig. 5.33 A suture cart is an example of Gestalt Principle of Common Region, where like colors are grouped together in close proximity

O.R. SCHEDULE

O.R.	TIME	PATIENT	PAT. RM.	SURGEON	ATTENDING	ANES. TYPE		REMARKS	ANES- THETEST	R.N.	SCRUB	CIRC.
3	09:15	ELIOT BURNS	4163	DR. HANNA	DR. KWONG	GEN	ACL	LATEX ALLERGY	E. Peters	E. Williams	B. Higgs	N. Shelby
2	10:30	TIMOTHY HAMILTON	3815	DR. MOREE	DR. NOLAN	GEN	T+A	DENTURE	M. Caputo	C. Woods	N. Viale	M. Shea
4	11:00	CHARLES BURTON	1654	DR. JONES	DR. KWONG	GEN	TOTAL KNEE		S. Peters	E. Williams	B.Higgs	N. Shelby
3	12:15	PETER OWENS	3344	DR. DANIELS	DR. BAKER	GEN	AAA		L. Jones	A. Logan	N. Viale	B. Marks
2	1:30	AMY RYAN	7546	DR. MOREE	DR. NOLAN	GEN	T + A		G. Scott	B. Suns	M. Smith	O. Deep
1	2:45	ALEX MURPHY	6541	DR. KING	DR. GORPON	LOC	CARPAL TUNNEL		M. Caputo	E. Williams	D. Stanton	M. Shea
2	5:45	LACE ADAMS	5467	DR. HANNA	DR. BAKER	GEN	ACL	LATEX ALLERGY	S. Peters	B. Suns	M. Smith	N. Shelby
1	6:15	KAREN MAYER	4221	DR. DANIELS	DR. GORPON	GEN	AAA		L. Jones	A. Logan	D. Stanton	M. Shea

Fig. 5.34 Operating room schedule board

Familiarity

The principle of familiarity suggests that we categorize visual stimuli according to past experience. If, in our experience, two objects go together, they become strongly associated in our minds and can be perceived as one object. For instance, In the

English language, the letters "s" and "h" often co-occur. This pair could begin to be grouped into just one unit in our minds during reading.

5.17 Information Structure

Our brains try to identify structure in noisy environments and visual scenes. Consequently, structured information is easier to perceive than unstructured information. Consider the example in Fig. 5.34, which shows a common tool used to communicate an Operating Room (OR) schedule. Technicians use a marker to write procedure details on a color-coded magnetic strip. The strips can be moved around and reused. Anesthetist, RN, Scrub Techs, and Circulator names are labeled on magnets and used to show who will assist with the procedure.

Visual Hierarchies

One powerful method for providing structure is through visual hierarchy, that enables people to understand the structure of information in a product, and guide their attention to the information or functionality they seek. For example, Table 5.1 shows the structure of a SOAP note that documents high-risk patient information. SOAPs must include specific subjective information provided by the patient, objective physiological information, assessment based on both subjective and objective data, and the plan moving forward. The acronym organizes the information. Medical professionals are taught to write structured SOAP notes to ensure that the most

Table 5.1 Structure of a SOAP note

S	Subjective
	● Identify patient
	● Describe symptoms, feelings, medical and surgical history, and progress
	● May use first person quotes or third person observations (e.g., "I don't have an appetite" or "the patient reported")
O	Objective
	● Physical exam
	● Vitals, lab results, imaging, etc.
	● Clinical observations related to medical assessment
A	Assessment
	● Description of clinical condition
	● Narrative of circumstance using subjective and objective information
	● Identify issues and differential diagnoses
P	Plan
	● Outline a treatment plan
	● Describe next steps regarding evaluation and/or treatment
	● Include timeline and quantities

important information is included in the note. However, in practice, notes are not always easy to read because some Electronic Health Records (EHR) systems do not afford the ability to format text, line spacing, bold font, use of bullets, and so on. In these situations, notes might look something like Note #1 below. When EHR systems do afford text formatting, however, HCPs can create notes with more structure such as in Note #2. Take a look at both notes. Which would be easier to read?

Note #1:

Michelle is a 66 y/o woman who presents with a rash that began 2 weeks ago. She first noticed the rash a couple of days after working in her garden. "I first noticed the rash on my back and it spread to my belly". For 2 days she had no associated symptoms; however, on the third day she began feeling "a burning pain that hurts more when my clothes touch it". She has been experiencing worsening pain in the evening making it difficult to sleep. She took "Tylenol a couple of times and tried applying vitamin E lotion" with minimal relief.

Michelle reported having chickenpox as a child. She has a history of hypertension and her medications include: Nifedipine XL 30 mg daily and Metoprolol 50 mg BID. Michelle is an older woman, energetic and in no distress. VS: BP 130/80, P 60, RR 12, T 98. Skin exam revealed an erythematous rash of grouped vesicles with clear fluid. The rash extends from the midline of the back anterior to the left side of the T12/L1 dermatome region. Likely herpes zoster, given location, distribution and associated pain. Less likely, but due to recent gardening another possibility is contact dermatitis. Michelle will take Tylenol every 4–6 h for pain relief. Michelle was informed that she may experience pain after the rash resolves. Michelle was told that she should return to the clinic if the rash does not resolve, becomes more severe, or if she develops a fever.

Note #2:

Subjective

Patient: Michelle is a 66 y/o woman who presents with a rash that began 2 weeks ago.

Symptoms: She first noticed the rash a couple of days after working in her garden. "I first noticed the rash on my back and it spread to my belly". For 2 days she had no associated symptoms; however, on the third day she began feeling "a burning pain that hurts more when my clothes touch it". She has been experiencing worsening pain in the evening making it difficult to sleep.

Progress: She took "Tylenol a couple of times and tried applying vitamin E lotion" with minimal relief.

History: Michelle reported having chickenpox as a child. She has a history of hypertension and her medications include:

- Nifedipine XL 30 mg daily.
- Metoprolol 50 mg BID.

Objective

Physical Exam: Skin exam revealed an erythematous rash of grouped vesicles with clear fluid. The rash extends from the midline of the back anterior to the left side of the T12/L1 dermatome region. Michelle is an older woman, energetic and in no distress.

Vitals:

- BP 130/80
- P 60
- RR 12
- T 98.7

Assessment

New onset painful rash in a dermatomal distribution. Likely Herpes Zoster: Given location, distribution, and associated pain. Less likely, but due to recent gardening another possibility is contact dermatitis.

Plan

Treatment: Tylenol every 4–6 h for pain relief.

Further Evaluation: Michelle was informed that she may experience pain after the rash resolves. Michelle was told that she should return to the clinic if the rash does not resolve, becomes more severe, or if she develops a fever.

Structured material is easier to scan and read. Imagine that you are a healthcare worker, and you are responsible for entering the serial number of an implanted device into a computer. You are busy, the surgeon is talking to you, and you have been paying attention to blood loss. Which of the numbers below would you prefer to enter? Which would be easier to read? Which would be easier to remember? You would probably choose the bottom, more structured, option.

BD73929841XC

BD7-3929-841-XC

5.18   Design Advice Based on Visual Perception

So far, we have covered how visual perception works, but now we think it is helpful to bring the discussion back to design. Since we perceive the world by using bottom-up and top-down processes, good design supports them both. Supporting bottom-up processes means providing stimuli of adequate size, contrast, brightness, and appropriate color. Facilitating top-down processes means providing structure, and making use of context and expectancies. This requires familiar stimuli, appropriate item placement, intelligent grouping, understandable nomenclature, and consistency. Below, we provide more specifics for achieving both of these goals.

Item Placement and Grouping

- Place important pieces of information where the user is most likely to look.

Fig. 5.35 This stovetop has poor mapping between control (knobs) and display (burners). (Image by STILLFX/shutterstock.com)

Fig. 5.36 A poor mapping between control (knobs) and display (burners). (Image by Paolo Bendandi/unsplash.com)

- Place controls, displays, and other items according to their importance (most important items should be larger, and easier to see and reach).
- Organize items by frequency of use, with more frequently used items closer, more prominent, and easier to reach.
- Group items by relatedness. Those items that are similar or that co-occur in the completion of tasks should be grouped together.

Fig. 5.37 Image that
shows clear mapping
between knobs and burners
on a stove

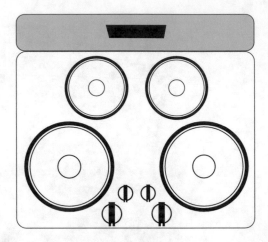

- Place controls close to the things they control and their related displays. This is called proximity-compatibility (Wickens & Carswell, 1995).
- Provide a mapping between controls and the thing they control. For example, Engineering 101 classes often feature the following example (this one borrowed from our colleague Mike Darnell, www.baddesigns.com). Students are asked what is wrong with the design of this stovetop? Eventually they conclude that there are two problems: (1) the controls are at the back so you sometimes have to reach over scalding hot pots and pans to use them; and (2) you cannot determine which knob controls which burner. You have to figure this out by trial and error (Fig. 5.35).

The example in Fig. 5.36 is even worse. Neither of these configurations provide a clear mapping between the control (the knobs) and the thing being controlled (the burners).

A simple and elegant solution is shown below in Fig. 5.37. This design has mapped the controls to the burners, and uses smaller controls for the smaller burners and larger controls for the larger burners. Nothing is left to chance.

Consistency

- Aim for consistency in color, font, nomenclature, item placement, grouping, and information architecture within your product. Also, aim for consistency within your brand, so that people know how to use multiple models manufactured by your company.
- Adhere to industry standards to prevent the user from needing to relearn products.

Fig. 5.38 Stop signs are
good examples of the use
of redundant coding

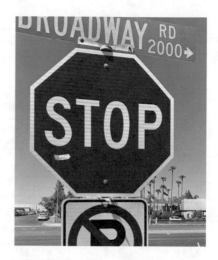

Adhere to User Expectancies

- Use familiar stimuli (words, fonts, symbols, icons)
- Avoid technical jargon
- Use whole words rather than abbreviations
- Usability test icons and nomenclature

Redundant Coding

- Use redundant coding. This is when the same message is expressed more than once in alternative physical forms. For example, we can recognize the Stop sign (Fig. 5.38) because it: (1) reads "Stop"; (2) is red; (3) is octagonal; (4) has a white border around it; (5) is usually in predictable places such as street corners; and (6) even has a standard height. These different ways of identifying the same thing make it easy to recognize.
- Whenever possible, include labels on your icons.

Make Text Legible

- Make sure text is large enough for the viewing distance. The farther away the user is from the text, the larger the text needs to be. Use the following equation to calculate the required character height for a specific reading distance (Association for the Advancement of Medical Instrumentation, 2018).

$$\text{Character height (inches)} = \text{Distance (minutes of arc)} / (57.3 \times 60)$$

Table 5.2 Recommended character height and font sizes according to reading distance[a] (Courtesy of Association for the Advancement of Medical Instrumentation)

Character height (in.)	Reading distance (in.)	Visual angle (minutes of arc)		Font size (points)
0.112	16	24	Upper size limit	8
0.168	24	24		12
0.251	36	24		18
0.838	120	24		60.5
1.257	180	24		91
0.102	16	22	Preferred (upper bound)	7.5
0.154	24	22		11
0.230	36	22		16.5
0.768	120	22		55.5
1.152	180	22		83
0.093	16	20	Preferred (lower bound)	6.5
0.140	24	20		10
0.209	36	20		15
0.698	120	20		50.5
1.047	180	20		75.5
0.084	16	18	Adequate (upper bound)	6
0.125	24	18		9
0.188	36	18		13.5
0.628	120	18		45.5
0.942	180	18		68
0.074	16	16	Adequate (lower bound)	5.5
0.112	24	16		8
0.168	36	16		12
0.558	120	16		40.5
0.838	180	16		60.5
0.056	16	12	Minimum threshold	4
0.084	24	12		6
0.126	36	12		9
0.419	120	12		30.5
0.628	180	12		45.5

[a]Assumptions: (a) normal vision, (b) contrast ratio > 7:1, (c) luminance >35 candelas per square meter (cd/m^2).

Smallest lower-case letter height. Although overall character size (width as well as height) is the essential characteristic, it is assumed that the width of readable characters is proportional to height (see 19.4.1.3). Consequently, character height is a common convention for specifying character size.

Font size is the distance from the highest ascender to the lowest descender of any character in the font set.

- Use the following equation to calculate the font size for a particular character height (where 1 point = 1/72.3 in. when measured from the highest ascender to the lowest descender of any character).

$$\text{Font size}(\text{points}) = \text{Character height}(\text{inches}) / 0.013837$$

- Use Table 5.2 (from Association for the Advancement of Medical Instrumentation, 2018) to determine the recommended character sizes for various reading distances (under favorable lighting conditions). For reading distances not listed, character heights and font sizes can be determined by using 1 point = 1/72.27 in. or 0.014 in.

- Proper text size also depends on whether the text is a title or other key element, a critical element or a static noncritical element (Lee et al., 2017). Titles and key elements should be at least 0.50°. Dynamic (moving) or critical elements should be at least 0.33°. And, static or non-critical elements should be at least 0.27°.
- Use familiar fonts and avoid fonts with ornamentation.
- Use sans serif fonts (e.g., Arial, Verdana and Tahoma) for digital displays and on-product labeling, but use serif fonts for paper (e.g., Times and New York), such as instructions for use.
- Avoid all capitals in long strings of words. There are two reasons for this. The first is the orthographic theory, which suggests that a wide variety of shapes in words makes it easier to recognize words based on their overall shape. The second is the familiarity theory. That is, we are used to seeing mixed case text, and we recognize familiar things faster than unfamiliar.
- Avoid unnecessary similarity because it can cause confusion. Similarity is the ratio of similar features to different features. For example, AJB648 is more similar to AJB658 than 48 is to 58. Even though both have only one digit difference, the percentage of similar digits is higher in the longer string. Where possible delete unnecessary similar features and highlight different features.
- Avoid confusion between similar characters, especially when mixing text and numbers as in codes. Table 5.3 provides some examples of confusable characters (Association for the Advancement of Medical Instrumentation, 2018).

Table 5.3   Frequently confused characters

T and Y	S and 5	I and L	X and K	1 and 1
O and Q	O and 0	C and G	D and B	H, M and N
K and R	2 and Z	B and R		

Fig. 5.39 Recognizable error symbols

Contrast

- Contrast ratios—Use contrast ratios between 3:1 to 7:1. The higher the contrast ratio, the easier text is to see (Lee et al., 2017).

Make Sure Errors Capture the User's Attention

- To make sure errors attract attention, flash, bounce, or wiggle an object briefly (Weinschenk, 2011).
- Place the error indicator or message near the error itself.
- Use recognizable error symbols or messages. (Fig. 5.39).

Color

Advice for the effective use of color includes:

- Design for monochrome first. This makes sure that you are making good use of other design strategies such as spacing, grouping, and visual hierarchy.
- Use color sparingly by limiting the number of colors and the amount of color used on your product. For example, use no more than three colors for screen displays.
- Use color to get attention.
- Use distinctive colors in your designs. Avoid subtle distinctions in shade.
- Make sure colors on displays vary in luminance as well as hue.
- Design for color deficiency.
- Use color redundantly with other design cues to emphasize its meaning. Spacing, size differences, grouping and other strategies can communicate your message effectively as well.
- Use red for errors.
- Adhere to culture meanings of color.
- Draw attention to warnings.
- Be alert to problems with color pairings.

- Use color changes to indicate status changes.

Resources

- Goldstein, E. B., & Brockmole, J. (2016). *Sensation and perception*. Boston, MA: Cengage Learning.
- Johnson, J. (2014). *Designing with the mind in mind: Simple guide to understanding user interface guidelines*. Boston, MA: Morgan Kaufman.
- Ware, C. (2008). *Visual thinking for design*. Boston, MA: Morgan Kaufman.

References

Anstis, S. M. (1974). A chart demonstrating variations in acuity with retinal position. *Vision Research, 14*, 589–592.

Association for the Advancement of Medical Instrumentation. (2018). *ANSI/AAMI HE75:2009/ (R)2018 human factors engineering – Design of medical devices*. Fairfax VA: Association for the Advancement of Medical Instrumentation.

Goldstein, E. B., & Brockmole, J. (2016). *Sensation and perception*. Boston, MA: Cengage Learning.

Grill-Spector, K., & Malach, R. (2004). The human visual cortex. *Annual Review of Neuroscience, 27*, 649–677.

International Standards Organization. (2008). ISO 9241-303:2008. Ergonomics of human-system interaction — Part 303: Requirements for electronic visual displays. Available at.

Johnson, J. (2014). *Designing with the mind in mind: Simple guide to understanding user interface guidelines*. Boston, MA: Morgan Kaufman.

Koffka, K. (1935). *Principles of gestalt psychology*. New York: Brace and Company.

Lee, J. D., Wickens, C. D., Liu, Y., & Boyle, L. N. (2017). *Designing for people: An introduction to human factors engineering*. Scotts Valley, CA: CreateSpace.

Levitin, D. J. (2014). *The organized mind: Thinking straight in the age of information overload*. New York: Plume.

Llorente, L., Barbero, S., Cano, D., Dorronsoro, C., & Marcos, S. (2004). Myopic versus hyperopic eyes: Axial length, corneal shape and optical aberrations. *Journal of Vision, 4*(4), 5.

O'Regan, J. K. (1992). Solving the "real" mysteries of visual perception: The world as an outside memory. *Canadian Journal of Psychology/Revue Canadienne de Psychologie, 46*(3), 461.

Raichle, M. E., MacLeod, A. M., Snyder, A. Z., Powers, W. J., Gusnard, D. A., & Shulman, G. L. (2001). A default mode of brain function. *Proceedings of the National Academy of Sciences, 98*(2), 676–682.

Thornton, T. L., & Gilden, D. L. (2007). Parallel and serial processes in visual search. *Psychological Review, 114*(1), 71.

United States Census Bureau. (2020). Retrieved on August, 2, 2020, from https://www.census.gov/popclock/.

Ware, C. (2008). *Visual thinking for design*. Boston, MA: Morgan Kaufman.

Weinschenk, S. (2011). *100 things every designer needs to know about people*. Indianapolis, IN: New Riders.

Wickens, C. D., & Carswell, C. M. (1995). The proximity compatibility principle: Its psychological foundation and relevance to display design. *Human Factors, 37*(3), 473–494.

Wurtz, R. H., & Kandel, E. R. (2000). Central visual pathways. *Principles of Neural Science, 4*, 523–545.

Young, F. A., & Biersdorf, W. R. (1954). Pupillary contraction and dilation in light and darkness. *Journal of Comparative and Physiological Psychology, 47*(3), 264–268.

Chapter 6
Hearing

6.1 Introduction

It's strange to think that the concept of sound—a phenomenon so interwoven with human experience—is such a finicky thing to describe. One problem is that, although it is made up of measurable, quantifiable energy, it is simultaneously experiential and unique to each person and environment, making it extraordinarily complex.

Another struggle is that sound is everywhere. In the healthcare environment, it comes from the chatter of people talking in the hallway, blips and beeps from patient monitors, and even the subtle whirring of the HVAC system. Each sound moves at 343 m/s and fills a three-dimensional space (see Fig. 6.1). Yet, our auditory systems effortlessly and instantaneously parse these noises to tell a sensible story about what is happening around us. It stitches together a sonic picture of what is happening, with whom, and in some cases, even why. This happens every moment of every day. You don't tell it to do this for you. It just runs. And, the vast majority of time—it works without a hitch.

This chapter covers what sound is from both a quantitative and experiential standpoint. Afterward, it discusses the basics of our auditory system, as well as its strengths and weaknesses with respect to medical device design. The conclusion addresses what happens when your auditory system strays from "normal hearing," and ventures into the realm of abnormalities, impairments, and disorders.

6.2 What Is Sound?

Sound can be thought of as "acoustic waves" traveling through the air, compressing and expanding. When they have enough energy behind them, they continue to pass through the air, acting on it like a toppling domino passing along its energy to the next in line. Along the way to your ears, however, some aspects of these waves will

© Springer Nature Switzerland AG 2021
R. J. Branaghan et al., *Humanizing Healthcare – Human Factors for Medical Device Design*, https://doi.org/10.1007/978-3-030-64433-8_6

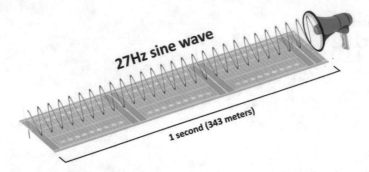

Fig. 6.1 Illustration of speed of sound—a single burst of sound travels slightly more than three (American) football fields per second

change. For example, over a very long distance, the waves with the highest frequencies will be absorbed by things in the environment, such as walls, furniture, and people. Or, they may simply be snuffed out by the air's slight but steady resistance to being displaced. If the distance between you and the source of these waves is far enough, none of them will reach you.

Keep in mind, however, that these waves aren't just a handful of vibrations. Depending on the type of sound produced, there may be hundreds or thousands of waves oscillating at different rates, intensities, onsets, and decay rates. Each wave affects each other. And, each wave is also affected by the environment and medium through which it's moving. Yet in this chaos of movement, your auditory system makes sense of the world around you.

The Building Blocks of Sound

Sound is quantifiable insomuch as we can measure its components. The measurable parts include intensity, frequency, and timbre (pronounced as "TAM" + "BURR"). Time is important as well (see Fig. 6.2). It mediates all the other components, making a sound come across as abrasive or smooth. It's also responsible for making parts of human speech comprehensible. For the sake of simplicity, however, this book will not cover these temporal elements on their own. Instead, they will be discussed in relation to the other elements described below in this chapter where appropriate.

Intensity

Sound needs energy to move through the air. For speakers inside a medical device, this energy is provided by an electrical current that stimulates a magnet at the base of the speaker. The magnet essentially "pushes and pulls" a cone (and other

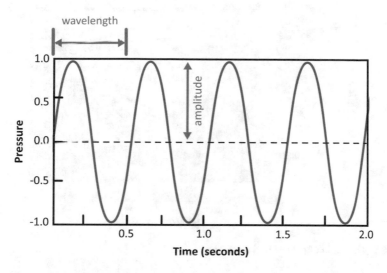

Fig. 6.2 Amplitude and wavelength of a sound wave

materials) attached to it at a rate that matches the rapid electrical pulsing. The cone, in turn, literally pushes air back and forth as it moves, creating the mechanical energy we hear as sound.

Sound energy is measured in decibels (dB). The decibel gets its name from the formative years of telecommunications, when miles upon miles of cabling were being laid for telegraph and telephone services around the world. Bell Labs coined the term Bel—in honor of Alexander Graham Bell (Martin, 1929). A decibel is the equivalent of 1/10th of a Bel (i.e., "deci" + "bel"). The initial intention of a decibel was to measure the amount of signal loss one could expect per (approximately) one mile of telephone cabling. In the field of psychoacoustics, however, the same underlying idea applies to sound (i.e., amount of signal loss). However, instead of wires and linear distance the focus is on people and three-dimensional space.

Since sound intensity involves the pushing and pulling of air all around us, our ears are placed under mechanical stresses. These stresses cause pain when they reach certain sound pressure limits. Generally, pain will occur after being exposed to sounds at or above 120 dB for more than 200 ms. Figure 6.3 shows how the threshold for pain for an average person relates to sound frequency.

Note that an MRI machine is loud enough to produce pain for patients. For this reason, patients should be given ear plugs (or similar) to protect their hearing during a scan. Otherwise, they run the risk of short-term hearing damage lasting for at least a few days. Furthermore, sounds at this intensity can cause you to experience stress and panic. This can mean the difference having a patient complete their examination or stopping part way through. Hearing protection (e.g., ear plugs, muffs) is usually rated to "remove" a specific amount of sound intensity. Hearing protection comes in a variety of styles, sizes, and protection levels (see Fig. 6.4). Each type should

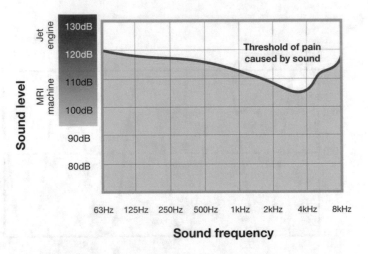

Fig. 6.3 Threshold of pain caused by sound

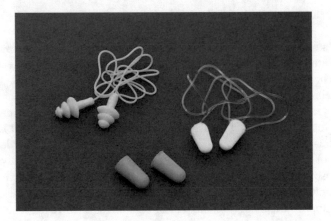

Fig. 6.4 Different styles of ear plugs. (Image by artfotoxyz/shutterstock.com)

indicate its Noise Reduction Rating (NRR)—the amount of decibels it reduces sound intensity.

Sound intensity diminishes as the listener moves farther away from the sound source. One reason is that sound is not directional; it disperses from the source in a three-dimensional space—up, down, forward, backward, etc. Because sound moves in all directions, it essentially spreads its energy over a bigger area.

In healthcare, the interaction between sound intensity and space can mean the difference between hearing a medical device alert (or not). For example, imagine two nurses listening to a patient monitor alarming in an ICU, such as shown in Fig. 6.5. The first nurse is positioned 10 m from the monitor and the second nurse is twice as far (20 m) from it. Since sound occupies a three-dimensional space on its way to these nurses, we know that a good amount of sound energy (intensity) will

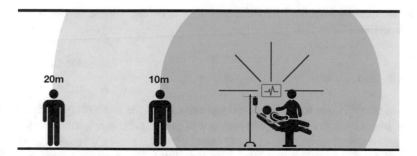

Fig. 6.5 Nurses at 10 and 20 m away from an alarm

Fig. 6.6 Sound meter

be dispersed into the air between the first and second nurses. The question is: how much sound energy will be "lost" in transit?

The answer depends on a few factors, such as the shape and size of the room, how the speaker is designed, and the position of the speaker itself inside this room. However, the underlying rule is that sound intensity (I) is equivalent ($\asymp$) to radius squared (r^2):

$$I \asymp 1 / r^2$$

As a result, the nurse at 20 m would receive a sound intensity approximately four times less than the nurse at 10 m, despite being only twice as far away.

Of course, there are other variables at play too. For example, hard surfaces such as the ceiling and floor will reflect some acoustic energy. When these reflections rapidly build on one another, reverberation occurs. Reverberation refers to multiple, reflected sounds that decay over time. These factors make it difficult to calculate the exact sound intensity reaching a person a priori. For exact measurements in specific locations, however, you can use a sound level meter, such as the one shown in Fig. 6.6.

A sound level meter measures ambient sound intensity. For example, the observation room at Research Collective has an ambient sound level of 41.4 dBA, despite the room being "quiet." An average use-environment will always have at least some ambient sounds, including subtle things like a ticking clock, or the hum of an HVAC system.

Intensity vs. Loudness

It seems like splitting hairs, but sound intensity is not the same thing as "loudness." Sound intensity is the quantifiable measure of sound pressure, whereas loudness is our psychological experience of sound intensity. Each person's perception of loudness will be different. For example, an older adult with advanced hearing loss will perceive an 80 dB sound as less "loud" than a younger adult with average hearing. Yet beyond these physiological differences, there are other factors that affect how loud you perceive a sound. Although a complete discussion of this topic is beyond the scope of this book, we'll discuss a few of the important details when designing a medical device.

Decibels vs. Loudness

People often expect decibels and perceived loudness to follow a linear, 1:1 relationship. They assume that a 10% increase in decibels would create a 10% increase in perceived loudness. Part of this misconception is due to volume controls that follow a steady, consistent increase in loudness as we turn them up.

In truth, decibels follow a logarithmic function—not a linear one. As a result, our perception of loudness doubles with every 10 dB increase in intensity. For example, a 50 dB sound is twice as loud as a 40 dB sound. Importantly, however, this doubling phenomenon is only true among sounds above 40 dB. When the sound is below 40 dB, our perception of loudness actually requires a smaller range of sound intensity for you to experience a doubling (or halving) of loudness.

Equal-Loudness Contour

Generally speaking, high frequency sounds are perceived as louder than low frequency sounds, even when they are at the same intensity. For example, a 40 dB sound at 3000 Hz, sounds much louder than a 40 dB sound at 30 Hz. Frequencies between about 2–4 kHz are perceived as the loudest. Then, frequencies higher than 4 kHz experience a gradual decrease in loudness and get very quiet again as frequencies approach the upper limit of our hearing capabilities (i.e., 20 kHz).

Figure 6.7 illustrates these relationships as the equal-loudness contour. The standard unit of measurement of the subjective experience of sound intensity is the phon (Hartmann, 2004). The baseline for comparison is our perception of loudness of a 1 kHz pure tone (i.e., sine wave) in increments of 1 dB. Thus, a 1 kHz sound played at 1 dB is equal to 1 phon, 2 dB is 2 phons, etc. All other frequency perceptions are compared against this reference point.

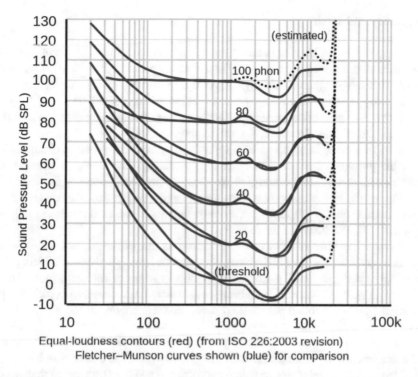

Equal-loudness contours (red) (from ISO 226:2003 revision)
Fletcher–Munson curves shown (blue) for comparison

Fig. 6.7 Equal loudness contours The *standard for perceived loudness is defined by ISO 226:2003 revision.* (Image by Lindosland/English Wikipedia/Public Domain)

The other important part of equal-loudness contours is that humans have a lower threshold at which sound is too quiet to be perceived. Frequencies between about 2 and 4 kHz have a very low threshold, meaning that you hear those frequencies with hardly any sound intensity behind them. But very low frequencies and very high frequencies have much higher thresholds. For example, the sound intensity for the lowest frequency of human hearing (20 Hz) must be intensified to just over 60 dB to even detect it. As a result, that 20 Hz frequency must be amplified about 64 times louder, compared to a 1 kHz frequency.

Sound Duration and Complexity

Loudness is also influenced by how long a sound is played, with the tipping point at about 200 ms (i.e., 1/5th of second). Sounds presented for less than 200 ms are perceived as quieter than longer sounds. This difference is subtle, but important to keep in mind when designing auditory components in medical devices, such as auditory feedback after interacting with a control, or even building layers of sounds to create an alarm.

Frequency and Pitch

The range of human hearing is remarkable; an average adult with normal hearing can hear sounds as deep and low as 20 Hz and as crisp and high as 20 kHz (Ashihara, 2007). This range expands slightly for younger children, and gradually decreases with age. By the time the average person reaches the age of 60–65 years, frequencies above 3–4 kHz often become difficult to hear without amplification (Wiley, Chappell, Carmichael, Nondahl, & Cruickshanks, 2008).

Sound frequency is based on how quickly or slowly air pressure changes. The faster the pressure changes, the higher the frequency. The slower it changes, the lower the frequency. Importantly, these changes are rapid—to say the least. For example, while the lowest point of the human hearing range (20 Hz) creates a pulse of air pressure change 20 times per second, the upper end of our hearing pulses a staggering 20,000 times per second.

This massive difference in cycles per second is one of the reasons why low frequencies travel further than high frequency sounds. And, it's also one of things that makes them better at "getting around" obstacles in your environment, such as passing through walls, doors, and semiabsorbent surfaces (e.g., carpets, people). Simply put, high frequency sound waves shed more of their energy in the air due to the faster rate of oscillations.

To understand this better, imagine you followed a sound wave's action closely, and each time the wave hit a peak or a trough, you punched a punching bag. Since lower frequency sound waves move up and down at a slower rate (time) than higher frequency sounds, you would throw far fewer punches when mimicking low frequencies (see Fig. 6.8). You would fatigue much faster when punching at the faster rate too. The same concept holds true with sound waves—they naturally lose energy as they move through the air. The faster the waveform changes direction, the faster it dissipates energy.

Frequency vs. Pitch

Similar to the difference between sound "intensity" and "loudness," there is also a difference between "frequency" and "pitch." As it applies to acoustics, frequency is a constant, quantifiable phenomenon. On the other hand, pitch is our psychological

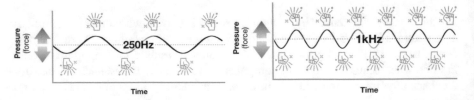

Fig. 6.8 Comparison of 250 Hz and 1 kHz sine waves—less "energy" is lost in a 250 Hz frequency compared to a 1 kHz frequency

experience (i.e., perception) of that frequency. As such, there are several things that can make two people perceive different pitches, even when presented with the exact same sound frequency. This book will only cover a few factors, due to the fact that this topic can get complicated in a hurry.

Sound Intensity and Pitch

Sound intensity can affect how we perceive pitch. Very loud sounds (above 90 dB) make us perceive low frequency sounds as lower, and high frequency sounds as higher than they actually are. However, when sounds have a low intensity, the exact opposite pattern is found—low frequencies sound higher, and high frequencies sound lower. As a general rule, sounds occurring at conversational levels (e.g., 60–70 dB) have little to no effect on how we perceive pitch. Likewise, sound frequencies around 2 kHz are also unaffected.

The Doppler Effect

The odds are good that you have experienced the Doppler effect, even if you don't know it by name. The Doppler effect is the changing of pitch as the sound producing object moves toward as well as away from you (or as you move from it). The examples you've likely witnessed are the siren of an ambulance as it wizzes by your vehicle, or the jingle of an ice cream truck as it passes through your neighborhood.

The Doppler effect reflects the difference between the sound wave size and movement happening simultaneously. This combination effectively "compresses" the sound waves as the sound source moves toward you, and "stretches" the sound waves out as the sound source moves away from you. In reality though, the true sound frequency is somewhere between these lows and highs. Our auditory system makes sense of this compressing and stretching in the form of raising and lowering the pitch, respectively.

Noisy Signals

Our auditory system can only sort through so many frequencies happening at once before it gets overwhelmed. This point of confusion happens rapidly if multiple sounds co-occur with slight differences in frequencies (e.g., sounds at 440, 450, 462 Hz). One reason for this is the phases of each sound wave are slightly out of sync with each other. This causes parts of the wave to cancel out.

Most sounds—for example, a person speaking, a beep of a medical device—are not overwhelming or disorienting. One reason for their clarity is that they possess a clear fundamental frequency, the most pronounced frequency in the assortment of tones. This is almost always a very low frequency. For example, the fundamental frequency of an average male speaker is between about 85–180 Hz (Fitch, 2000;

Pisanski et al., 2014; Titze, 1989), even though their speech also contains an assortment of higher frequencies as well.

Sounds that lack a clear fundamental frequency sound noisy and are difficult to pinpoint. In addition to lacking a fundamental frequency, noisy sounds have little to no rhyme or reason to the frequencies that combine together to make it. It's simply a collection of sonic odds and ends. For example, radio static is an assortment of atmospheric noises, stray radio frequencies, and thermal energy.

Timbre

Every sound has a characteristic tone to it. It's what makes you instantly recognize a good friend or family member's voice, or whether you are hearing a heartbeat through a stethoscope or an ultrasound machine. Amazingly, it's also a component of how we subconsciously recognize when the source of a sound is near or far from us. Timbre is the quality of a sound; the sonic fingerprint that makes it unique. It also happens to be the proverbial "junk drawer" of psychoacoustics. Meaning, it encapsulates all of the sound elements that cannot be neatly described in terms of "intensity" and "frequency" alone (McAdams & Bregman, 1979).

Timbre is based in part on the presence of multiple frequencies occurring in tandem, ranging from just a few to several hundred (or more). Each frequency is produced at a precise intensity level within this tapestry of sound. The Acoustical Society of America Standards Secretariat (1994) adds that other factors such as sound pressure levels and timing—down to mere microseconds (μs)—play a significant role in timbre as well. These timing details are especially important, as they define the envelope of a sound—that is, it's speed of onset and decay. Particularly in speech, envelope doesn't just apply to the start of individual words, but also to each words' various parts and pieces (i.e., phonemes).

Some sounds have a more specific timbre than others. For example, a metal tuning fork produces an indistinct, sterile sound that resonates at a specific frequency. This is done by design to provide musicians with a "clean" sound to use as a reference point while tuning an instrument. A sound with only one frequency to it is known as a pure tone.

A pure tone—also known as a sine tone—are rare occurrences in nature (Stainsby & Cross, 2009). They must be man-made through analog or digital means. Pure tones follow a predictable, smooth pattern of oscillation (see Fig. 6.9). What's more, there are no other frequencies produced that effectively "cancel out" parts of this wave, masking parts of the sound. A tuning fork comes close to a pure tone, but still possesses a few extra frequencies that effectively rules it out of that category (more on this in a moment though).

Both waves in Fig. 6.9 have the same fundamental frequency (f_0) of 500 Hz, but they will sound much different due to the extra frequencies (harmonics, partials, etc.) in the complex tone.

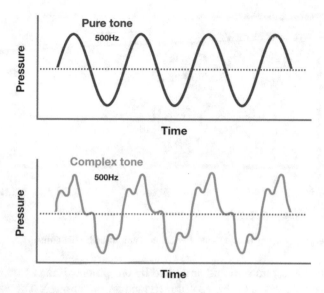

Fig. 6.9 Comparison of a pure tone and complex tone

6.3   How Do We Hear Sound?

Sound is inherently a mechanical thing. That is, it propagates by way of physically "manipulating" the medium through which it is propelled. Our brains, on the other hand, are not mechanical things. Rather, they are complex structures based on electrical and chemical processes. So, the question is: how does a mechanical "sound" get detected by an electrochemical "receiver"? The answer comes in the form of complicated handoff of energy from components both in and outside the ear.

Outer Ear

The visible outside of our ear is called the Pinna (plural pinnae). Its primary responsibility is to funnel and amplify incoming sound toward the auditory canal to aid in hearing. Depending on the sound frequencies reaching the ear, pinnae will amplify that sound by approximately 10–15 dB.

The pinnae are shaped to help detect frequencies used in speech, between 1.5–7 kHz (Fig. 6.10). These frequencies are also important in determining a sound's direction (i.e., localization) as well as what (or who) produced it.

Note that while humans cannot talk or sing as high as 6–7 kHz, parts of our speech (phonemes) create sounds that do. For example, the "th" phoneme sound in the word, "therapy," can register as high as (about) 7 kHz due to the way the sound is produced with our teeth and tongue position.

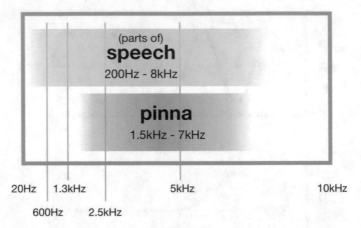

Fig. 6.10 Speech frequency overlap with frequencies amplified by the pinnae

Once sound is centralized and amplified by our pinnae, it is passed through the ear canal and focused onto the eardrum (tympanic membrane). The eardrum is a sensitive, thin, cone-shaped membrane that divides your ear canal, marking the boundary between the outer ear and the middle ear. Its main function is to relay the (mechanical) acoustic waves going through the air to vibration that gets passed along through to the tiny bones inside the middle ear.

Middle Ear

Vibrations passed across the eardrum cause small but precise movements in three bones in the middle ear: the malleus, incus, and stapes. These three bones, referred to as (auditory) ossicles, are the smallest bones in the body. The largest of these three bones—the malleus—is a mere 7–8 mm in length in an average human (Noussios, Chouridis, Kostretzis, & Natsis, 2016). On the other hand, the smallest bone—the stapes—is tiny enough to fit on top of a #2 pencil eraser. These three, tiny bones work in concert, transferring the vibrational energy from the eardrum to the oval window at the threshold of the inner ear. Importantly, the ossicles amplify the auditory signal even more—in some cases, over ten times higher than the signal strength that hit the eardrum in the first place (Moore & Dalley, 2018).

Inner Ear

The sound signal communicated via the ossicles and oval window is transmitted into the fluid-filled region of the inner ear known as the cochlea (see Fig. 6.11). The cochlea is a spiral-shaped tube rolled up on itself like—for lack of a better

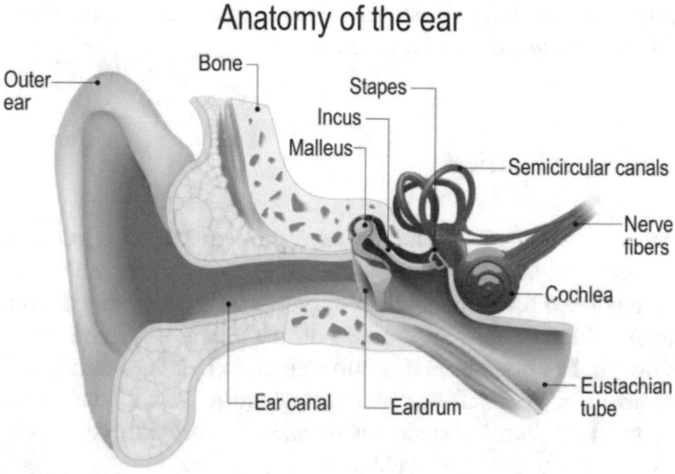

Anatomy of the ear

Fig. 6.11 Overview of the ear. (Image by Designua/shutterstock.com)

analogy—a cinnamon roll. This is the first point at which the "mechanical" sound signal begins to change into a "neural" one. Though, there are a couple more steps before the "electrical" component begins.

One important part of the cochlea is the basilar membrane. This long, narrow, tubular structure nested inside the cochlea contains thousands of hair-like fibers called stereocilia (Stainsby & Cross, 2009). Various segments of the basilar membrane are more or less sensitive to specific sound frequencies. High frequencies (about 6–20 kHz) are picked up by the area closest to the oval window at the start of the cochlea. Middle and low frequencies are registered increasingly deeper into the cochlea, respectively. These areas of high sensitivity to specific frequencies along the basilar membrane are called the characteristic frequency.

One reason that high frequencies get captured at the start of the basilar membrane is that it is comparatively thin and stiff in this region. As you move deeper into the cochlea, the basilar membrane becomes thicker and looser. About 4–5 times looser, in fact (Oghalai, 2004). This is similar to the strings on a musical instrument—thicker, longer strings will produce "low notes" compared to thinner, shorter strings when all the strings are tightened to the same amount of tension.

As specific areas of the basilar membrane resonate with a specific sound frequency, the stereocilia in that region trigger the release of neurotransmitters that interact with thousands of neurons along the auditory nerve. Sound frequencies are interpreted by our brains based on where the stereocilia activate different regions of the auditory nerve, as well as the patterns of stereocilia activation itself. These activation points along the basilar membrane are important because there is a limit to how many points we can discern in close proximity to each other. For example, if two frequencies are too similar to each other—such as a tone 900 Hz and a second at 905 Hz—they stimulate a significant overlap along the basilar membrane and stereocilia. The auditory nerve is not sensitive enough to detect that there are in fact two, distinct frequencies (or more) present.

6.4   Sound Localization

Humans are binaural, meaning that we have two sources (ears) to take in sound. Various physiological adaptations enable us to hear more efficiently. For example, the pinnae are pointed forward and have curves to help determine the direction of a sound. Sounds behind or above us bounce off the pinnae differently than sounds in front or below us. The brain uses this information to localize sound, so that sounds to your left arrive at your left ear sooner, and are a little louder to that ear than sounds on your right. Further, since our pinnae face forward, you can hear sounds in front of you better than sounds behind you.

Our brains automatically process sound location based on the difference between when and how a sound hits each ear. These differences include things like the sound's intensity, frequency spectrum, and timing details (Blauert, 1997). For example, if the sound is coming directly from your left side, it will take just slightly longer for that sound to continue traveling and reach your right ear. We are talking fractions of a single millisecond—microseconds (μs). And, since our heads are solid objects, it will also mask parts of the remaining sound passing by, giving it an "acoustic shadow" by the time it hits our other ear (Akeroyd, 2014). However, when the sound source is directly in front of us, the sound will hit both of our ears at basically the same exact time. The technical term for this concept is the Interaural Time Difference (ITD).

A related concept—Interaural level difference (ILD)—is based on the same idea, but deals with sound level/intensity instead of frequencies. Generally speaking, ILD and sound localization correlate positively—meaning, as ILD increases, so does our ability to localize a sound source. This means that we will experience the greatest reduction in sound level from our left ear to our right ear when we are dealing with higher sound frequencies. The reason for this is that our heads physically "filter out" a lot of the high frequencies passing from left to right. Lower frequencies are less affected, and thus, continue to the next ear largely unscathed. This effect is most pronounced when the sound source is coming directly from our left or right side (i.e., 90°). While still present, the relationship between frequency and ILD is less significant as the sound source moves in front of us (i.e., 0°) or behind us (i.e., 180°). Figure 6.12 shows how ILD sound changes as a function of both the frequency and where the sound is positioned along the user's horizontal auditory field (i.e., azimuth).

One reason for accurate sound localization is to direct visual attention (Akeroyd, 2014). In other words, we use our ears as a type of "alarm" system for our eyes. This makes great sense from an evolutionary perspective—our hearing tells us generally where a threatening sound is coming from, and our eyes handle the fine tuning

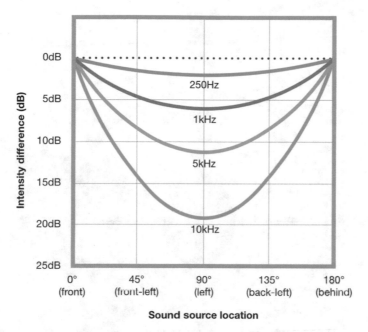

Fig. 6.12 Interaural level difference

afterward. Ostensibly, the more accurate and faster that first effort is, the more likely that organism is to avoid a threat and survive.

Humans do surprisingly well at localizing sounds directly in front of them, as well as slightly off axis up to about 45° from center. This is partially due to the forward angle of our pinnae. By comparison, humans are relatively poor at localizing sounds coming from behind or above us. Similarly, people tend to be more accurate and faster at localizing high frequency sounds (greater than 1 kHz), rather than low frequencies (less than 1 kHz) (Sanders & McCormick, 1987).

These differences are called the Minimal Audible Angle (MAA). MAA refers to the smallest amount of change in location/angle that a person can accurately detect in a horizontal listing plane (i.e., azimuth). Generally speaking, a small MAA is a good thing. It means that our auditory system is able to pare down our environment to a small area from which the sound originated. Then, this area can be visually assessed for further accuracy using the high acuity portion of our eyes known as the fovea, which covers about 5° of visual angle. Not surprisingly, MAA is smallest when the sound source is directly in front of us (0° azimuth). The angle grows increasingly wider as the sound source moves left or right from this center point (Moore, 2012).

Figure 6.13 is an interpretation from an example discussed in Wang and Brown (2006). It shows that the smallest detectable change in angular position between two sound sources is about 1° when those two sounds are played directly in front of someone with average hearing capabilities. However, as these two sound sources are moved toward the person's right side (about 75°), their difference can only be

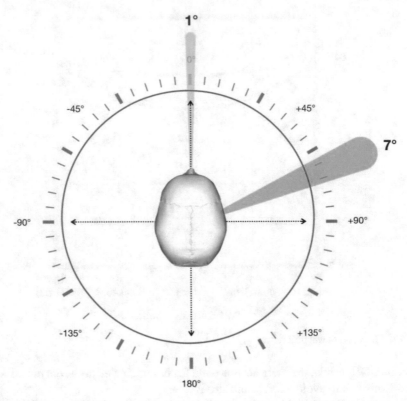

Fig. 6.13 Minimal audible angle (MAA) is smallest near the center of azimuth, and widens as it moves off axis

detected if they are separated by at least 7°. Keep in mind that most laboratory studies of MAA use simple sounds (e.g., sine waves) as the auditory stimuli. This helps verify that the effects observed in the lab are free of potential confounding variables. In the real-world, though, simple sounds are far and few between. Complex sounds are not as "clean" as sine waves, and may lead to worse MAA outcomes, especially in the peripheral regions of the azimuth. The values pointed out in Fig. 6.13 are essentially a "best case" scenario. Outcomes may be worse in a busy, noisy environment such as a hospital ICU or PACU.

Pitch is a vital aspect of identifying the location of a sound as well. As a rule of thumb, people are better at localizing a sound source when that sound covers a broader range or "band" of frequencies, rather than a narrow band or a single sound frequency. This is one reason why a "low battery" smoke alarm warning ("chirp") is difficult to localize in a large room or long hallway. The single, +3 kHz simple tone does not have enough complexity to localize. A similar issue exists among ambulance sirens. Drivers in the United States, for example, often have difficulty localizing the sirens of an ambulance because they use a narrow frequency band ranging from 500 to 1.8 kHz (Withington & Chapman, 1996).

Our ability to use sound to locate a source improves when the frequencies are similar to a human voice (about 300 Hz to 3 kHz). Yost and Zhong (2014) found that sound source localization was most accurate when stimuli were between 1 and 2 octaves apart from each other. Sound localization accuracy decreased when sound frequencies were too close to one another (i.e., less than half an octave apart).

6.5   Hearing Impairments and Disorders

Hearing impairments are common. For example, approximately 10% of all middle-aged adults in the United States experience some form of hearing loss or impairment that affects how they communicate with others (Czaja, Boot, Charness, & Rogers, 2019). One large-scale study suggests that these estimates may be even higher, when accounting for hearing loss affecting at least one ear among those older than 12 years old (Lin, Niparko, & Ferrucci, 2011).

Older adults are especially susceptible to hearing impairments due to natural aging. The prevalence increases to nearly 40% by the time adults reach 65 years of age (Hoffman, Dobie, Losonczy, Themann, & Flamme, 2017). The same study reported that males are twice as likely to have hearing loss that affects their perception of speech compared to females (Hoffman et al., 2017). Though hearing-related impairments have decreased slightly over the past decade, researchers estimate that the prevalence will rise again over the coming years as the median age of the population in many parts of the world continues to increase (Lin et al., 2011).

Beyond age, however, hearing impairment is more prevalent among certain populations with underlying health conditions. For example, those with diabetes are about 2.1 times more likely to have hearing impairment as a result of their disease (Horikawa et al., 2013). This relationship is consistent even when accounting for age. In other words, younger adults with diabetes were consistently about twice as likely to have a hearing impairment when compared to other younger adults (Horikawa et al., 2013).

Hearing impairments and disorders come in a few different forms, including sensorineural hearing loss, conductive hearing loss, auditory processing disorders, and mixed hearing loss. The following subsections discuss a few characteristics and dilemmas related to each one as they pertain to medical device design.

Sensorineural Hearing Loss (SNHL)

SNHLs are a neurological type of hearing loss affecting structures within the inner ear. The cause of SNHL includes acute or chronic damage to small, sensitive things like the cochlea, stereocilia, or auditory nerves. In most cases, the damage is brought about through loud and/or long sound exposures.

One of the most common forms of sensorineural hearing loss among healthcare workers is Noise Induced Hearing Loss (NIHL). Like many hearing impairments, NIHL affects the higher frequency range (i.e., above 4kHZ). Interestingly, this is true regardless of what frequencies of sound a person with NIHL is exposed to. The two critical factors in NIHL are how intense the sound exposure is, and how long the person is exposed to those sounds. Very loud sounds can take mere seconds to induce temporary or permanent NIHL, whereas exposure to moderately loud sounds (e.g., 90 dB) for sustained periods can produce the same deleterious effects.

A study conducted by Konkani and Oakley (2012) revealed that some hospital environments are noisier than others, leading some providers to work in conditions that exceed the National Institute for Occupational Safety and Health (NIOSH) recommendations for safe noise exposure. That study found that neonatal intensive care units (NICUs) had an average noise level of approximately 85 dB, with a peak level of nearly 140 dB (Konkani & Oakley, 2012; Spencer & Pennington, 2015). Spencer and Pennington (2015) argues that the long-term effect of this noise exposure may cause nurses to develop suboptimal hearing, which in turn, could lead to missing critical alarms and events signaled through sound.

Another form of SNHL—referred to as presbycusis—also involves the reduction of perceivable frequencies above approximately 4 kHz (Hoffmann et al., 2016). Unlike NIHL however, presbycusis is a natural phenomenon associated with advancing age and deterioration of certain middle and inner ear components, as well as nerve pathways leading to the brain (Cheslock & De Jesus, 2020). One result of

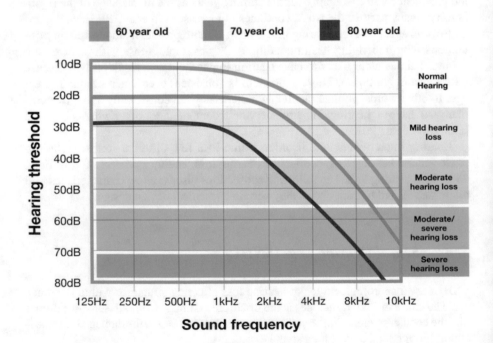

Fig. 6.14 Presbycusis is an age-related hearing impairment that affects high sound frequencies

presbycusis is a soundscape in which everything appears slightly muffled, similar to the experience of wearing large headphones with no music.

At about the age of 60 years, the highest frequencies of your hearing begin to diminish (see Fig. 6.14). These frequencies can be recovered to some extent via devices such as hearing aids. By the time you reach 80 years old, about half of the frequencies used in conversational speech degrade. Raising your voice to an older adult with hearing loss does not help with comprehension, because it also amplifies many of the low sound frequencies that are not affected by presbycusis.

High frequency impairment can diminish speech comprehension. English speech consists of approximately 44 phonemes—the smallest contrastive units of sound specific to language (Harley, 2017). Each phoneme has a unique frequency range as well as envelope. For example, most vowel phonemes occupy the lower frequency range, while about half of the consonant phonemes register in the higher frequencies. Interestingly, it's the consonant phonemes that tell us the most about the "meaning" of words. Phonemes near the top-most range of sound frequencies fall into the category of "voiceless consonants"—sounds that you produce without the use of vocal cords.

This is one reason why words like "sat" and "fat" can be misunderstood over the phone or in areas with lots of background noise. The other reason is that many voiceless consonant sounds are naturally low volume. For example, the "th" sound like in the word "therapy" is both high frequency (i.e., about 6 kHz) and occurs at near whisper intensity levels (i.e., about 20–30 dB). Try as we might, we simply cannot yell or shout certain phonemes like we can with others. This is one of the reasons why simply talking louder or yelling at someone with certain SNHLs won't help them understand you; the voiced phonemes get louder, but the problematic, voiceless ones remain quiet. In fact, yelling can actually have the opposite effect on

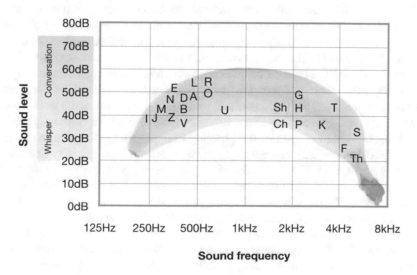

Fig. 6.15 The "banana" shape of speech

speech comprehension altogether, as it distorts key facial features that we implicitly use to aid in speech comprehension.

This spread of phonemes over the hearing frequency spectrum and intensity levels is informally referred to as the "banana shape of speech." Moving left to right, experiment by speaking each phoneme on its own. Listen to each one's natural pitch and loudness (Fig. 6.15).

Deterioration of higher sound registers can also reduce an individual's ability to interact with technology. For instance, elderly individuals with presbycusis might struggle hearing the phone ring, the stove's timer, turn signals, or even the low-battery beep from the smoke detector (Gates & Mills, 2005). This is important when designing medical systems.

People with many types of SNHLs benefit from sound augmenting technologies such as hearing aids. Or, in more advanced or profound cases of hearing loss, binaural cochlear implants may be a viable option. These devices selectively increase about a dozen or so frequency bands to normalize sound perception as much as possible. However, hearing aid technologies to-date lack the nuance and granularity of natural hearing (Balling, Townend, Stiefenhofer, & Switalski, 2020).

Conductive Hearing Loss (CHL)

CHLs are the result of a physical separation or occlusion at some point along the ear's system which inhibits the transfer of sound waves. The cause of CHL can include trauma to the pinnae (e.g., perichondrial hematoma), a foreign body or buildup of wax inside the ear canal, or even scar tissue on the eardrum following a rupture. Less frequently, factors such as autoimmune disorders or physical abnormalities in the ear cause CHLs. For example, abnormal bone growth in the middle ear region—known as otosclerosis—will impede at least one of the three ossicles from moving freely (Upala, Rattanawong, Vutthikraivit, & Sanguankeo, 2017).

The symptoms of CHLs vary between mild and severe, depending on a variety of factors. For example, a total ear cannula occlusion by a foreign object would significantly lower the amount of sound intensity perceived, as well as filter out most of the high frequencies. The resulting effect would be similar to what the world sounds like when you have an earplug in one of your ears. Other severe cases may produce symptoms of vertigo or tinnitus (i.e., ringing in the ears). On the other hand, mild cases of CHL might only cause a slight impairment of sound intensity or frequencies. As a result, it's difficult to categorically create design strategies for users with CHLs; the underpinnings of each person's condition are simply too different.

An example of CHL is tinnitus (from Latin "tinnire" meaning, "to ring") is ringing in the ear is caused by persistent, loud sounds rapidly changing the sound pressure that reaches your ear. These swings in pressure eventually damage the tiny hairs in your basilar membranes. While this affliction is usually temporary for younger adults, tinnitus affects many elderly individuals on a much longer, or even permanent basis (Fig. 6.16).

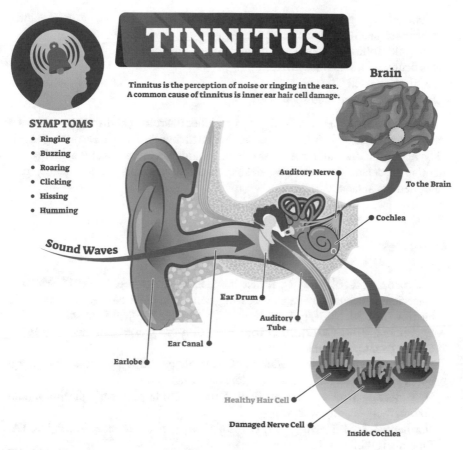

Fig. 6.16 Tinnitus. (Image by VectorMine/shutterstock.com)

Auditory Processing Disorder (APD)

APD is a neurological disorder that compromises key auditory mechanisms, such as pattern recognition, temporal resolution, or the ordering of sounds. Importantly, those with APD have normal hearing thresholds, as defined by standardized pure tone hearing tests (Heine & Slone, 2019). Like other types of hearing impairments, men tend to have a higher prevalence rate of APD compared to women (Golding, Carter, Mitchell, & Hood, 2004).

Those with APD may have difficulty assigning meaning to phonemes, words, or entire sentences while conversing. This is especially pronounced in busy environments with lots of noise and people speaking at once, or even in relatively quiet places that happen to have lots of reverberation or echo. The critical factor is how much acoustic stimuli are present in the person's environment. According to Heine and Slone (2019), APD can produce the following types of behaviors issues among those affected by it:

- Frequent request to "repeat" what was said
- Misunderstanding statements or questions
- Difficulty following complex (i.e., multi-part) directions
- Difficulty with sound localization
- Easily distracted
- Intolerance toward loud sounds

In some cases, APD is part and parcel of other neurological disorders, such as a stroke, a traumatic brain injury, or advancing Alzheimer's disease. Designing medical devices for users with these types of comorbidities requires careful planning and attention to detail, since effective design must address auditory and perceptual-motor deficits in tandem.

Resources

- Atcherson, S. R., Nagaraj, N. K., Kennett, S. E., & Levisee, M. (2015, August). Overview of central auditory processing deficits in older adults. In seminars in hearing (Vol. 36, No. 3, p. 150). New York: Thieme Medical Publishers.
- Eggermont, J. J. (2017). *Hearing loss: Causes, prevention, and treatment.* Cambridge, MA: Academic Press.
- Fact Sheets. American Academy of Audiology: https://www.audiology.org/ publications-resources/consumer-information/fact-sheets.
- How Hearing Works. Hearing Health Foundation: https://hearinghealthfoundation.org/how-hearing-works.
- Northern, J. L., & Downs, M. P. (2002). *Hearing in children.* Philadelphia, PA: Lippincott Williams & Wilkins.

References

Acoustical Society of America Standards Secretariat. (1994). Acoustical terminology ANSI S1.1-1994 (ASA 111-1994). American National Standard. ANSI/Acoustical Society of America.
Akeroyd, M. A. (2014). An overview of the major phenomena of the localization of sound sources by normal-hearing, hearing-impaired, and aided listeners. *Trends in Hearing, 18,* 233121651456044.
Atcherson, S. R., Nagaraj, N. K., Kennett, S. E., & Levisee, M. (2015, August). Overview of central auditory processing deficits in older adults. In *Seminars in hearing* (Vol. 36, No. 3, p. 150). New York: Thieme Medical Publishers.
Balling, L. W., Townend, O., Stiefenhofer, G., & Switalski, W. (2020). Reducing hearing aid delay for optimal sound quality: A new paradigm in processing. *Hearing Review, 27*(4), 20–26.
Blauert, J. (1997). *Spatial hearing: The psychophysics of human sound localization.* Cambridge: MIT Press.
Cheslock, M., & De Jesus, O. (2020). Presbycusis. In *In StatPearls [Internet].* Treasure Island, FL: StatPearls Publishing.

Czaja, S. J., Boot, W. R., Charness, N., & Rogers, W. A. (2019). *Designing for older adults: Principles and creative human factors approaches*. Boca Raton, FL: CRC Press.

Eggermont, J. J. (2017). *Hearing loss: Causes, prevention, and treatment*. Cambridge, MA: Academic Press.

Fitch, W. T. (2000). The evolution of speech: A comparative review. *Trends in Cognitive Sciences, 4*(7), 258–267.

Fletcher, H., & Munson, W. A. (1933). Loudness, its definition, measurement and calculation. *Bell System Technical Journal, 12*(4), 377–430.

Gates, G. A., & Mills, J. H. (2005). Presbycusis. *Lancet, 366*, 1111–1120.

Golding, M., Carter, N., Mitchell, P., & Hood, L. J. (2004). Prevalence of central auditory processing (CAP) abnormality in an older Australian population: The Blue Mountains Hearing Study. *Journal of the American Academy of Audiology, 15*(9), 633–642.

Harley, H. (2017). *English words: A linguistic introduction*. Hoboken, NJ: John Wiley & Sons.

Hartmann, W. M. (2004). *Signals, sound, and sensation*. New York: Springer Science & Business Media.

Hoffman, H. J., Dobie, R. A., Losonczy, K. G., Themann, C. L., & Flamme, G. A. (2017). Declining prevalence of hearing loss in US adults aged 20 to 69 years. *JAMA Otolaryngology—Head & Neck Surgery, 143*(3), 274–285.

Hoffmann, T. J., Keats, B. J., Yoshikawa, N., Schaefer, C., Risch, N., & Lustig, L. R. (2016). A large genome-wide association study of age-related hearing impairment using electronic health records. *PLoS Genetics, 12*(10), e1006371.

Horikawa, C., Kodama, S., Tanaka, S., Fujihara, K., Hirasawa, R., Yachi, Y., et al. (2013). Diabetes and risk of hearing impairment in adults: A meta-analysis. *The Journal of Clinical Endocrinology & Metabolism, 98*(1), 51–58.

Konkani, A., & Oakley, B. (2012). Noise in hospital intensive care units: A critical review of a critical topic. *Journal of Critical Care, 27*(5), 1–9.

Lin, F. R., Niparko, J. K., & Ferrucci, L. (2011). Hearing loss prevalence in the United States. *Archives of Internal Medicine, 171*(20), 1851–1853.

Martin, W. H. (1929). Decibel—The name for the transmission unit. *The Bell System Technical Journal, 8*(1), 1–2.

McAdams, S., & Bregman, A. (1979). Hearing musical streams. *Computer Music Journal*, 26–60.

Moore, B. C. (2012). *An introduction to the psychology of hearing*. Leiden: Brill.

Moore, K. L., & Dalley, A. F. (2018). *Clinically oriented anatomy*. Gurugram: Wolters Kluwer India Pvt Ltd..

Northern, J. L., & Downs, M. P. (2002). Hearing in children. In *Lippincott Williams & Wilkins*. Philadelphia, PA.

Noussios, G., Chouridis, P., Kostretzis, L., & Natsis, K. (2016). Morphological and morphometrical study of the human ossicular chain: A review of the literature and a meta-analysis of experience over 50 years. *Journal of Clinical Medicine Research, 8*(2), 76.

Oghalai, J. S. (2004). The cochlear amplifier: Augmentation of the traveling wave within the inner ear. *Current Opinion in Otolaryngology & Head and Neck Surgery, 12*(5), 431.

Pisanski, K., Fraccaro, P. J., Tigue, C. C., O'Connor, J. J., Röder, S., Andrews, P. W., et al. (2014). Vocal indicators of body size in men and women: A meta-analysis. *Animal Behaviour, 95*, 89–99.

Sanders, M. S., & McCormick, E. J. (1987). *Human factors in engineering and design* (6th ed.). New York: McGraw-Hill.

Spencer, C., & Pennington, K. (2015). Nurses with undiagnosed hearing loss: Implications for practice. *OJIN: The Online Journal of Issues in Nursing, 20*(1).

Stainsby, T., & Cross, I. (2009). The perception of pitch. *The Oxford Handbook of Music Psychology*, 47–58.

Titze, I. R. (1989). Physiologic and acoustic differences between male and female voices. *The Journal of the Acoustical Society of America, 85*(4), 1699–1707.

Upala, S., Rattanawong, P., Vutthikraivit, W., & Sanguankeo, A. (2017). Significant association between osteoporosis and hearing loss: A systematic review and meta-analysis. *Brazilian Journal of Otorhinolaryngology, 83*(6), 646–652.

Wang, D., & Brown, G. J. (2006). *Computational auditory scene analysis: Principles, algorithms, and applications*. Hoboken, NJ: Wiley-IEEE Press.

Wiley, T. L., Chappell, R., Carmichael, L., Nondahl, D. M., & Cruickshanks, K. J. (2008). Changes in hearing thresholds over 10 years in older adults. *Journal of the American Academy of Audiology, 19*(4), 281–292.

Withington, D. J., & Chapman, A. C. (1996). Where's that siren? *Science and Public Affairs, 2,* 59–61.

Yost, W. A. (2016). Sound source localization identification accuracy: Level and duration dependencies. *The Journal of the Acoustical Society of America, 140*(1), EL14–EL19.

Yost, W. A., & Zhong, X. (2014). Sound source localization identification accuracy: Bandwidth dependencies. *The Journal of the Acoustical Society of America, 136*(5), 2737–2746.

Chapter 7
Cognition

For 300,000 years, we *Homo sapiens*, the most advanced species on the planet, existed mainly as hunter-gatherers. We had evolved a nervous system that responded automatically to sudden threats, such as the crack of a branch or the darkening of an entrance (Levitin, 2014). These conditions were very different from our modern world. For most of our evolutionary history, a typical human would rarely travel more than a few dozen miles and meet only a few hundred people in an entire lifetime. It was rare for them to move at a pace faster than 12 miles per hour (5-min mile), and that pace could be maintained only for a short distance. These conditions were very different from our modern lives.

Now of course, we have a 24-h news cycle, millions of technological gadgets, and are exposed to thousands of emails, newspapers, magazines, blogs, text messages, postal mail, advertisements, signs, and labels. These all exploit a technology that did not even exist until recently in our evolutionary history—reading. If we go grocery shopping, we are confronted with about 50,000 different products, brightly labeled with cartoon characters, names, and colors, all vying for our most precious resource—attention. On our way to work, we can encounter thousands of people, and our cars travel over 75 miles per hour (admit it), as we commute. If you commute on an airplane, you can reach speeds ten times that.

Not only has our pace of life changed, the pace of technological progress has changed by several orders of magnitude. This is a new development as well. For most of our evolutionary history, technology did not change much from decade to decade or even over millennia (Ornstein & Ehrlich, 2018). Advances did, however, begin about 15,000 years ago, when the world's population was about 5 million (Wright, 2000). Some people began to farm and domesticate animals, and the pace of technological progress accelerated. By today's standards, the rate of progress was less than breakneck; it took humans another 10,000 years to start using written language, mainly to keep track of transactions. Over time, however, the pace took off. For example, the population grew from 600 million to 1 billion in the 250-year period between 1600 and 1850. The population doubled again in the 43-year period between 1945 and 1988. There are now more than 7.5 billion people on earth, and

© Springer Nature Switzerland AG 2021
R. J. Branaghan et al., *Humanizing Healthcare – Human Factors for Medical Device Design*, https://doi.org/10.1007/978-3-030-64433-8_7

next month the world population will grow by more than the number of humans that lived on the planet just 100,000 years ago (Levitin, 2014).

This degree of growth and change causes real challenges for our hunter-gatherer minds. In fact, it is important to remember that our brains are based not only on our hunter-gatherer homo-sapiens ancestors, but also on hominids who walked upright 6 million years ago, and lived much like our hunter-gatherer ancestors. It is no wonder that progress is dizzying these days; space vehicles, televisions, personal computers, and the Internet were all invented in the last 50–70 years, just 1/100,000th of time since hominids first walked upright. This technology, and the way in which life has changed, can be a mismatch for our hunter-gatherer brains, creating anxiety and sometimes error. Our evolutionary history is evident in what we are good at (our capabilities) and what we are not good at (our limitations) today. The design challenge is clear that provides life improving and lifesaving devices while accommodating a brain that evolved under vastly different circumstances.

7.1   Cognitive Resources

Cognition refers to perceiving, thinking, remembering, reasoning, and problem solving (Neisser, 1967). Our ability to do these things is impressive but limited. For example, we can only pay attention to a small amount of information at once, and it's difficult to think of two things at once. In fact, attempting to do more than this depletes our mental energy or what is often referred to as our cognitive resources.

We encounter information in the world much faster than we can process it. For example, the processing capacity of the conscious mind is about 120 bits per second (Clarke & Sokoloff, 1999), representing a sort of speed limit for information we can pay conscious attention to at any moment. Of course, there is far more processing happening subconsciously, below the threshold of our awareness, and this impacts how we feel. Nonetheless, there are clear limits on our conscious processing, and exceeding those limits diminishes our performance. What's worse, all this information processing, including distinguishing the inconsequential from the important, makes us tired. Manufacturing, dispersing, and reuptaking brain chemicals takes energy. Neurons are living cells with a metabolism; they need oxygen, glucose, and rest to survive, and when they work hard, we get fatigued (Gailliot & Baumeister, 2007). To give you an idea of how metabolically expensive our brains are, the entire brain weighs three pounds (1.4 kg) and so is only a small percentage of an adult's total body weight, typically 2%. But it consumes 20% of all the energy the body uses.

To survive, neural communication needs to be rapid, and neurons communicate with each other hundreds of times per second. So, even an hour of relaxing or mind wandering requires about 11 calories. Light mental activity such as reading, which requires significant use of the prefrontal cortex, consumes about 42 calories per hour. Absorbing new information by learning requires about 65 calories, mostly for synaptic transmission; that is, connecting neurons to one another and, in turn, connecting thoughts and ideas to one another (Clarke & Sokoloff, 1999).

Evidence of this metabolic cost comes from the finding that eating or drinking glucose improves performance on mentally demanding tasks. For instance, when one half of research participants eat sugary food, and the other half do not, the sugary food group solves difficult problems significantly better. The additional glucose appears to feed the neural networks involved in solving the problem (Hoyland, Lawton, & Dye, 2008).

The limitations on our cognitive resources present serious challenges for design. How do we attract attention when everything else is trying to attract attention? How do we get people to interact successfully with our device, yet still have cognitive resources remaining to interact with the patient? How do we present data in a way that informs rather than overwhelms? Let's begin to answer some of those questions by examining our most precious, and probably limited, resource—attention.

7.2   Attention

For context, we show an adaptation of Lee, Wickens, Liu, and Boyle's (2017) information processing model again in Fig. 7.1. Chapter 5 already discusses the sensory register and perception, so we will not revisit that here. Instead, we start with attention, and move on later to working memory and long-term memory.

Attention, defined as the narrow focus of consciousness, is required to perceive stimuli, to focus on important information, and to filter out distractions. What we

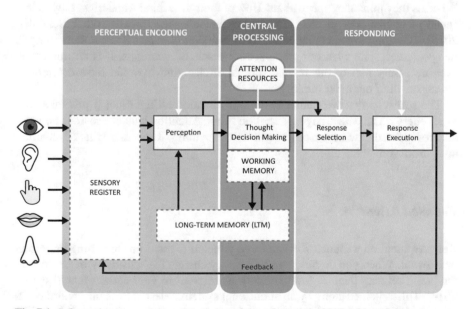

Fig. 7.1 Information processing model. (Adapted from Lee et al. (2017) with permission)

perceive, what we recognize and make sense of, is dictated to a large degree by what we pay attention to. If you do not attend to something, you will not perceive it.

As mentioned in Chap. 5, some conspicuous stimuli, such as loud sounds, bright lights, or things moving quickly (that might indicate a predator), grab our attention automatically. In fact, our brains get used to (habituate to) current situations, and any change from the current situations attracts our attention. This is especially true of changes in the environment or things in the environment that stand out from all of the other things.

Figure 7.1 highlights a central point: Attention is required for almost every mental activity. We need it to perceive stimuli in the environment, hold information in memory, make judgments, solve problems, make decisions, choose alternatives, and execute actions. Attention does so much, yet there is so little of it to go around that we must be selective about what we pay attention to. Selective attention means attending to, and perceiving, one thing while ignoring others. This is the type of attention we experience at noisy parties, where you can focus on conversation even though there are other conversations happening at the same time. This is called the cocktail party effect. In fact, one of the reasons we are exhausted after a cocktail party, besides the booze, is that selecting what to pay attention to and what to ignore is hard work, requiring lots of cognitive resources.

On the other side of the attentional spectrum, there are times when we do not pay attention to anything in particular, letting thoughts wander in and out of consciousness in a seemingly random fashion. This is a sort of default situation for attention, called the mind-wandering mode. Mind wandering is critically important for our ability to recover from intense cognitive work, as well as for creativity and problem solving. People tend to underestimate how much time they spend mind-wandering. Whereas they think they spend about 10% of their time mind wandering, they in fact spend more like 30% of their time in this default state (Smallwood & Schooler, 2015). The thing to remember about mind-wandering is that it is not something to be avoided, despite what our grade school teachers' warnings. It is an important, even necessary part of our mental life. Shortchanging it would have deleterious effects on all of our other thinking.

The reality is that we cannot focus our attention all the time; it just takes too much energy. The phrase paying attention is apt. Attention is such a scarce and valuable resource, that when we pay attention to one thing, it necessarily takes attention away from something else.

Focused Attention

But, pay attention we must. You can force yourself to focus on something important or relevant. When you do this, you effectively change the sensitivity of neurons in the brain, refining them to fire in situations relevant to your current task (Levitin, 2014). This is helped along by an attentional system, referred to as the central executive, which prevents you from being distracted when you're engaged in a task. The central executive helps by limiting what will enter your consciousness so that you can focus on what you're doing uninterrupted. In fact, the central executive operates in direct opposition to the mind wandering mode. When one activates, the other deactivates.

We often have numerous things vying for our limited attention, leading us to attempt to process multiple sources of information at the same time. This, however, does not work. We simply cannot process all the information coming at us at once. Instead, we prioritize the input as best we can, using selective attention. We determine what to pay attention based on four factors (Lee et al., 2017):

- Salience—Salient stimuli are distinct, loud, large, etc., so they capture our attention. Certain stimuli tend to stand out because they are unlike others in their surroundings or because they have been important for survival in our ancestral environment.
- Effort—It is easier to pay attention to things that do not require much effort. If the stimulus is directly in front of our face, it requires little effort to attend to it, and is usually perceived. Further, we are frugal with our energy, so we are more likely to scan short distances rather than long ones (Lee et al., 2017), especially when we are fatigued.
- Expectancy—We pay attention to places where we expect to find valuable information. This is one of the reasons that user interface consistency is so important. If devices are consistent, people will establish accurate expectancies for where to find valuable information, and will focus their attention there.
- Value—You only pay attention to what you need for the task at hand, and to information that helps you complete that task.

Multitasking

If you are in the job market, you are likely to encounter job ads looking for people who can multitask. Multitasking entails processing multiple things or tasks at the same time. If you are a human being, do not apply for that job, because you are not qualified. It is true that many of us think we are good at multitasking, but actually we are deluding ourselves. And, in case you were wondering, young people do not multitask any better than older people (Carrier, Rosen, Cheever, & Lim, 2015). At any one time, our attention can be either selective or divided. When it is selective, it forces us to pay attention to only one task or source of information. When it is divided, it tries to allocate extremely limited attentional resources between tasks. The tasks then interfere with each other, and we perform poorly on both of them. This causes us to work slower, to make more errors, or both. When we think we are multitasking, we are just switching from one task to another rapidly, and usually ineffectively. As a rule, we would be better off conducting each task in succession.

But it gets worse; there is a cognitive and metabolic cost every time we switch tasks. Shifting attention from one activity to another causes the brain to consume substantial amounts of glucose. Repeated shifting causes the brain to burn through fuel so quickly that we feel exhausted and disoriented after even a short time. We literally deplete nutrients in the brain, leading to decreased cognitive and physical performance, and increased anxiety. Making matters worse, multitasking requires repeated decision making about which task to focus on and when. Just making these minor decisions is metabolically expensive—small decisions use as much energy as big ones. This leads to a depleted state in which, after making lots of insignificant decisions, we might end up making truly bad decisions about something important (Levitin, 2014).

Moreover, multitasking increases the production of the stress hormone cortisol as well as the fight-or-flight hormone adrenaline. These can overstimulate our brains and cause mental fog or scrambled thinking. If we do it too often, it can even make us dizzy. So, forget multitasking, once you are on a task, it is best if you can stay on that task.

Sustained Attention

Because our nervous systems are finely tuned to detect changes, we are far less adept at paying attention to the same thing that doesn't change for long periods of time. For example, vigilance tasks, like reading many diagnostic images one after another for long periods of time, can potentially reduce performance. In laboratory studies of vigilance, both speed and accuracy tend to follow a curve like the one in Fig. 7.2, with performance decrements over the first half hour or so. Again, our evolutionary history works against us here. In our history we rarely needed to stare

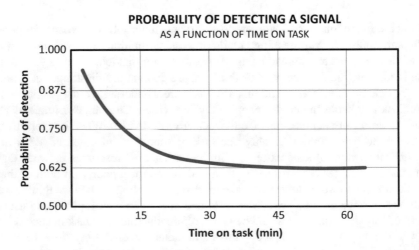

Fig. 7.2 Common performance on vigilance tasks

at one thing for hours on end. Doing so, would put us at risk of being eaten by predators approaching from the opposite direction.

7.3 Memory

Memory is a system for retaining information and bringing it to mind when needed, enabling us to use previous information about stimuli, images, events, ideas, and skills after the original information is no longer present (Goldstein & Brockmole, 2016). In general terms it involves three processes (Baddeley, Eysenck, & Anderson, 2020)

- Encoding—Converting information into a form that is usable by memory. This requires transducing electromagnetic energy, sound waves, pressure, etc. into neural impulses processed by our nervous system. Generally, this demands attention. Failure to encode material is one of the most common problems in memory.
- Storage—Retaining information in memory.
- Retrieval—bringing stored information to mind—Retrieval is influenced by interference and stress. For example, have you ever been put on the spot when giving a presentation, and cannot recall what you wanted to say? Often, we know the correct response, but we cannot quite retrieve it—until later once we have relaxed.

We will also discuss the structure of memory and what each type of memory is for, beginning with working memory.

Working Memory

Working memory (WM) contains the contents of our awareness, combining our most recently perceived information with relevant contents of Long-Term Memory (LTM). This combination enables conscious cognition such as thinking, calculation, judgment, comparison, and prediction. What you are thinking about right now is contained in WM. In fact, in many ways, WM is what we think about when we think about thinking. Sounds pretty important, right? It is, but despite its importance, WM has some serious limitations. Specifically, it can only hold a small amount of information for a brief period of time, and that information tends to be fragile.

Capacity of WM

It is difficult to know the exact capacity of WM. George Miller (1956) inspected several studies, and concluded that the average person could hold seven plus or minus two items in WM. Follow-on studies and reviews (Baddeley, 1994; Cowan, 2009, 2010) estimate that WM capacity may be closer to four items.

You might ask, what is an item? If you just met a person named Timothy, and he is in your WM. Is that one item named Timothy or is it seven items, one item for each letter? For instance, the sequence of letters, "M B and I" are probably not meaningful to you, so each letter represents its own item in WM. Those letters would fill three of four (or seven if you believe Miller was correct) available slots in your WM. But what if we reordered the sequence to I B M. Is that still three items? Well, if you are familiar with this large technology company, it is only one. You could add in HP, AT&T, and BP to round out your four items. Similarly, to medical practitioners, another reordering to B M I (body mass index) might also have meaning.

Combining what seems to be separate items like this into a meaningful whole is called chunking. Our WM holds about four chunks. Chunking is the process of grouping individual, but related, pieces of information into one item, enabling us to expand the capacity of WM.

Chunking is powerful. Here is an example. Ericsson and Staszewski (1989) asked a student to participate in a research study on memory. As expected, when the student started, he had an average digit span of about 7. Over the next several months, in 1-h sessions, the experimenters read digits to the participant at a rate of one per second, and then asked him to repeat the digits back to them. Over time, the participant got better and better and, after 320 one-hour sessions, could memorize up to 79 digits. This is an impressive feat, and Ericsson and Staszewski suspected that he must be chunking the numbers, but they were not sure how. The answer was simple; this student just happened to be a distance runner, and runners are often obsessed with times. So, imagine the first numbers were 2 9 2 0 4 5. The student might recognize that this time is only 2 s slower than the women's world record in 10,000 meters of 29:18.45. If the next bunch of numbers were 4 0 8 2 4, he might recognize that this was close to his mile time (4:08.24) last week. The key was to chunk numbers into times that were either good or bad for various distances. They were no longer digits, they were filled with meaning (distance, good or bad, men's or women's race, etc.). Adding the meaning resulted in chunks.

Categorization has a similar effect as chunking, providing the means to remember more, by placing things in related categories. This in fact is why categorization is so integral to information architecture (see Chap. 12). People have a natural tendency to form hierarchical categories to make sense of the world and to increase our WM. In fact, a portion of our prefrontal cortex is dedicated to this task of partitioning long events into chunks (Levitin, 2014). In this way, we create hierarchies with little effort.

Duration of WM

Despite chunking, WM mainly provides temporary storage. Its contents are transient, decaying quickly unless rehearsed or otherwise paid attention. For example, Peterson and Peterson (1959), gave participants triplets of letters, like XPJ or BTP, to recall, and found that participants' ability to remember the letters got worse the longer they waited between learning and testing. These results suggested that short-term memory (what WM used to be called) lasts up to 18 s. Using different methods, contemporary scientists however paint a bleaker picture, suggesting that WM lasts as little as 2 s (Fig. 7.3).

A famous cognitive model developed by Card, Moran, and Newell (1980) indicates a half-life (the time after which recall is reduced by half) of WM items to be 7 s for three chunks of information and 70 s for just one chunk.

Because contents in WM decay quickly we remember short worlds better than long ones. That is our ability to recall a word gets worse as the time required to say the word gets longer. People can remember only as many words as it takes to say in roughly 2 s. This is one of the reasons that healthcare providers use the term EEG rather than electroencephalogram or PET rather than positron emission tomography. Think about this the next time you choose names for features or functions in your user interface. Keep the names as short as possible without introducing confusion. Of course, we say that like it's easy. Design is difficult.

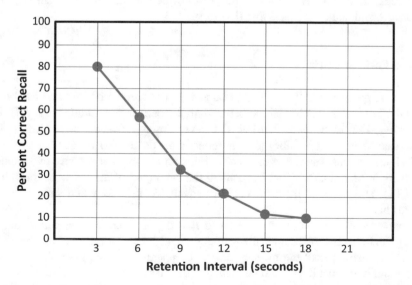

Fig. 7.3 Duration of short-term memory. (Data from Peterson, L. R., & Peterson, M. J. (1959). Short-Term Retention of Individual Verbal Items. Journal of Experimental Psychology, 58, pp. 193–198)

WM and Attention

Material in WM is fragile, requires attention, and gets disrupted by other activities and interruptions. This means that maintaining information in WM requires a great deal of mental effort, which depends on a limited supply of attentional resources. If these resources are being used on another task, WM contents decay rapidly. We all have experienced forgetting between screens or even on the same screens when integrating or comparing information that is far away. Information. It gets worse of course when performing complex tasks with lots of information, and when we are distracted.

Phonological Similarity Effect

Information in WM is stored in various ways. For instance, if we see the word, "Cat" we store a visual representation (image of a cat and image of the word, cat), a semantic representation (knowledge about cats, like they chase mice and have bad attitudes), and a phonetic representation (the sound of the word cat). Interestingly, the phonetic representation is particularly strong. We tend to say the word, "cat" to ourselves sub vocally and it sticks around in WM. As a result, if we are asked to remember a list of words that happened to include the word, we sometimes confuse it with rhyming words like hat, sat or pat. Similarly, when recalling letters, we are more likely to confuse the letter P for E than E for F or R for P. The phonetic representation is stronger than the visual. This is called the phonological similarity effect, and suggests that items and labels on your device should have different sounding names. Avoid names or labels that rhyme.

Serial Position Effect

If you are given a long list to recall, you are likely to remember the first few words and the last few words in that list better than the ones in the middle (Fig. 7.4). The heightened ability to remember items at the beginning is called the primacy effect, whereas the enhanced ability to remember items at the end is called the recency effect. In fact, this was one of the first findings suggesting that we have two memory systems, WM and LTM. That is, items in the beginning of the list were transferred to LTM, and those at the end were still available in WM when saying the list words back.

Interestingly, follow-on work showed that if you delay asking participants to recall the items by 30 seconds, the primacy effect (memory for words at the beginning) stays strong, but the recency effect disappears. This suggests that the most important items in a list should be placed at the beginning.

The specifics of these studies are intended to provide background and to round out your understanding, but the main points are that:

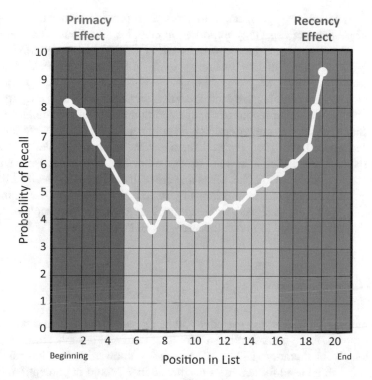

Fig. 7.4 Serial position effect. (Data from Murdock Jr., B. B. (1962). The serial position effect of free recall. Journal of experimental psychology, 64(5), 482)

- WM stores very little information at a time
- You can effectively increase the capacity of WM by chunking
- Material in WM decays quickly
- Focusing attention on WM contents keeps them active longer
- As soon as you take attention away from those contents (e.g., when interrupted), they are likely to be lost
- Newly added items vie for our limited attention, and displaces old information
- Working memory capacity decreases under stress

Prospective Memory

Prospective memory is the memory to do something in the future. Often, we forget to do things we plan to do. For example, we might leave the house without our wallet, or we might forget to take our medication at night. Worse, we might forget to pre-clean an endoscope before it is sent to sterile processing (Jolly, Hildebrand, & Branaghan, 2013). Errors in prospective memory are common in everyday life but can be particularly problematic in healthcare. Reminders can be built into systems that enable people to remember to do things.

One way to improve prospective memory is to provide a checklist (Gawande, 2010). Checklists are especially important in complex situations with many tasks, subtasks, interruptions and time constraints. It is ideal to have two individuals use the checklist together, with one reading the items and the second checking the items or completing the activities.

Aviation and spaceflight have used checklists for decades. In fact, they are largely responsible for the stellar record of commercial flight in developed countries. In the United States, checklists are required on all commercial flights. Despite this, medical professionals have often resisted them. HCPs seem more likely to employ checklists after being reminded that airplane pilots and astronauts use them. They also seem more likely to use checklists that are well designed. Such checklists take advantage of grouping (chunking) items by stage or by role. They use familiar but short terms. They are designed to be as quick as possible to complete. In other words, they are designed to match the strengths and weaknesses of WM.

Long-Term Memory

Some information in working memory will be rehearsed enough, and become well enough known, to be transferred to a very long term and durable store of memory called long-term memory (LTM). LTM includes knowledge of facts, skills, and events, and can be used for activities conducted in WM and in perception. It is different from WM in that it has a very high capacity, and very long duration; information in LTM can potentially last forever.

Contextual Memory, Recognition and Recall

Because LTM's capacity is so large, because it stores so much information, it can be difficult to retrieve the specific piece of information you want. Some pieces of information, however, are easier to retrieve than others. For example, it is simple to remember information that we use frequently or that has been used recently. If you use the same device several times a day, every day, you will retrieve the information you need easily. On the other hand, if you were trained on a medical device 3 months ago, and have not used it since, you will find retrieval difficult. This is a problem for emergency procedures, which are rarely required. Their infrequency makes them less likely to be correctly recalled when needed. Relying on LTM in this situation would be a mistake. As a result, many emergency procedures use checklists rather than relying on the user's memory.

LTM houses our knowledge, arranging it according to meaning so that related topics are closer to each other than unrelated topics. In reality, of course, their neural representations are not physically closer in the brain, but their neural connections are stronger. Because of this reliance on associations, we remember better when we are reminded by something related—something in context.

An example of this is recognition memory. When we recognize something we simply need to judge if the information presented to us matches information stored in memory. This is like taking a multiple-choice test, rather than a fill in the blank or, even worse, an essay test. Related items serve as cues to help us remember.

This reliance on context and associations is responsible for the phenomenon of state dependent memory. People tend to remember more information under the same physical and mental conditions in which they learned them. Godden and Baddeley (1975) conducted a fascinating experiment demonstrating this. They asked research participants to memorize a list of words either underwater while scuba diving or on land. Later, they tested those participants to recall those words either on land or underwater. They found that if people learned the words underwater, they recalled them better while underwater. Similarly, if they learned the words on land, they recalled them better when they were on land. This is often referred to as state dependent (or context dependent) memory. The cues that were present during the encoding (memorizing) phase were helpful as cues in the retrieval phase. Then, when we try to remember the experience, those seemingly irrelevant stimuli can serve as retrieval cues. This is why it is best for students to take tests in the same room where they learned their lessons. Various parts of the environment, even ones you would never expect, serve as retrieval cues. This also provides evidence for the value of simulation training and usability testing in well simulated environments.

As an aside, context-dependent memory works for physiological states too. For example, state-dependent memory occurs in response to drugs. When participants are asked to learn word lists after smoking marijuana or smoking a placebo, they remember more words when tested under the same conditions. It appears then that contextual cues can be either external in our physical environment or internal in our physiological environment.

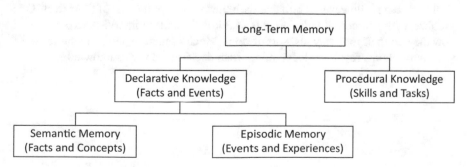

Fig. 7.5 Structure of long-term memory

Structure of Long-Term Memory

To get along in the world, we need to know a lot. We need to know facts like the security code on the office door, events like when we talked with our boss last time, and what was discussed. We need to maintain skills like reading, writing and maybe computing. We need to remember locations like where our office is located or where the "Save" function resides on a computer screen. Although these things all require memory or knowledge (LTM), they seem like very different kinds of knowledge. Cognitive scientists have tried to understand the organization of this knowledge for quite a long time. For instance, cognitive scientists often divide LTM into several processes, illustrated in Fig. 7.5. First, LTM is divided into either declarative knowledge or procedural knowledge, discussed next.

Declarative vs. Procedural Knowledge

Declarative knowledge involves facts and events. You might, for example, remember that Abraham Lincoln was the 16th president of the United States even though you do not recall exactly when you learned that fact. You might also have a vivid memory of your sixteenth birthday; you remember a specific episode in your life. These are both examples of declarative knowledge, knowledge of facts or events rather than skills or procedures. They are called declarative because they are explicit and can be verbalized. You can tell someone else about this knowledge. There are two types of declarative knowledge; the factual knowledge of Abraham Lincoln is called semantic memory, and the event knowledge of your birthday is called episodic memory.

Procedural knowledge is different; it includes knowledge of skills and tasks. It cannot be taught by verbalizing, but must be taught by demonstration and learned through practice, trial and error. Procedural knowledge includes skills that are largely subconscious and non-verbalizable; you cannot tell someone a skill. You do not teach someone to ride a bicycle by telling them. They need to try it for themselves, and learn it through trial, error, and feel. It's going to get messy. Once the skill is mastered, however, it remains with the learner for the rest of their lives. One problem with procedural knowledge is that it is difficult to interview experts about how they do their job; they just sort of do it. Much of their expertise is the result of procedural, nonverbal knowledge rather than declarative, verbal knowledge.

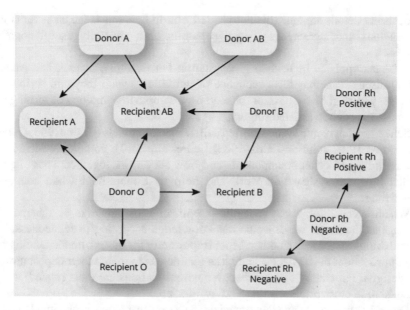

Fig. 7.6 Semantic network of blood typing derived using Pathfinder network algorithm. (Courtesy of Professor Roger W. Schvaneveldt)

Organization of Semantic Memory

As mentioned above, declarative knowledge is divided further into semantic and episodic memory; that is into memory of facts and memory of events, respectively. The critical thing to remember about semantic memory is that it is organized according to meaning. We discuss a few of these organizational schemes below. For example, we can think of semantic memory as a network in which concepts are depicted as nodes, and relationships among those concepts are depicted as links. When we think of one concept we naturally tend to think of related (linked) concepts. This phenomenon is called priming, and the mechanism that makes it work is referred to as spreading activation. An example of a semantic network is depicted in the Pathfinder (Schvaneveldt, 1988) network in Fig. 7.6. This network depicts an expert's knowledge of blood typing. You can see that two factors (Rh and blood type) need to be combined to make the right choice about donors and recipients. For example, Rh-negative donors can donate to Rh-negative and Rh-positive recipients, whereas Rh-positive donors can only donate to Rh-positive recipients. Another portion of the network donors with the blood type O can donate to patients with blood types, A, B, AB or O. And, people with blood type AB can receive blood from people with blood types A, B, AB, or O.

Sometimes it is helpful to think of knowledge as organized around a central concept or topic, often referred to as a schema. When we encounter new information, we attempt to fit it into our existing schema, or set of expectations. Importantly, details that are consistent with our schemas are more likely to be retained, whereas inconsistent details are more likely to be left out. In fact, this is the basis for confirmation bias and stereotypes—negative results of a generally useful cognitive process.

Schemas apply to more than just systems. For example, Schank and Abelson (1977) discuss schemas for a sequence of activities, which they call scripts. For instance, you might have a script for sending a sample to pathology or turning on a robot used in robotic assisted surgery.

Schemas of dynamic systems, often called mental models (Craik, 1943; Gentner & Stevens, 1983; Johnson-Laird, 1983; Lee et al., 2017), commonly include our understanding of system components and how they work together. These help us predict how the system will work and help us troubleshoot when something goes wrong.

Whether you think about semantic organization as networks or schemas, the message is the same; we organize our knowledge according to relatedness, with related things closer to each other. More importantly for design, understanding how users think about their systems and tasks enables us to match their expectations in our designs. For example, if we discover what functions are most related, we can place them in close proximity on the user interface,

One other type of knowledge organization, cognitive maps, is absolutely critical to design. Cognitive maps represent spatial information like the layout of a surgical suite or even the layout of a computer display. Because of our hunter-gatherer roots, requiring us to remember where food and water could be found, we have evolved specialized cells in the hippocampus for remembering the spatial location of things. This is the exact same structure, by the way, that enables squirrels to locate buried nuts several months after burying them.

Getting back to people, one famous study demonstrates the importance of the hippocampus and location memory. To pass a qualifying exam, London taxi drivers

Fig. 7.7 The hippocampus in London taxi drivers are larger than those in other people of comparable age and education ("London" by mister.johnson is licensed under CC BY-NC-ND 2.0)

often study for 3 or 4 years to memorize routes through the city. London is difficult to learn since it has many dead ends, one-way routes and so on. Neuroscientists found that the hippocampus in London taxi drivers was larger than those in other people of comparable age and education (Maguire et al., 2000). The drivers' hippocampus had increased in volume due to the location information they needed to track. Today, place memory tells us where to look on a display, and is the main reason that items on displays should be placed consistently. Inconsistent placement frustrates our eons old place memory (Fig. 7.7).

Categorization

To avoid being overwhelmed with the information in the world, people spontaneously group items and events into categories (Mervis & Rosch, 1981). This frees us up from needing to remember every single item, effectively dividing long-term memory retrieval into stages. We first remember the category, and then if more information is needed, we remember the item within the category. This promotes cognitive economy (Conrad, 1972; Rosch, 1999), freeing us from having to make decisions that deplete our energy and cognitive resources. Importantly, categories are often hierarchical, and people have improved memory for recalling items in organized hierarchies rather than random lists.

We spontaneously assign items to the same category if they are more related than unrelated. Items can be related either because they are similar in some way, or because they co-occur. For example, two items that are used in the completion of some task or activity might be grouped together in the same category. We humans seem to have a natural need to categorize, which is capitalized on by business. For example, The Container Store, California Closets, Tupperware, and many other companies have built impressive businesses catering to just this need (Levitin, 2014).

Knowledge in the World vs. Knowledge in the Head

So far, when we have discussed memory or knowledge, we have referred to knowledge that resides in our minds. This includes knowledge about, facts, events, and skills. We mentioned that the knowledge available in working memory is severely limited, fleeting, and requires attention to maintain. Norman (2013) calls this "knowledge in the head," because people need to store and retrieve it from their own brains. Insightfully, he points out that, if we could provide cues to what products do and how they work, we could reduce the load on the user's memory, making products easier to learn, more efficient to use, easier to remember and more satisfying; in a nutshell, more usable (Nielsen, 1994). He refers to this strategy as providing "knowledge in the world."

In conversations with the ecological psychologist J. J. Gibson, famous for his theory of direct perception, Norman became convinced that often behavior can be guided by cues, hints, and guidance that exists in the environment. If these cues and hints did not naturally exist on the product, the designer would need to design them in to tell the user what each thing is, what it does, and how it does it—making it easier for users to do the right thing than the wrong thing.

As an example, consider the following situation in which little if any knowledge in the world was provided. In 2006, 24 years old Robin Rodgers was 35 weeks pregnant, and began vomiting and losing weight. As a result, her doctor admitted her to the hospital so that she could be fed through a tube until she gave birth to her daughter. Tragically, instead of inserting the tube through Robin's nose and into her stomach, the nurse accidentally attached the liquid-food bag to a tube that entered a vein. The New York Times article describing the incident likened this to pouring concrete down a drain. Ms. Rodgers and her baby died from the incident.

This was simply one of hundreds of deaths or serious injuries caused by tube mix-ups. Patients are often attached to numerous clear plastic tubes that are largely interchangeable. These tubes deliver a variety of substances including food, fluids, medication, and blood. Since the tubes feed into veins, arteries, the stomach, lungs or bladders, it is critical that the right substances be paired with the right tubes. Not surprisingly, nurses connect and disconnect dozens of tubes each day. Indeed, tubes meant to inflate blood-pressure cuffs sometimes get connected to intravenous lines, and intravenous fluids sometimes get connected to tubes intended to deliver oxygen, leading to suffocation (Harris, 2010).

When we discuss this case with students their first reaction is to fire, convict, or provide more training for the nurse; it's hard to know which is most punitive. After some reflection though, they begin to discuss design solutions. For example, you could provide color coding so that you only attach tubes with the same color. In fact, shape coding accomplishes the same thing, but also prevents two different shapes from being connected. Some suggest providing tubes with different textures. They discuss providing icons, labels and even electronic ID. They attempt to make it easier and more likely for the user to do the right thing, and harder to do the wrong thing. Effectively, they want to place knowledge in the world, just as Norman

Fig. 7.8 A smartphone with imperceptible affordances. (Image by Mungkhood Studio/ shutterstock.com)

advised. Knowledge in the world requires no learning, relies on recognition rather than recall, and is always present. However, it must be available at the right time, in the right place and in the right situation. It must be there when you need it.

According to Norman (2013), good design provides discoverability and understanding. Discoverability is the ability to determine what actions are possible and where and how to perform them. Understanding is comprehending how the product is used, as well as the function of the various controls and settings. These result from applying six important psychological concepts:

- Affordances—an affordance is a relationship between an object and the ability of a person to use that object. For example, a desk can be used to place papers on, so they afford paper placing. A doorknob affords turning. These affordances are *relationships* between the user and the object, rather than attributes of the thing itself. For instance, stairs afford climbing for many people, but they do not for people in wheelchairs. Further, some affordances are visible (or perceptible in general) and some are not. Many actuators on sleek consumer electronics afford pressing to turn on, but there is no indication of this relationship at all. These devices have affordances that are not perceptible (Fig. 7.8).

- Signifiers—Affordances exist even if they are not perceptible, but for designers it is critical to highlight the affordance and how it works (Norman, 2013). The highlight is called a signifier. A signifier indicates the existence of an affordance, and the way to make the affordance work. A drawer pull is an example of a signifier. So is a "push" label on an electronic device. As Norman points out, "Affordances determine what actions are possible. Signifiers communicate where the action should take place."

- Constraints—As described in the tubes example, constraints inhibit certain operations. Physical constraints rely on properties of the physical world for their operation, so no special training is needed. Physical constraints are best if they are easy to see and interpret. When this is the case, the set of actions is restricted before anything is done. Otherwise, a physical constraint prevents a wrong action from occurring only after it has been tried (Norman, 2013).

Forcing functions are especially strong constraints that prevent inappropriate behavior. Interlocks, one type of forcing function, demand that operations take place in a proper sequence. In cars with automatic transmissions, for example, you cannot get out of Park, unless you also press the brake pedal. Some electronic health records systems (EHR) will not allow an order to be finalized unless allergy information, patient weight, and patient height are entered to help avoid dosing errors and adverse medication reaction dues to pre-existing allergies.

Importantly, not all constraints are physical or procedural. For example, standardization is a type of cultural constraint. When no other solution appears possible, simply design everything the same way, so people only have to learn how to use the device once.

- Mapping—Devices are easier to use when the set of possible actions is visible and when the controls and displays exploit natural mappings. Mapping makes the relationships among things, for example, controls and displays, visible. The most famous example of this is the relationship between knobs on a stove top and the burners they control (see Fig. 5.35). Natural mapping makes use of spatial analogies to improve understanding. For example, to turn up volume, move the control up. Groupings and proximity are important principles from Gestalt psychology that map controls to function. Another design recommendation is that related controls should be grouped together.

- Feedback—Feedback communicates the results of an action. It must be immediate: even a delay of a tenth of a second can be disconcerting. Feedback must also be informative. It needs to tell us that something has happened, and what specifically has happened.

Conceptual models—As mentioned, we automatically categorize information and develop models of how things work. The problem is that the designers, who know the product best, have their model for how the product works, and the only way to convey that model to the user is through the design itself (what is sometimes called the system image). It is important for the designer to aid the user in developing the appropriate mental model. This can be done through metaphors and analogies as in the desktop metaphor on personal computers. It can also be done by organizing features and functions in a way that matches the user's mental model, through appropriate organization of functions and intuitive nomenclature.

This chapter has covered a great deal of material, which we suppose is appropriate for our discussion of mental life. The next section provides some tips for designers, based on the material we covered.

7.4   Tips for Designers

- Do everything possible to reduce memory load.
- Keep the number of things people need to retain in their memories, as well as the time required to retain it, as low as possible (Lee et al., 2017).
- Avoid long strings of digits, letters, characters, or combinations of them.
- Keep numbers and letters separate, where possible. Mixing numbers and letters in one code or string makes them difficult to remember.
- Provide placeholders for sequential tasks (Leer, et al., 2017). Provide visual reminders of what steps have been completed and which still need to be done.
- Make the names of features and functions as short as possible while making them identifiable.
- If pieces of information must be compared, present the information side by side.
- Increase the visibility, legibility and audibility of your labels, messages and displays. Some ways to do this are increasing text size and contrast, or ensuring that sounds are substantially louder and different from ambient noise.
- Avoid negatives. An instruction such as do not do X is easily misremembered as do X, because negation requires an extra cognitive step.

- Use familiar stimuli because they already exist in LTM and are expected. This makes them easier to recognize, remember and pay attention to. Also, include mixed case text, especially for longer strings, because that is what people are most familiar with.
- Be consistent in placement and grouping of items. This enables users to better guess the location of the information they seek.
- Use meaningful icons that people can recognize. Use standard icons when they exist, and make sure to usability test all icons.
- Use words rather than acronyms or abbreviations.
- Us a small vocabulary, because it is easier to learn, easier to make guesses and items are less likely to be confused with each other.
- Usability is especially important when people try to multitask; make sure to conduct usability testing as people try to multitask (Weinschenk, 2011).
- Make sure important labels, messages, and warnings are conspicuous. Make them different from their surroundings, larger, more intense, a different color, a different shape, and maybe a different sound from their surroundings.
- Use redundant coding by displaying the same information in two different ways, for example visually and auditorily. This helps improve recognition. Remember the Stop sign example; it uses color, shape, label size and location coding. Each part helps us recognize that it is a Stop sign.
- Remember that peoples' attention wanders up to 30% of the time, so they will not focus for long. Also, they will tend to get interrupted. Assume that users' minds will wander often.
- Provide feedback, and the ability to find their place again if users' minds do wander. It will be important to ensemble them to pick up where they left off.
- Provide placeholders for tasks with sequential steps. This prevents users from getting lost in their work.
- Ask yourself, if I were interrupted, would I remember what I was doing (Weinschenk, 2011).
- Remember that people work in distracting environments.
- Decrease the burden on working memory. Realize that people will forget, due to the limits of WM. This requires us to minimize working memory load at every turn.
- Avoid requiring people to recall, rather than recognize, information.
- Where reasonable, provide external memory aids such as reminders and checklists.
- Place controls, displays, labels and other interface elements where people expect them.
- Be consistent. Use consistent labeling, information placement, syntax and sensible defaults.
- Organize items according to relatedness.
- Use short words and short sentences. Eliminate extraneous words and images whenever possible.

- Provide progressive disclosure. Especially for tasks performed rarely or intermittently, guide the user through the task, breaking the task into a series of bite sized chunks.
- Since words that rhyme get confused in Working Memory, make sure your commands and labels do not sound too similar to each other.
- Redundancy helps Working Memory, so provide labels on icons. Present items as pictures as well as words.
- Make use of chunking. For example, chunk long numbers and characters into chunks of three to four. Chunk stimuli into meaningful characters. Also, keep numbers and letters separate.
- Spell out acronyms.
- Avoid requiring users to remember things from one place to another, or even one screen to another.
- If people need to compare figures or outputs, enable them to make comparisons side-by-side.
- Provide auditory and visual information together. That is, enable the user to both see and hear information displayed. This makes it easier for them to understand and remember.
- Avoid relying on the user's memory. Users get distracted by other tasks and people. And they often get interrupted. Attempt to present the information users need, when they need it, and in a format that is useful.
- Design for people with less than ideal memory capabilities. Younger adults remember more than children and elderly. For young designers, this means that your users may not perform as well as expected, given your own capable memory.
- Rely on recall rather than recognition memory. We evolved to recognize things and situations quickly. We did not evolve to recall arbitrary facts.
- Provide knowledge in the world by including signifiers such as labels and indentations for pushing.
- Remind people of what to do, when to do it, and how to do it.
- Use icons with labels, and use concrete, easy to visualize labels when possible. Remember long-term memory is better for pictures than for words, and better for visualizable words than abstract words.
- Structure your user interface as logically and consistently as possible. Organized knowledge is learned faster and retrieved faster than unorganized material.
- Minimize the need to rely only on reading in your user interfaces and your instructions for use. Reading and writing have existed for only about 10,000 years, a very short blip in our evolutionary history. And for the vast majority of that time they were practiced by only a select few. It takes years to learn how to read and write.
- Use familiar words. Familiar and frequently used words are recalled better and easier than less familiar and words that are less frequently used.
- Provide mnemonics wherever possible. For example, the mnemonic MUDPILES (Medschool Tutors, 2020), described below is commonly used to remember the causes of increased anion gap metabolic acidosis:

M—Methanol
U—Uremia (chronic kidney failure)
D—Diabetic ketoacidosis
P—Paracetamol—a stabilizer in medications
I—Infection, iron, isoniazid, inborn errors of metabolism
L—Lactic acidosis
E—Ethylene glycol
S—Salicylates

- Match the mental model of your users or of experts. People have mental models that dictate expectations of how a new device works. Aim to identify and characterize those mental models so they can be accommodated in your design (Association for the Advancement of Medical Instrumentation, 2009).
- Make use of natural mappings.
- Avoid using jargon. This can be unfamiliar and difficult for users to recall.
- Where possible, design to adhere to industry standards. This removes much of the burden from the user's long-term memory. If the device operates like others, it does not require additional learning and remembering.
- Provide memory aids for tasks that will not be frequently conducted, and tools that will not be used often. These memory aids can be in the form of posters, checklists, on-device help, and so on.
- Consider dividing big tasks into sequences of small tasks, each requiring one decision. This is often called progressive disclosure. Automatic teller machines (ATMs) provide a good example of this. They guide users through a series of simple decisions for transactions. This way customers do not need to recall exactly how the ATM works. Instead the ATM provides simple choices, based on recognition rather than recall. This type of design is also ideal for devices that will be used intermittently or infrequently.
- Where possible, standardize controls, displays, and procedures. Standardizing enables people to develop expectations and habits that they can use over and over again. Imagine the trouble if half of our cars had the brake on the left and half had the brake on the right.
- Use memory aids. This is especially important when a task will be done infrequently or when doing things just right is critical.
- Use concrete, visualizable words rather than words that are more abstract.
- Spend a great deal of effort on organizing information.
- Avoid technical jargon.

Resources

- Baddeley, A. D., Eysenck, M. W., & Anderson, M. C. (2020). *Memory*. New York: Routledge.

- Gawande, A. (2010). The checklist manifesto: How to get things right. In *Henry Holt and Company*. New York.
- Goldstein, E. B. (2019). *Cognitive psychology: Connecting mind, research and everyday experience* (5th ed.). Boston, MA: Cengage.

References

Association for the Advancement of Medical Instrumentation. (2009). *ANSI/AAMI HE75-2009: Human factors engineering—Design of medical devices*. Arlington, VA: Association for the Advancement of Medical Instrumentation.

Baddeley, A. (1994). The magical number seven: Still magic after all these years? *Psychological Review, 101*(2), 353–356.

Baddeley, A. D., Eysenck, M. W., & Anderson, M. C. (2020). *Memory*. New York: Routledge.

Bartlett, F. C. (1932). *Remembering. A study in experimental psychology*. Cambridge: Cambridge University Press.

Card, S. K., Moran, T. P., & Newell, A. (1980). The keystroke-level model for user performance time with interactive systems. *Communications of the ACM, 23*(7), 396–410.

Carrier, L. M., Rosen, L. D., Cheever, N. A., & Lim, A. F. (2015). Causes, effects, and practicalities of everyday multitasking. *Developmental Review, 35*, 64–78.

Clarke, D. D., & Sokoloff, L. (1999). Regulation of cerebral metabolic rate. *Basic Neurochemistry: Molecular, Cellular and Medical Aspects, 6*.

Conrad, C. (1972). Cognitive economy in semantic memory. *Journal of Experimental Psychology, 92*(2), 149–154.

Cowan, N. (2009). Capacity limits and consciousness. In T. Bayne, A. Cleeremans, & P. Wilken (Eds.), *Oxford companion to consciousness* (pp. 127–130). Oxford: Oxford University Press.

Cowan, N. (2010). The magical mystery four: How is working memory capacity limited, and why? *Current Directions in Psychological Science, 19*(1), 51–57.

Craik, K. J. W. (1943). *The nature of explanation*. Cambridge: Cambridge University Press.

Engle, R. W. (2002). Working memory capacity as executive attention. *Current Directions in Psychological Science, 11*(1), 19–23.

Ericsson, K. A., & Staszewski, J. J. (1989). Skilled memory and expertise: Mechanisms of exceptional performance. *Complex Information Processing: The Impact of Herbert A. Simon, 2*, 235–267.

Gailliot, M. T., & Baumeister, R. F. (2007). The physiology of willpower: Linking blood glucose to self-control. *Personality and Social Psychology Review, 11*(4), 303–327.

Gawande, A. (2010). *The checklist manifesto: How to get things right*. New York: Henry Holt and Company.

Gentner, D., & Stevens, A. L. (Eds.). (1983). *Mental models*. London: Psychology Press.

Godden, D. R., & Baddeley, A. D. (1975). Context-dependent memory in two natural environments: On land and underwater. *British Journal of Psychology, 66*(3), 325–331.

Goldstein, E. B., & Brockmole, J. (2016). *Sensation and perception*. Boston, MA: Cengage Learning.

Harris, G. (2010). Diabetes drug maker hid test data, files indicate. *New York Times*, 13.

Hoyland, A., Lawton, C. L., & Dye, L. (2008). Acute effects of macronutrient manipulations on cognitive test performance in healthy young adults: A systematic research review. *Neuroscience & Biobehavioral Reviews, 32*(1), 72–85.

Johnson-Laird, P. N. (1983). *Mental models: Towards a cognitive science of language, inference, and consciousness*. Harvard: Harvard University Press.

Jolly, J. D., Hildebrand, E. A., & Branaghan, R. J. (2013). Better instructions for use to improve reusable medical equipment (RME) sterility. *Human Factors, 55*(2), 397–410.

Lee, J. D., Wickens, C. D., Liu, Y., & Boyle, L. N. (2017). *Designing for people: An introduction to human factors engineering*. Scotts Valley, CA: CreateSpace.

Levitin, D. J. (2014). *The organized mind: Thinking straight in the age of information overload*. New York: Plume.

Maguire, E. A., Gadian, D. G., Johnsrude, I. S., Good, C. D., Ashburner, J., Frackowiak, R. S., & Frith, C. D. (2000). Navigation-related structural change in the hippocampi of taxi drivers. *Proceedings of the National Academy of Sciences, 97*(8), 4398–4403.

Medschool Tutors. (2020). Retrieved July 12, 2020, from https://www.medschooltutors.com/blog/the-true-utility-of-mnemonics-for-your-usmle-ste-prep.

Miller, G. A. (1956). The magical number seven, plus or minus two: Some limits on our capacity for processing information. *Psychological Review, 63*(2), 81.

Murdock, B. B., Jr. (1962). The serial position effect of free recall. *Journal of Experimental Psychology, 64*(5), 482.

Neisser, U. (1967). *Cognitive psychology*. Upper Saddle River, NJ: Prentice-Hall.

Norman, D. (2013). *The design of everyday things*. New York: Basic Books.

Ornstein, R., & Ehrlich, P. (2018). *New world new mind*. ISHK.

Peterson, L., & Peterson, M. J. (1959). Short-term retention of individual verbal items. *Journal of Experimental Psychology, 58*(3), 193.

Rosch, E. (1999). In E. Margolis & S. Laurence (Eds.), *Principles of categorization*. Cambridge, MA: MIT Press.

Schank, R. C., & Abelson, R. P. (1977). *Scripts. Plans, Goals and Understanding*. Mahwah, NJ: Lawrence A. Erlbaum.

Smallwood, J., & Schooler, J. W. (2015). The science of mind wandering: Empirically navigating the stream of consciousness. *Annual Review of Psychology, 66*, 487–518.

Tulving, E. (1972). Episodic and semantic memory. In W. Donaldson & E. Tulving (Eds.), *Organization of memory* (pp. 381–403). Cambridge, MA: Academic Press.

Weinschenk, S. (2011). *100 things every designer needs to know about people*. Indianapolis, IN: New Riders.

Wittgenstein, L. (1953). *Philosophical investigations: The German text, with a revised English translation*. Hoboken, NJ: Blackwell.

Wright, R. (2000). *Non-zero: The logic of human destiny*. London: Vintage.

Chapter 8
Use-Error

8.1 Introduction

Try going a day without making an error; heck, try going an hour. Unless you are asleep, you won't be able to do it. Even in casual conversation, we make plenty of errors; we choose the wrong word, stumble over our own speech, and misunderstand our conversational partner. However, because conversation involves a partnership between people attempting to understand each other, most conversations end in success. The errors are hardly noticed, they are quickly remedied, and the conversation continues unabated. In our work, we have reviewed thousands of hours of interview and usability test transcripts (pretty glamorous, we know). Most transcripts, even of eminent medical practitioners, are filled with run-on sentences and ambiguous pronouns, and are punctuated by "ums," "uhs," and "ya know what I means." People handle the messy ambiguity just fine.

Machines, however, do not handle ambiguity well at all; they are less forgiving than people, and this can result in tragic errors. For example, the FDA estimates that up to one-third of device failures that result in suboptimal treatment, injuries or death, are failures of device use rather than failures of the device itself (Association for the Advancement of Medical Instrumentation, 2018). They wrote that "user error, caused by designs that are either overly complex or contrary to users' intuitive expectations for operation, is one of the most persistent and critical problems encountered by FDA," and that HFE issues are "inherent in virtually every device related error." In other words, the weak link is neither the person or the device, but the interaction of the person with the device. And the result is use-error.

Use-errors are undesirable or unexpected events resulting from the interaction between a user and a device. We choose the term, *use-error* (International Electrotechnical Commission, 2004) rather than user error, because the blame should not be attributed to the user, but rather to the user-device system. Often, if you fix how the device presents information, organizes content, or receives input,

© Springer Nature Switzerland AG 2021
R. J. Branaghan et al., *Humanizing Healthcare – Human Factors for Medical Device Design*, https://doi.org/10.1007/978-3-030-64433-8_8

you eliminate the use-error altogether. Indeed, it is far more effective to improve the device than to try to somehow improve the user.

In medical devices, use-error occurs when a user's action—or lack of action— results in an unintended outcome, possibly causing an adverse event (AE; Barg-Walkow, Walsh, & Rogers, 2012). As a reminder, adverse events are injuries caused by medical care and can result in life-critical situations. Not all AEs are caused by use-errors and not all use-errors result in AEs. For instance, a patient may develop a hospital acquired infection even though there was no use-error. Similarly, a medical technician may press the wrong button on a device, recognize the problem and quickly remedy the issue before it ever affects the patient.

Healthcare providers are human—no more, no less; and like all humans they are imperfect. They have good days and bad. They get interrupted, fatigued, and sleepy just like anyone else. No degree of shaming and blaming will change this. As Reason (1997) points out, errors are unintentional, so it is pointless for management to try to control what people did not intend to do in the first place. Blame does not improve safety but encourages people to hide the errors instead—to sweep them under the rug. A culture of shame and blame predictably reduces the likelihood of reporting problems; and we cannot learn from errors we do not know about. For example, a University of Texas study found that only five adverse drug reactions were reported per year in a 900-bed hospital. Yet, when they examined the actual medical records, they found an average of 240 adverse reactions per year. This represents a reporting rate of only 2%. That is, the actual rate error was 50 times higher than the reported rate (Wachter, 2012).

These use-errors are not intentional, nor are they due to a lack of vigilance. You cannot blame this one on the millennials. Instead, many factors contribute to each use-error, including multitasking, caring for sick patients, unfamiliar equipment, inconsistent designs, noisy environments, lack of sleep, and fatigue. These are the parameters designers need to work within. Ultimately, the design of medical devices can either increase use-errors or reduce them.

8.2   What Is the Cause of All of These Use-Errors?

Too often, people blame the healthcare provider or caregiver who administered the patient care, assuming that they were careless, poorly trained, or not paying attention. This is not the case. Use-errors are not the user's fault. Most errors are not committed by a few bad apples, but are made by hard-working, well-intentioned, conscientious, healthcare providers, patients, and caregivers. Fundamentally, these errors have two things in common: they involve using some type of medical device and they involve human beings.

Not only is blaming the user inappropriate, it actually makes things worse. Errors will not be prevented by admonishing people to be more careful or by penalizing them for their mistakes. Instead we need to focus on anticipating errors, and preventing them or catching them before they cause harm. Rather than focusing on the mistakes users make, HFE examines the human-device interaction from many per-spectives. The following sections outline several factors that contribute to the use-error epidemic in the United States.

Size and Complexity

Healthcare is probably the most complex industry in the world (Leape, 2004). It is also enormous. In the US alone, healthcare employs 17 million people in at least 75 different roles, including physicians, nurses, physicians' assistants, therapists, tech-nicians, pharmacists, reprocessing technicians, procurement specialists, and many more (Bureau of Labor Statistics, 2018). This represents about 7% of American adults, and doesn't even count office staff, technical support, and janitorial staff. Additionally, the industry is growing; in the 12 months preceding March, 2020, United States healthcare employment grew by 368,000 (Bureau of Labor Statistics, 2020).

Size and complexity are exacerbated by the sheer variety of products used, the variability among people who use medical devices, the environments of use, and even the tasks that are performed. Amazingly, there can even be significant variation within the same person depending on the amount of attention they give to a device, the time of day, their mood, the work shift, and other variables.

With all these moving parts, there is little consistency between intended user, use-environment, and the tasks a medical device strives to complete. Because no two use-cases are the same, manufacturers must design devices that work in several places, times, and by several different people. With all these considerations, it's easy to miss something, and it's even easier to manufacture a product that doesn't work *for* the user quite as efficiently as it could.

Emphasizing Technology Over the User

Engineers are trained to be experts in technology and systems rather than to be experts at human behavior. Their academic curricula include calculus, differential equations, physics, computer programming, mechanical engineering, and electrical engineering. In fact, due to accreditation requirements, engineering students rarely have the flexibility to take nontechnical courses. Engineering school deans com-plain that there is little room in their curriculum for nontechnical electives. Many programs do not even require a course in introductory psychology, even though these students will usually be designing products for human use. As a result of these curriculum restrictions, engineers learn little about human perception, attention, memory, judgment, decision making, or emotion. They don't know how to consider the strengths and limitations of the human user in their designs.

Of course, the product would not exist without the tenacious focus on technology. Technology delivers the necessary functionality. But, technology is complicated, requiring specialized knowledge and skill to design and develop. It is unfortunate, but engineers spend so much time, brainpower, and resources solving technical challenges, that there is little remaining time to think about the end users.

Feature Creep

"Feature creep" (Norman, 2013) is the tendency to add functions to a product, whether they are needed or not. This phenomenon is often the result of market challengers competing in a feature-based arms race. To manufacturers' detriment, extraneous features complicate the design, hiding its main purpose and value, and muddle what could be an otherwise elegant solution. This is well known, and stems from failing to prioritize user needs. The fact is that simpler devices, without bells and whistles, tend to be easier to learn, more efficient to use, more memorable, and even more satisfying. Not surprisingly, they are also likely to be safer (Branaghan, Hildebrand, & Foster, 2020).

Most products are, and should be, designed to do one or two things, with some variations, options, or parameters. Yet, it is tempting to include every possible technical function to add more value to the product. People fail to realize that, after a certain point, adding features actually diminishes value. That point is different for every product, and requires customer research to identify, but success depends on understanding the "Goldilocks" point, where the number of features and functions is "just right."

Assuming Users Will Become Experts

Designers and developers often assume incorrectly that users will become experts with their device (Association for the Advancement of Medical Instrumentation, 2018). At best, a user might master a feature that is critical to their work and/or one that they use frequently. The other features are likely to be overlooked until needed, perhaps in an emergency. At that point, when needed urgently, the user is effectively a novice with that functionality. Further, due to the variety of products, manufacturers and models, many users may, at that moment, be interacting with the product for the very first time. At the very least, they may experience your product only intermittently. This is not an ideal way for healthcare professionals to interact with potentially life-saving devices.

Relying on Training

Training is treated as the "safety net" of design. Often, however, users fail to receive proper training before interacting with a device. Users often lead busy lives, and are unlikely to spend their free time attending training sessions. Even if they attend, there is no guarantee that they will pay adequate attention. The reliance on training is particularly problematic for new or substitute employees because they may not have been in attendance on training day. More importantly, since memory decays quickly, even users who did receive high quality training are unlikely to recall the details. As always, this problem is especially acute for intermittent users.

Underestimating Environmental Challenges

Medical devices are used in a variety of environments, frequently in unison with many other medical devices. Designers and developers often fail to appreciate these interactions and environments. They work as if users will be able to dedicate time and attention to that singular device. Of course, the opposite is true: practitioners should dedicate full time and attention to their patients, not to a device.

Healthcare environments can be loud, confusing, and teeming with other people (who might be vying for the user's attention). Conditions like these diminish a user's capacity to learn, remember, and make good decisions. Further, devices can migrate to environments which were not considered during product development. For example, a device that was designed for use in the hospital or other clinical environment could also be used at home. As a result, users may operate the device in environments with far less light, far more noise, and far more motion than expected. This is especially problematic because home devices may be used by people with compromised physical, perceptual, or cognitive conditions (not to mention appropriate training!).

Failing to Design for the "Worst Case Scenario"

When investigating reliability and durability, engineers often conduct drop tests— dropping the device from a pre-specified height. Other tests are similar; devices are exposed to extreme heat or are shipped across the country several times to see what breaks in transit. These tests simulate the worst-case scenario for the device. The idea is this: even though the device would never experience these conditions, it's good to know if the device can survive them.

Such drop-testing is needed for the user interface as well (Association for the Advancement of Medical Instrumentation, 2018). Usability testing is a common and robust way to evaluate how a device is used; usability tests can be conducted during normal product development and can help form the product's final shape and design. The testing could implement the following conditions:

- Complicated task(s)
- Untrained users
- Time pressure
- Stressful situation
- Poor lighting
- High vibration
- Moving environments (such as an ambulance or a helicopter; Association for the Advancement of Medical Instrumentation, 2009)
- Loud environment
- Busy and constrained environments

Such testing enables designers to plan for the worst case, thus improving the ease of use of the device, and preventing use-errors in the long run. Too often, however, usability tests are conducted under ideal conditions for the user, in which they only have to focus on one device, and they have plenty of time to do so. This is not the case in the real world, and it is certainly not the worst-case scenario.

Failing to Expect Use-Errors

People make mistakes—it's as simple as that. As we've discussed, they make these errors for a variety of reasons: they may be fatigued, distracted, unfamiliar with the task or device, or inadequately trained. They may be quickly moving from task to task, they may have failed to read the IFU, or they may have forgotten their training. Use-errors are inevitable, so manufacturers should wholly expect them to happen.

Underestimating User Diversity

Designers often assume that users are like them, which of course is not true. In reality, there is wide variation in the capabilities, characteristics, and limitations among users, but we tend to underestimate that diversity. Not only are people different from each other, but *what* is different depends on the situation, what day it is, or even what part of the day it is.

Fatigue, mood, shift, and stress all play a role. Some devices have well-targeted users, cardiac electrophysiologists, for example. Others have more varied audiences, such as patients with diabetes who inject insulin at home. These patients can be young, old, or somewhere in between. They can be administering medication at a clean kitchen table, or in their bedroom amongst toys and dirty socks. Over the counter (OTC) devices, in fact, can be used by just about anyone, anytime, anywhere.

This variability is enough to make you throw your hands up, design the device the way you would like to design it, and let the users adapt. Unfortunately, the ability for users to adapt is also variable; some do it fairly well, whereas plenty of others struggle mightily.

Expecting People to Multitask

As a rule, the human brain evolved to pay attention to one thing at a time, and paying attention to one thing, necessarily means ignoring other things. Today, we are faced with many competing stimuli. Since we cannot multitask, we are forced to switch tasks frequently, turning our attention from one task to another and back again. This attention switching is one of the most metabolically expensive things we can do. This is one of the reasons you crave peanut M&Ms. when you are, say, writing a book.

Rapidly switching our attention not only fatigues us (which has HFE considerations in its own right), but it also causes interference. That is, if you try to do two tasks at once, the performance on at least one task will suffer. Interacting with medical devices is no different; healthcare workers are extremely busy, tending to several important tasks at once. Devices need to be so easy to use that they require very few attentional resources. This is a great, and difficult, design challenge.

Overestimating User Capabilities and Motivation

Healthcare providers are often expected to behave like super-people, never making mistakes, no matter how difficult the situation. Because of this, they are less likely to report medical errors due to concern for their job, legal liability, or their reputation. Designers fall victim to this unfair assumption as well; unwitting designers expect people to have acute eyesight, perfect hearing, superhuman attention, faultless memory, and logical judgement. Of course, none of that is true. People are limited in each regard, and nowadays the limitations are more severe than in the past.

Failing to Involve Users Early in Design

Too often, manufacturers lack knowledge of their device's users, their tasks, and the environments in which they work. Since medical devices are designed for people, these people should be involved in the design of the product. Early in the design, users can help manufacturers understand their tasks, hassles, needs, and preferences, which in turn ensures the right problems are being addressed. There is no sense in solving a problem nobody has. Without users providing feedback, medical device creators fly blind, risking missed deadlines, expensive late-stage development changes, failed launches, or recalls.

Business decisions aside, failing to invest in user research and usability testing is a leading cause of use-errors. Without putting in the effort and resources early on, the human component of the user-interface interaction simply isn't considered. At that point, manufacturers either identify usability issues too late to make changes, or don't find out about them until after the product has launched.

Excessive Reliance on Thought Leaders

Thought leaders, accomplished physicians and surgeons, and those from key accounts, can provide great help during product design and development. This is sometimes referred to as "eminence-based device development" rather than "evidence-based device development." They are, after all, among the most knowledgeable people regarding that topic. However, these same leaders are far from representative of end users. They are, by definition, special. For example, they may be more knowledgeable or sophisticated than your average user or they may prefer to see information presented in a different, more intricate format. They may be overly familiar with how things used to be done, and prefer the old approach. Likewise, they may opt for advanced methods that are not helpful for users early in their career. Although making use of thought leaders can be helpful, it should only be one component of the overall understanding of user needs, goals, and characteristics. Use-errors can be caused by too much emphasis on important people and not enough on actual end users.

Lack of Focus on Human Factors

Product developers never intend to design bad products—products that people avoid, dislike, or that cause use-errors. The problem is that they rarely understand human capabilities and limitations, which means that they design for users who they assume exist, rather than the real thing.

Product development organizations employ extraordinary talented people, including experts in mechanical and electrical engineering, manufacturing, marketing, accounting, and software development. On the other hand, they rarely employ enough experts in human behavior, human performance, human decision-making and human capabilities and limitations; they are often lacking in HFE. Adding this focus would reduce use-errors,

Human factors engineers have a long history of studying and reducing risk and error. Much of the original work came from other high-risk domains like the military, nuclear power, aviation and driver safety. Recently however, with the recognition of error in healthcare, HFE has placed more focus on medical devices. The next section explores use-error further, and discusses a useful taxonomy for understanding and categorizing types of errors. There are three varieties of use-error categorized by their cause (Fig. 8.1): slips, lapses, and mistakes.

8.3   Slips

Slips occur when you intend to do one thing but accidentally do something else (Norman, 2013). Usually, when there is adequate feedback about the result of your actions, you detect the problem and are able to fix it. Sometimes however, when there is not enough feedback, the error might not be detected (Norman, 2013). There are a few different varieties of slips, which we discuss below.

Capture Slip

The first type is the capture slip, which happens when the activity you intend to do and the one you accidentally do are similar and even share steps. In a capture slip, the accidental activity is more familiar, more common, or has been conducted more recently than the one you intended to do. After doing the identical part, you proceed to complete the more familiar or frequent parts rather than the intended ones (Norman, 2013). As an example, have you ever decided to go to the store, only to start driving to work instead? We go to work a lot; in fact, you probably just went there yesterday. You probably go to the store less often. Both activities—going to the store and going to work—involve some of the same steps, including leaving the house, starting the car, and backing out of the driveway. In this case, however, the more frequent and recent activity of going to work captures what you intended to

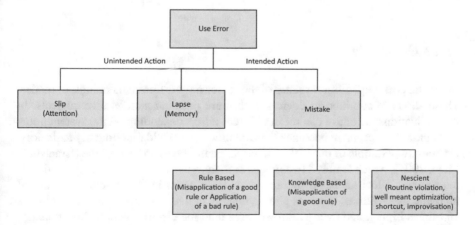

Fig. 8.1 Use-errors

do. On the other hand, because there is usually adequate feedback, we eventually recognize the error, curse our absentmindedness, and turn around to head to the store.

Description Similarity Slip

In a description similarity slip, you act on an item that is similar to the target. This happens when the description of the target is sufficiently vague. Here is an example. One of us recently went for a blood draw. While making polite discussion, the technician wiped one vein with alcohol, and then proceeded to draw blood from a different adjacent vein, one she had not wiped. It was the correct action, but the wrong target. In her defense, she sees hundreds of veins every day, and they probably begin to look alike. Thankfully, there were no ill effects, but there could have been. Similarly, healthcare providers see plenty of controls and displays from medical devices all day long almost every day, so it is easy to confuse one for another.

In a famous similarity slip from World War II, experienced pilots of military planes frequently retracted their landing gear after touching down on the runway. Let that sink in: they pulled up the landing gear once the plane was on the ground! Alphonse Chapanis, a US Army Lieutenant and experimental psychologist, was asked to investigate the problem. He found that in these aircraft the levers controlling the landing gear were identical to, and placed right next to, the levers controlling the flaps. As a result, when the pilots reached over to engage the flaps (the right intent), they retracted the landing gear (the wrong action).

In considering how to fix this problem, Chapanis knew that moving the controls farther apart, in an entire fleet of aircraft at a time of war, was cost prohibitive. He could, however, change their shape. So, he simply attached a small wedge to the control for the flaps and a rubber disc to the control for the landing gear. Now people could feel the difference in the controls, which eliminated the so-called human error (Vicente, 2010).

Mode Error Slip

In a mode error slip, different states of the device have different meanings, or modes. These slips are common in devices with more possible actions than controls. In these situations, a control can mean different things depending on the system's current mode. This problem becomes more common as we add functionality to devices. One famous example of this is the Therac-25 radiation therapy machine (Israelski & Muto, 2004), which featured two modes of therapy: electron beam therapy and photon therapy. Electron beam therapy provides a constricted low-current beam of high-energy electrons, whereas photon therapy delivers X-rays with a beam that is 100-times higher, and over a narrower area than the electron beam. Unfortunately, in this incident the technician chose the right sequence of actions, but the system was in the wrong mode. This slip delivered a lethal dose of radiation and weeks later the patient died from radiation poisoning.

You can imagine how this happened: Have you ever typed away on your keyboard only to look up and realize you have been typing all capital letters? That was a mode error; the problem with the Therac-25 is that it involved radiation rather than emails.

Mode errors are especially common when the device fails to make the mode visible. These facts lead to two pieces of design advice: (1) Avoid modes whenever possible. Your design mantra should be, "one control for each function and one function for each control"; and (2) If you must use modes, make it abundantly clear which mode is in effect.

8.4 Lapses

Lapses are memory breakdowns in which the user fails to do what they intended or does not evaluate the results of their actions. This leads to failing to follow steps of a procedure, or conducting the same step twice. Lapses can often be caused by interruptions or delays between the start and end of a task sequence. After interruptions, the user must remember exactly where they were in their task. Otherwise, they are likely to skip steps or even redo steps.

Memory-lapse slips are difficult to detect precisely because there is nothing to see. With a memory slip, the required action is not performed. When no action is done, there is nothing to detect. It is only when the lack of action allows some unwanted event to occur that there is hope of detecting a memory-lapse slip.

Lapses can be reduced by decreasing the number of task steps, providing reminders of where you are in the sequence, or the use of a forcing function. A classic example of a forcing function to reduce lapses is the automatic teller machine that requires you to remove your card before receiving your cash.

8.5 Mistakes

In contrast to slips, in which the user intends to do one thing but accidentally does another, mistakes occur when users mistakenly intend to do the wrong thing. Their actions are correct for their intention, but their intention is just plain wrong. Here we discuss two types of mistakes: rule-based and knowledge-based.

In a rule-based mistake, the user diagnoses the situation correctly, but then chooses the wrong response. Users apply the wrong if-then rule. For example, imagine the colossally incorrect if-then rule, "if the patient's blood glucose is low, administer insulin." The "then" part of this rule is not only wrong, but dangerous, and would produce a rule-based mistake.

In a knowledge-based mistake, the problem is misdiagnosed because of errone-ous or incomplete knowledge. The user fails to understand the situation or is unaware of any rules to apply to the situation. In this case, they need to "figure out" what to do. And sometimes, under these novel conditions, the user makes the wrong decision. Unfortunately, mistakes are difficult to detect because there is seldom any-thing that can signal an inappropriate intention.

8.6   Root Cause Analysis

One problem with errors and accidents is that they rarely have just one cause. An error resulting in harm was probably preceded by several problems. If even one of those problems had not transpired, there would have been no accident. Root cause analysis (RCA) analyzes errors until the underlying cause is found (Norman, 2013). Using RCA, we continuously ask questions until the reason for the failure is revealed. If it's discovered that a human is involved in the error, as mentioned previ-ously, it's important to remember not to blame the user. Rather, we investigate why the person made the error, what contributed to it, and what can be done to prevent it.

One tool we can use to identify underlying causes is called the "Five Whys" which requires us to continue asking" why" until we can identify the reason(s) for the incident. For example, if an infusion pump was programmed to administer 10 times the dose of a drug, we do not stop and conclude that the user made the error. Instead we ask exactly *why* the user made the error. Perhaps the user pressed a digit twice when they meant to only press it once. Then, we must ask *why* this happened. This might reveal that the keys on the keypad are too sensitive. Then, we might ask *why* the user did not notice this, which might reveal that the font on the display does not show the decimal very well, and so on. The beauty of this approach is that the conclusion is not to attempt to fix the user (which hardly ever works), but instead to fix the device, which is more effective. Despite the name, you won't always need to ask exactly five whys. Sometimes you may need to ask more; rarely you will need to ask fewer. Five, on the other hand, is a good rule of thumb.

8.7   Hindsight Bias

One of the challenges of learning from error is hindsight bias (Fischhoff, 1975), which suggests that events, which were unpredictable before they occur, appear obvious and predictable after they occur. For example, psychologist Baruch Fischhoff presented people with various situations and asked them to predict what would happen. Participants were not very good at this, predicting correctly only at chance level. Fischoff then presented a second group the same situations, but included the actual outcomes. He then asked participants in this group to estimate how likely each outcome was. When the outcome was known, they were estimated not only to be plausible, but to be likely. The participants felt like they knew it all along. That, in fact, is how hindsight works; everything is obvious in retrospect. Foresight, on the other hand, is a different matter altogether; there are rarely obvious clues or signs when an error or accident is unfolding.

8.8 Designing for Error

Swiss Cheese Model

Harm (adverse event) usually has multiple contributors, with no single one qualifying as the root cause. Reason (2000) describes this through the analogy of multiple slices of swiss cheese, each slice with several holes. Each slice of cheese represents a barrier to harm. In medical device terms, we think of these as efforts at mitigating harm. For example, one slice might represent a standardized operating procedure (SOP), whereas another represents the activity of double-checking drug labels. No barrier to harm is perfect, so the imperfections are represented as holes in the slices. If one failure occurs and makes it through one slice, it usually hits a barrier at the next. Consequently, harm can happen only if holes in all slices of cheese are aligned just so. This metaphor suggests of course that you can reduce the likelihood of harm by adding more slices (barriers to harm or lines of defense) or by improving each of the barriers (reducing the number of holes or making the holes smaller) (Fig. 8.2).

Our job in HFE is to make it difficult to err, yet easy to discover and recover from errors. One simple error should not be able to harm. Below we list a few recommendations, which are elaborated in the next section.

* Understand the causes of error and design to minimize those causes.
* Do sensibility checks—does the action pass the "common sense" test?
* Make it possible to reverse or undo actions.
* Make it easier for people to discover the errors that do occur, and make them easier to correct.

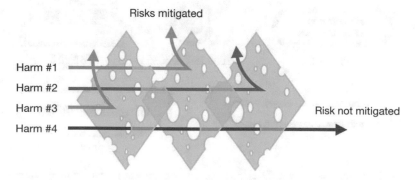

Fig. 8.2 The Swiss Cheese Model of safety incidents

Constraints

Error prevention often involves constraining the types of actions that can be conducted. This can be done through the use of shape and size. For example, when filling your car with gas, it is impossible to fill a gasoline engine with diesel because the size of the nozzle will not fit. Many medical devices often make use of constraints that make it impossible to connect the wrong things. An example from electronics is illustrated in Fig. 8.3. Different connectors using different communications protocols cannot be connected to the wrong port.

Undo

Perhaps the most powerful tool to minimize the impact of errors is the Undo command in modern electronic systems, reversing the operations performed by the previous command, wherever possible. The best systems have multiple levels of undoing, so it is possible to undo an entire sequence of actions.

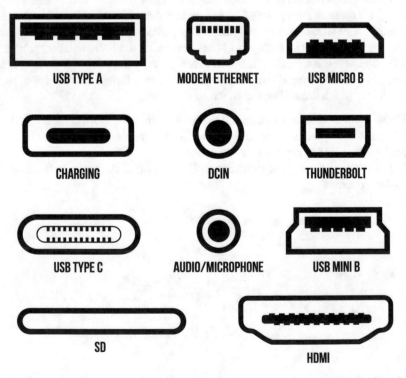

Fig. 8.3 Connectors have size and shape constraints that prevent the wrong things from being connected. (Image by Mikhail Grachikov/shutterstock.com)

Sensibility Checks

Another approach to reducing errors is to check that input is sensible. For example, some infusion pumps will not begin infusing a medication volume, rate, or dosage that is outside normal ranges. Sometimes the user can override this, but the idea is to make the user consider whether these values are actually correct.

8.9 Regulatory Considerations

Considering and managing use-error are an increasingly important part of the design process for medical devices. All of the major regulatory bodies that govern the sale and marketing of medical devices have released guidelines for reducing risk associated with use-errors. It is the responsibility of the device manufacturer to assess the potential harm that may result from use-errors. Device manufacturers must examine use-errors that could lead to serious harm and mitigate them to the extent possible. The process of identifying potential use-errors, creating mitigation solutions, and demonstrating the effectiveness of the mitigations will be described in detail in Chap. 9.

Resources

- Bogner, M. S. (Ed.). (2018). *Human error in medicine.* Boca Raton, FL: CRC Press.
- Dekker, S. (2004). *Ten questions about human error: A new view of human factors and system safety.* Boca Raton, FL: CRC Press.
- Peters, G. A., & Peters, B. J. (2007). *Medical error and patient safety: Human factors in medicine.* Boca Raton, FL: CRC Press.
- Reason, J. (1990). *Human error.* Cambridge: Cambridge University Press.
- Wachter, R. M. (2012b). *Understanding patient safety* (2nd ed.). New York: McGraw-Hill Medical.
- Woods, D. D., Dekker, S., Cook, R., Johannesen, L., & Sarter, N. (2010). *Behind human error.* Surrey: Ashgate.

References

Association for the Advancement of Medical Instrumentation. (2009). *ANSI/AAMI HE75-2009: Human factors engineering—Design of medical devices.* Arlington, VA: Association for the Advancement of Medical Instrumentation.

Association for the Advancement of Medical Instrumentation. (2018). *ANSI/AAMI HE75:2009/ (R)2018 human factors engineering—Design of medical devices*. Fairfax, VA: Association for the Advancement of Medical Instrumentation.

Barg-Walkow, L. H., Walsh, D. R., & Rogers, W. A. (2012, September). Understanding use errors for medical devices: Analysis of the MAUDE database. In *Proceedings of the human factors and ergonomics society annual meeting* (Vol. 56, No. 1, pp. 872–876). Sage, CA/Los Angeles, CA: SAGE Publications..

Branaghan, R. J., Hildebrand, E. A., & Foster, L. B. (2020). Designing for medical device safety. In A. Sethumadhavan & F. Sasangohar (Eds.), *Design for health* (pp. 3–29). Cambridge, MA: Academic Press.

Bureau of Labor Statistics. (2018). *Occupational employment statistics*. Retrieved August 5, 2020, from https://www.bls.gov/oes/current/oes_stru.htm.

Bureau of Labor Statistics. (2020). *Employment situation summary*. Retrieved August 5, 2020, from https://www.bls.gov/news.release/empsit.nr0.htm.

Fischhoff, B. (1975). Hindsight is not equal to foresight: The effect of outcome knowledge on judgment under uncertainty. *Journal of Experimental Psychology: Human Perception and Performance, 1*(3), 288.

International Electrotechnical Commission. (2004). IEC 60601 Part 1–6. Medical electrical equipment: General requirements for safety—Collateral standard: Usability.

Israelski, E. W., & Muto, W. H. (2004). Human factors risk management as a way to improve medical device safety: A case study of the therac 25 radiation therapy system. *The Joint Commission Journal on Quality and Safety, 30*(12), 689–695.

Leape, L. L. (2004). Human factors meets health care: The ultimate challenge. *Ergonomics in Design, 12*(3), 6–12.

Norman, D. (2013). *The design of everyday things*. New York: Basic Books.

Reason, J. (2000). Human error: Models and management. *BMJ, 320*(7237), 768–770.

U.S. Food and Drug Administration. (2002). *General principles of software validation: Final guidance for industry and FDA staff*. Washington, DC: U.S. Department of Health and Human Services Food and Drug Administration, Center for Devices and Radiological Health, Office of Device Evaluation.

Wachter, R. M. (2012). *Understanding patient safety* (2nd ed.). New York: McGraw-Hill Medical.

Chapter 9
Human Factors Regulations for Medical Devices

As you might imagine, exceptional amounts of time, money, and effort are required to develop a medical device that accurately and reliably delivers clinical benefit. These endeavors are worth it, of course; advancements in medical devices, including combination products[1] and in vitro diagnostic devices (IVD)[2], have improved the quality of life of millions of people and saved countless lives.

Due to the life-altering and risk-laden nature of medical devices, we cannot sensibly allow them to be distributed to the public haphazardly and without proper vetting. And so, to help ensure the safety of medical devices on the market, governments around the world regulate how devices are developed, manufactured, labeled, and sold.

The regulation of medical devices has a history that began in the eighteenth century, during a time that could be succinctly described as exploitative. The Industrial Revolution introduced new large-scale manufacturing processes for food, clothing, and medicine. The invention of the canning process allowed foods to last long enough to be shipped around the world, and sit on store shelves for extended periods. Meanwhile, "patent medicines" like Luden's Throat Drops and Dr. Morse's Indian Root Pills, were promoted as a cure for a variety of ailments, and started being sold in catalogs.

This industrialization put a new veil between consumers and products (Hutt, 2018). Artful labels and hyped slogans touted the "benefits" of the product, but there was no way for the consumer to know what the product contained. Manufacturers abused this ambiguity. Using spices or additives, canners could mask the taste of expired meat and other substandard ingredients. Many patent medicines relied on large quantities of morphine or cocaine to give users a high instead of actually healing them. Towards the end of the nineteenth century, governments began to regulate

---

[1] Combination products are therapeutic or diagnostic products that combine a drug or biologic with a drug delivery device. Examples include prefilled syringes and inhalers.

[2] In vitro diagnostic devices (IVD) are tests done on samples such as blood or tissue that have been taken from the human body. Examples include pregnancy tests and antigen tests.

© Springer Nature Switzerland AG 2021
R. J. Branaghan et al., *Humanizing Healthcare – Human Factors for Medical Device Design*, https://doi.org/10.1007/978-3-030-64433-8_9

how food was prepared, marketed, and sold. Likewise, pharmaceuticals have been regulated for over 100 years to ensure they are both safe and effective. Medical devices were not formally regulated until the 1970s. Today, medical device regulations are intended to ensure that the therapeutic benefit received from a device outweighs the potential harm to the patient or user. There is an ever-expanding list of standards documents that describe how medical device manufacturers should go about developing their devices and properly documenting the process (Semler, 2019).

9.1 Human Factors Regulatory Guidelines

The primary goal of human factors and usability engineering for medical devices, combination products, and in vitro diagnostics (Medical Devices) is to ensure that the design of a device aids users in accomplishing its intended use while minimizing potential harm that could come through unintentional misuse. Providing evidence that these devices can be used safely has become an increasingly important part of a premarket application in the United States, Europe, and other markets. As such, most regulatory bodies have provided, or adopted, guidelines that describe how human factors and usability engineering should be implemented into the development of new devices. Table 9.1. lists the four human factors regulatory documents that manufacturers should adhere to depending on the type of device and the market in which the device will be sold.

Table 9.1 Human factors regulatory documents

Guidance or standard document	Publisher	Markets covered	Device types
IEC 62366-1 Medical devices—application of usability engineering to medical devices	International Organization for Standardization (ISO)	Worldwide	Medical devices and IVDs
Applying human factors and usability engineering to medical devices	Food and Drug Administration (FDA)	United States	Medical devices and IVDs
Human factors and usability engineering—Guidance for medical devices including drug-device combination products	Medicines and healthcare products regulatory agency (MHRA)	Great Britain	Medical devices, IVDs, and combination products
DRAFT[a]—human factors studies and related clinical study considerations in combination product design and development	Food and Drug Administration (FDA)	United States	Combination products

[a]DRAFT not for implementation. Contains non-binding recommendations

9.2   Human Factors Process for Medical Devices

As of 2020, while the vocabulary, report structure, and desired format vary, the regulatory documents listed above are similar, describing a nearly identical process for implementing HFE. Broadly speaking, the HFE process for medical devices can be summarized in three steps (Fig. 9.1):

1. Identify design requirements by characterizing device users, environments of use, device tasks, and any known use-related issues with similar products.
2. Design the device to reduce use-related risk and evaluate the design along the way through formative evaluations.
3. Confirm that potential harm related to unintentional misuse has been reduced sufficiently through validation/summative usability testing.

Each of these three steps have subcomponents that are vitally important to designing a safe and effective device, where effective means it can be used appropriately. Each must be documented in a human factors and usability engineering (HFE/UE) report within a regulatory submission. Let's examine these important steps.

Step 1: Identify Users, Environments, and Critical Tasks

Identify Device Users

Before starting to design any product, it is very important to understand who will be using it. It wouldn't make much sense, for example, if you built a system with doctors in mind when nurses will actually be the ones using it. Understanding the users of the product helps ensure the device is usable by those intended to use it (Lee, Wickens, Liu, & Boyle, 2017).

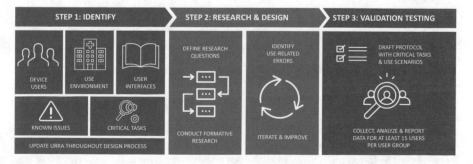

Fig. 9.1 Summary of medical human factors engineering process, a three-stop model

First, start by identifying all of the potential users of the device through a series of questions. It is notable that while some medical devices have just one intended user, others have several.

1. Who receives the device?
2. Who removes the device from its packaging?
3. Who sets up the device?
4. Who maintains the device?
5. Who uses the device for its intended purpose?
6. Who disposes of the device?
7. Who cleans or reprocesses the device?
8. Who teaches others to use the device?

Answering these questions provides a list of intended users of the device, which can be longer than one might expect. Let's consider a patient monitor as an example. Once shipped by the manufacturer, the patient monitor(s) would be received by the shipping folks at the hospital. They would likely move the boxes to the appropriate storage area within the facility. When the hospital is ready to transition to the new monitors, biomedical engineering staff would set them up for their colleagues, modifying specific settings according to the facility's policies and preferences. Once calibrated and set up to the facility's liking, staff such as physicians, nurses, and technicians may or may not be trained by an instructor on how to use the new patient monitors, but will certainly begin to use them in ways unique to their respective roles within the facility. Over time, several different people would be involved in maintaining the device: calibration checks, cleaning, or disinfection, to name a few.

Each distinct type of user could make up a "user group"[3]. In this example, even if you choose to exclude the shipping and receiving staff, we have identified the biomedical engineering staff, physicians, nurses, technicians, and trainers as potential user groups for the sample patient monitor.

Each user group can then be examined further, with the goal of understanding how their physical and cognitive characteristics may impact their experience with the device.

1. What are the physical characteristics of this user group?

   (a) Size
   (b) Strength
   (c) Dexterity
   (d) Coordination
   (e) Vision
   (f) Hearing
   (g) Tactile sensitivity

---

[3] IEC 62366-1 defines a user group as a subset of intended users who are differentiated from other intended users by factors that are likely to influence usability, such as age, culture, expertise or type of interaction with a medical device.

2. What are the cognitive characteristics of this user group?

   (a) Memory
   (b) Mental or emotional state while using the device
   (c) Literacy
   (d) Education
   (e) Training

3. Does this user group have medical conditions that might affect physical or cognitive abilities?
4. Does this user group have experience using similar types of devices?

Having a deep understanding of device users will help to guide the hundreds of design decisions that are made during product development and reduce the number of design problems that will be found in formative usability testing.

Device Use-Environment

Our surroundings affect, sometimes subtly, how we perceive, reason, and solve problems. Similar to identifying the characteristics of device users, having a deep understanding of the environments in which a device is used is necessary for good design. After identifying all of the environments in which the device may be used, a series of questions can be helpful to think through the characteristics of each environment.

1. What is the lighting level?

   (a) Is there natural or artificial lighting?
   (b) Is it bright, dim or variable?
   (c) Can the lighting level be changed if needed?

2. What is the noise level?

   (a) Are there noises from other devices?
   (b) Can the noise level be changed if needed?

3. Are there similar devices present that may affect selection of the proper device?
4. What is the physical size of the space?

   (a) Is the space clean or cluttered?

The characteristics of each environment of use should be taken into account during the medical device development process. Design decisions should then be made based on this understanding. Let us return to our patient monitor example. In hospitals and standalone clinics, the environments in which patient monitors are most commonly used, there are certain assumptions that can be made from a HFE perspective. Lighting conditions are expected to be anywhere from average to excellent. Meanwhile, noise levels could be anywhere from very quiet to very loud, depending on where the monitor itself is located within the facility. There may or may not be other (specialized) monitoring devices surrounding the patient and, similarly, there may or may not be multiple patient monitors located in the same room.

Based on this cursory analysis, there are several variables that should guide the patient monitor's design. First, because the lighting conditions are expected to be good, there might not be too much to worry about on that front. Second, because noise conditions could be quiet or loud, it might make sense to have adjustable volume controls. At absolute minimum, the patient monitor should be resounding when communicating something important, and should be hearable by hospital staff from multiple rooms away. Lastly, with other medical equipment competing for real estate and attention, the patient monitor's design would benefit from being unique (appearance and sound). In an emergency situation, the last thing you'd want to happen is for a nurse to run into the room and then have to figure out where the alarm is coming from.

Device User Interfaces

While identifying device users and device use-environments can benefit the design of the product, device manufacturers also need to ensure that they identify the entire device's user interface. A device's user interface includes all points of interaction between the user and the device. The user interface of a device could include:

- Hardware components like buttons, knobs, touchscreen, handles, and connectors
- Graphical user interface for software
- Visual and auditory alarms and indicators
- Packaging, labeling, instructions, and quick start guides
- Training materials and training procedure
- Online tutorials produced by the manufacturer
- Customer support (phone/chat/email)

All of these user interface elements should be identified and the design of each should be considered and evaluated through formative studies to ensure their support of safe and effective use. However, many medical device manufacturers don't realize that each of these items is actually considered to be part of the user interface.

The effect of this misconception is lack of attention to materials that are, in reality, very important. If we think about our sample patient monitor, the graphical user interface and physical device itself are the most straightforward portions of the user interface, but any auditory indicators, training checklists, training videos, instructions manuals, customer support lines, and online resources are part of the user interface as well. Many medical device manufacturers, even if they have followed a user-centered design model, pour resources into things like the graphical user interface and the training videos, leaving the instruction manual as a mere afterthought. We strongly recommend applying HFE principles to and evaluating each part of the user interface so that the entire human-system interface gets the attention it deserves.

Known Use-Related Issues

When designing a new medical device, it is important to know if there are any use-related issues that have occurred with devices similar to the one under development. If possible, the new device's design needs to learn from other manufacturers' mistakes. You'll look silly if your device has the same use-related issue another device has been dealing with for half-a-decade.

It is common for manufacturers to review internal complaint systems and reviews of their own devices when doing a known use problem analysis. However, investigations into known use-related issues usually mean studying use issues of other manufacturers' devices. The medical device industry provides a number of databases in which known-use-related issues may be documented. These databases include:

- FDA's Manufacturer and User Facility Device Experience (MAUDE) database
- FDA's MedSun: Medical Product Safety Network
- CDRH Medical Device Recalls
- FDA Safety Communications
- ECRI's Medical Device Safety Reports
- The Institute of Safe Medical Practices (ISMP's) Medication Safety Alert Newsletters

Table 9.2 Provides an example of a table that can be used to document known use-related issues and associated design mitigations.

Critical Tasks

When it comes to determining if a device can be used safely and effectively, not all tasks are created equal. Some tasks are given the distinction of being "critical tasks." Critical tasks are those that, if not performed correctly, could lead to harm to the patient or user. Many times, harm to the patient or user includes compromised medical care. It is important to refer to the definition in the regulatory document that governs the device being developed, as definitions vary slightly across the documents.

To identify critical tasks, an important first step is to identify all device tasks, regardless of their "criticality." A task analysis is an effective tool for doing this. With all the tasks identified, we can look at each task individually to determine the

Table 9.2 Summary of known use problems for (names of predicate devices or similar devices)

Source	Issue number	Summary of issue	Impact to user	Design mitigation
Internal customer complaint	1222	Customer was unable to connect the device to wifi for 3 days and therefore the physiological data could not be sent to the physician	Delay of therapy	If the user is not connected to wifi for more than 24 h then a customer support representative will call the user

potential use-errors that users may commit on each. Keep in mind the list of tasks may continue to evolve as the design is refined and as formative evaluations are performed.

Each possible use-error identified carries with it potential harm(s) and each harm has an associated severity score that is assigned through a risk assessment (often on a scale of 1–5). A systematic approach to identifying critical tasks, risk assessments use a combination of severity and probability to assign a singular risk score.

Often, the risk score can be reduced through engineering or design mitigations that reduce the probability of occurrence. For example, the probability of electric shock may be reduced by including a breaker that trips when a short is detected. Reducing the probability of electric shock may result in reduced risk, but the severity of harm that could occur remains the same. It is therefore important to note that probability scores are not factored into the determination of critical tasks. This process is shown in Table 9.3.

Step 2: Formative Research and Design Process

After the identification phase, device design and iterative evaluation of the device can commence. The goal of formative research is to formulate or improve a product's user interface by learning as much as possible about the intended users, the environments in which the product is used, and the device/user interactions. Formative research can be broken into two main types, generative research intended to provide insight into who the device users are, where they use the device and how they use it, and usability research to systematically evaluate the device design to ensure the device meets user's needs and can be used as intended.

Table 9.3 Device tasks and determination of critical tasks

Device tasks	Potential use-errors	Potential harm of use-errors	Severity of resulting harm	Critical task? (YES/NO)
Remove the instructions from packaging	User does not remove instructions from packaging User disposes of instructions User damages instructions	User does not review instructions prior to use	2 (minor)	NO
User places device on bicep	User does not place device on body User places device on location other than bicep	Inaccurate results	3 (severe)	YES

Formative Generative Research

Formative generative research is performed before beginning design of the user interface. Generative research methods include contextual inquiry, shadowing, and observation (see Chap. 2—Qualitative Human Factors Research Methods). The findings from generative research help to define the design characteristics of the device user interface. An example of generative formative research used to inform the design of the user interface for a total artificial heart is found in Branaghan, Hildebrand, and Foster (2020). Before initiating the design of a new Freedom driver, the pneumatic pump that powers the total artificial heart, SynCardia Systems performed contextual inquiry research, visiting patients in their homes to observe and discuss their experience living with a total artificial heart and using the Freedom driver (Holtzblatt & Beyer, 1997). A key finding from this research was that the current process of switching a patient from one Freedom driver to another was much too difficult and potentially life-threatening. The current driver's drivelines (tubes that connect to the total artificial heart) have two connectors that must be swapped simultaneously (Fig. 9.2).

Swapping drivers is a high-risk task often performed under incredibly stressful circumstances, often when the driver has malfunctioned and is no longer circulating the patient's blood. A patient connected to a failed driver may lose consciousness in a few seconds, and can die in a short amount of time. Under these life-or-death circumstances, the patient's caregiver must disconnect the patient from the failed

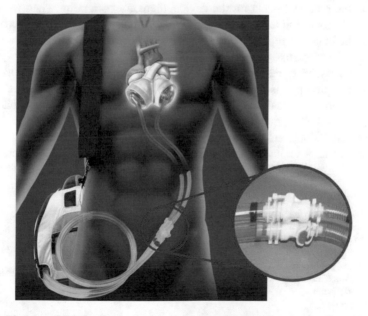

Fig. 9.2 Diagram of the SynCardia total artificial heart connected to the Freedom driver and a close-up view of the two connectors that must be disconnected when changing to a new driver. (Image courtesy of SynCardia Systems, LLC with permission)

driver and reconnect them to a functioning driver. Participants in the study described instances where they had to perform a driver swap in a car, a restaurant, or at a sporting event. With this increased understanding of the difficulties and potential harm associated with swapping drivers, SynCardia set out to design a new driveline connector, shown in Fig. 9.3, that dramatically reduces the time it takes to swap from one driver to another.

Formative Usability Research

As a product's user interface begins to take shape, it is important to perform formative usability research to determine if the product can be used as intended by representative users. When designing a new product there are all sorts of decisions that need to be made. Many of the design criteria may be determined by environmental or other constraints. For example, a device that needs to be moved regularly may need to be small enough to be carried or be placed on wheels if it is large. Once it has been decided that the device needs to be on wheels there are more decisions to be made. How many wheels should the device have? How large should the diameter of the wheels be? Should the wheels be hard or soft? Do the wheels need tires? Do the wheels need to lock? Again, we may look at where and how the device is used to make decisions about the characteristics of the wheels. When placed on a medical device, using the wrong wheels could be catastrophic (think of a life-sustaining device that is prone to tipping over). No matter how many meetings we hold to decide on the perfect wheels for the device, there comes a point when we need to put these wheels to the test to ensure that they actually perform as expected when placed in the hands of representative users in the device's intended environment of use. Queue the formative usability research.

Formative usability research is performed throughout design to help us decide between multiple options and ensure the decisions being made meet user's needs. To perform formative usability research, there are a number of methods that can be used based on development timelines, access to users, and components that are being evaluated. For example, heuristic evaluations (see Chap. 4) are a good tool for identifying general violations of usability design principles and can be completed with a small group of reviewers fairly quickly. Most commonly, usability testing is

Fig. 9.3 Early prototype of a new connector designed to reduce the time it takes to transfer a patient from one Freedom driver to another. (Image courtesy of SynCardia Systems, LLC with permission)

the method chosen when tangible prototypes or actual devices are available for evaluation (Wiklund, Kendler, & Strochlic, 2011). Usability studies implement a simulated use method where study participants (i.e., representative end users) perform realistic tasks with the device. Performing real tasks allows study personnel to observe the device/user interactions and identify elements of the user interface that help them use the device correctly, and any elements that may cause confusion or lead to use-errors. When we observe confusion or use-errors, we can talk with study participants to identify the root cause and consider ways to mitigate the issues through improvements to the design. Figure 9.4 shows a participant simulating use of an in vitro diagnostic used for blood analysis in a home environment.

From a regulatory perspective, there are a few things to note about formative usability research.

1. Formative usability studies are not officially mandated nor reviewed by any of the regulatory agencies. This gives the manufacturer freedom to evaluate the device design in the way they see fit throughout the design process, testing components of the user interface as prototypes and mock-ups become available. There are no requirements as to the number of participants included in formative usability studies. Often, as few as five to seven participants may provide sufficient data to determine if the design meets user's expectations, or needs improvement.

2. Components of the device user interface do not need to be finalized before performing formative usability studies. In fact, they should not be final designs. The whole concept of formative usability research is to evaluate whether or not the user interface design is appropriate, and to revise the design based on findings, as needed. Put prototypes in front of users as early and often as possible. Figure 9.5 shows a participant interacting with a 3D printed handheld device prototype and a prototype of the GUI mocked up on a tablet.

Fig. 9.4 Participant simulating use of an in vitro diagnostic in a home environment. (Image courtesy of Kahala Biosciences with permission)

Fig. 9.5 Participant of a formative usability study interacting with a 3D printed handheld device prototype and a prototype GUI mocked up on a tablet. (Image courtesy of MedicaSafe with permission)

3. Individual components may be evaluated independently. Rather than waiting for the device to be finalized, the instructions can be drafted and evaluated through a label comprehension study. Often, software can be designed and mocked up faster than hardware. Rather than waiting on the hardware to be complete, the GUI information architecture might be prototyped on a tablet and evaluated through a small usability study. Evaluating components independently, as they become available, can greatly speed up the design process.

4. Conduct at least one formative usability study as a prevalidation or presummative usability study. This means to design the study in the same way the validation/summative usability study will be designed and run it with three to five participants. The results of this prestudy will provide confidence going into the validation/summative usability study, or help to avoid a potentially unsuccessful study. Details of validation/summative usability study design are included in the following section.

5. Although formative research is not required, formative study results are included in summary form in the final HFE report. The summarized formative research results show reviewers that the final device was designed through a human-centered design process. Formative studies are a bit like showing your work on a math problem. If you get the right answer but don't show your work, the teacher is left to wonder if you used a calculator. Similarly, if a manufacturer shows positive results in their validation/summative validation study, reviewers are left to wonder how they designed such a usable product without evaluating it throughout the design process Experience shows these reports often receive more scrutiny from reviewers.

Step 3: Validation/Summative Usability Testing

The final step in the regulatory HFE process is validation usability testing (for FDA submissions) or summative usability testing (for EU and MHRA submissions). The purpose of validation/summative testing is to provide evidence that intended users can use the device safely and effectively in its intended environment of use. The usability method is no different than what is described in Chap. 4: Usability Assessment, but we will discuss the important considerations from a HFE regulatory perspective.

Preparing for Validation/Summative Usability Testing

Validation/summative usability testing can be compared to a university entrance exam, both carry a lot of pressure to get it right the first time. To prepare for the university entrance exam, a diligent student may take countless practice tests until they feel comfortable enough to register for the exam. Similarly, a company may perform several formative usability studies to gain confidence in the result they should expect in the validation/summative usability testing. As described in Step 2, at least one formative usability should be designed as a prevalidation/presummative study.

Participant Criteria for Validation/Summative Usability Testing

Validation/summative usability testing is performed with the users defined in Step 1 of the regulatory HFE process. For FDA submissions, a minimum of 15 participants from each identified user group need to be represented in the study (Faulkner, 2003). For non-FDA submissions, there is not a defined number of minimum participants, but 15 has become fairly widely accepted. There may be instances where more than 15 participants are requested, often the case for over-the-counter (OTC) devices where there is a wide range of characteristics within the population of OTC product users. Conversely, there are also instances where it is not feasible to obtain a sample of 15 users, often the case when dealing with rare conditions.

Simulated Use-Environment

Validation/summative usability testing should be performed in an environment that closely resembles the intended environment of use, especially elements that may affect device/user interactions. For example, a surgical tool used in operating rooms should be tested in an operating room environment with representative equipment, personnel, lighting, sounds, space, etc. It is impossible to know all of the environmental elements that may interact with device use, so providing as realistic an environment as possible is preferred. Figure 9.6 shows a simulated clinical environment used to evaluate usability of an infusion pump.

Market-Ready Devices

While formative evaluations may use early concepts, prototypes, or otherwise non-final materials, validation/summative usability testing is performed with final, production-quality components of the entire user interface. The final versions of the device's hardware, software, packaging, labeling, instructions for use, training, and customer support should be used.

Participant Training

If training is a part of the user interface, participants of validation/summative testing should also receive representative training prior to the evaluation activities. This includes allowing appropriate time for learning decay. Adequate time should be provided to the user to allow for learning decay before participating in the study. For products that would be used immediately following training, learning decay periods can be as short as 1 h. For devices that may not be used for 1 day or more after training, the learning decay period should be at least 24 h.

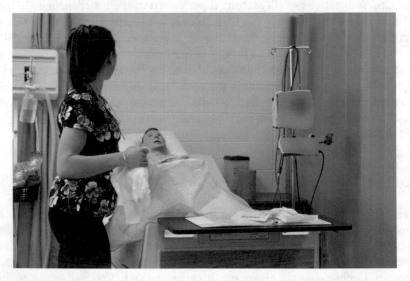

Fig. 9.6 Nurse participant in a simulated clinical environment. (Image by Samara Heisz5/shutterstock. com)

Tasks Included in Validation Testing

The goal of validation testing is to provide evidence that ALL critical tasks can be performed safely and effectively by intended users of the device. In order to test all critical tasks within the natural flow of device use; however, participants may need to perform many noncritical tasks as well. For example, there are often noncritical tasks such as plugging the device into the wall, powering it on, or logging in that a user must complete before encountering a critical task. In order to evaluate critical tasks within a realistic context of device use, participants should also perform those noncritical tasks and encounter the critical tasks when they occur within natural use of the device.

Data Collected

There are three types of data collected during a validation study: observational data, knowledge-task data, and open-ended interview data.

Observational Data

Most tasks can be put into simulated-use scenarios that allow study personnel to observe participants' performance of the tasks. Participants should be permitted to perform all tasks as they see fit without any interference or assistance from study personnel, but sometimes participants get stuck and need help to move forward for the study to continue. In these cases, the study moderator should allow the participant to work through the issue independently before offering assistance, recognizing that as soon as help is given, the task should be scored as a use-error. When it becomes clear that help is needed in order to proceed, the moderator should provide as little help as possible (e.g., "do the instructions offer any assistance to complete this task?") to keep the study moving forward. Figure 9.7 shows an interaction between a lay user participant and the study moderator.

Study personnel should document the participant's performance of each task and determine whether the task was performed correctly, correctly with difficulty, or incorrectly (use-error). As these determinations are subjective, success and failure criteria should be defined prior to conducting the study. An example of how task performance determinations might be defined is as follows:

- 2 = "Success"—Completed a task with little or no difficulty
- 1 = "Success with difficulty"—Completed a task with more than a little difficulty (e.g., confusion, multiple attempts, frustration)
- 0 = "Use-error"—Did not complete the task, completed it incorrectly, or required assistance.

Fig. 9.7 Study moderator interacting with a lay user participant

Knowledge Task Data

Some critical tasks cannot be evaluated through simulated use-scenarios. This is often the case for warnings and contraindications found in labeling. Another example is comprehension of an expiration date. As it may cause the participant to believe the study is "rigged" or "tricky," it would not be fair within the context of a simulated-use study to give a participant an expired device or drug and expect them to identify that it is expired. To evaluate comprehension of these knowledge tasks, the study moderator first asks the participant a question, and the participant reviews the device user interface and/or labeling in order to provide an answer. Knowledge and comprehension questions should be open-ended, neutrally worded, or require the participant to apply the information through a hypothetical scenario.

Open-Ended Interview Data

Observation of participant performance and knowledge task questions should be followed by questions to collect participants' subjective feedback about their experience using the product. Following up in this way may help study personnel uncover difficulties or confusion that may have gone undetected. After starting broadly, the study moderator should ask the participant to describe their experience performing tasks that were scored as use-errors or difficulties. The goal of this discussion is to identify potential root causes for the use-related issues experienced by the participant. Again, these questions should be open-ended and neutrally worded to reduce leading or biasing participant responses as much as possible. Questions that may be considered are:

- "What did you think of the device overall? How was your experience using the device?"
- "Did you have any trouble using it? If so, what kind of trouble did you have?"
- "Was there anything confusing? If so, what was confusing?"

Questions to ascertain root cause of use-related issues

- "Please tell me about your experience [performing a task where their use-error or difficulty was identified]. How did that go for you? What happened? How did that happen?"
Characteristics of validation/summative usability testing at a glance
Goal: Demonstrate the device can be used safely and effectively
Characteristics:

- Perform testing with a minimum of 15 representative participants, per user group
- Perform testing in an environment that simulates the intended environment of use
- Use only final, market-ready, device, packaging, labeling, etc.
- Provide realistic training, including learning decay (if applicable)
- Use simulated-use scenarios to allow participants to perform real device tasks
- Include all critical tasks
- Do not ask follow up questions until after all tasks have been completed

Analysis of Validation/Summative Usability Test Results

The most important part of the validation/summative usability test report is the analysis of use-related issues (use-errors or difficulties). If the study yields use-errors or difficulties on critical tasks, each issue should be analyzed independently to determine the user interface element, or elements, that may be responsible for the issue (i.e., the root cause). Critical task results are best documented in a table that describes the use-related issues observed, the follow up discussion with participants, the potential root causes of the issues, the potential opportunities to mitigate the issue through design, and whether or not redesign of the interface is necessary.

Tables 9.4 and 9.5 provide examples of a table that can be used to document the critical task results.

This table is what regulatory agency reviewers will analyze to determine if the device being tested can be used safely and effectively by representative users. The fewer results in the table, the better. However, having use-related issues identified in the table does not mean the device will not pass the regulatory review. The criteria for a successful validation/summative usability test is a bit of a moving target. There is no number or percentage of correct uses that can be set to determine if the usability test was successful or unsuccessful. The potential harm associated with each use-related issue must be examined to determine if a change to the device user

Table 9.4 Example analysis of critical task results

Critical task	Use-Errors (UE) and difficulties (D)	Observations and follow-up	Potential root cause(s) of use-errors and difficulties	Potential harm	Possible risk control measure	Redesign needed? (YES/NO)
Place device (surgical implant) in sterile field	(UE) one nurse participant dropped the device on the floor while attempting to place the device on the table in the sterile field	The participant immediately stated that she would not use the device dropped on the floor and asked for a new device which she correctly placed in the sterile field. During debrief, the participant said it was hard to get the implant to fall out of the packaging on its own, "it was stuck in there" and she did not want to touch it. She shook the packaging and the implant dislodged abruptly and fell onto the floor. She said she knew from her training as a nurse that any equipment dropped on the floor must be discarded and may not come in contact with the patient	The package is designed to allow users to release the implant when held upside down, but the implant was stuck in the package	Patient infection (serious)	Ensure manufactured packages do not hold the implant preventing it from falling out when held upside down	NO, the design is appropriate as observed in 14 of 15 instances of this task in the validation study. Manu-factured packages must meet the design requirements in regard to size and shape

Table 9.5 Example analysis of critical task results

Critical task	Use-errors (UE) and difficulties (D)	Observations and follow-up	Potential root cause(s) of use-errors and difficulties	Potential harm	Possible risk control measure	Redesign needed? (YES/NO)
Send blood pressure result to healthcare provider (home-use blood pressure cuff)	(UE) Three (3) participants were unable to send the blood pressure results to the healthcare provider	All three participants were confused by the label term "UPLOAD RESULT." All successfully navigated to the correct page in the GUI architecture, but did not understand the term upload. During debrief they said they were looking for a button to "share" or "send" the result. One participant said it would be helpful if there was an icon or image that represented a doctor in addition to the phrase "send result to doctor"	Use of unfamiliar terminology to intended users of the device	Delay in therapy (serious)	Consider rewording the label to "share result with doctor" or "send results to doctor" to use terms that are more familiar to the intended user	YES

interface is warranted to reduce residual risk. The greater the potential harm of a use-related issue, the more likely that a design change might be needed to mitigate that issue. Likewise, if a use-related issue was observed repeatedly in the usability testing, it may warrant a design change. In cases where a design change is warranted, the usability of the new design element should be evaluated through further testing. Note that only the new user interface element would need to be evaluated. For example, if one warning message in the instructions is reworded, only that new warning message needs to be tested in the follow up study. The parts of the interface that have not changed can be omitted. Top 10 Questions Related to Human Factors Regulations for Medical Devices.

How Many Use-Errors Will the FDA/Regulatory Agency Accept?

"It depends." It's an unsatisfying answer, but a number of variables play a role:

- Type of device and potential harm
- The severity of the potential consequences of use-errors
- The number of tasks that have use-errors
- The number of use-errors on a given task
- Whether a design change can be made to correct the problem
- Whether a design change is made to correct the problem
- The reviewer's background
- Whether the medical device manufacturer already attempted to submit before

Each use-error, difficulty or close call, requires an in-depth analysis to determine the root cause. From there, the medical device manufacturer will need to decide if the residual risk requires additional mitigation or explain why the residual risk is acceptable as is.

Do We Have to Evaluate Tasks That Aren't Critical? If So, Should Noncritical Task Results Be Included in the Report?

A validation/summative usability study should be primarily interested in evaluating performance of critical tasks. However, evaluating every possible task is advisable. Evaluating noncritical tasks is an opportunity to further understand the use of the medical device and to improve it. It is also prudent to have simulated-use data available for each task as regulators may determine that some tasks classified by the manufacturer as noncritical should be considered critical.

Noncritical task results are not typically provided in detail. Instead, a summary of noncritical task results is appended to the report. This allows the reviewer a high-level view, with the ability to probe further should he/she deem it necessary.

How Do We Define Critical Tasks?

The definition of a critical task varies slightly between regulatory groups, but basically a critical task is a user task which, if performed incorrectly or not performed at all, would or could cause harm to the patient or user. In this case, harm is defined to include compromised medical care.

Critical tasks should be determined from severity scores given in a risk analysis or failure modes and effects analysis (FMEA). While the risk analysis or FMEA traditionally gives each user task two scores, severity and probability, critical tasks are not concerned with probability. Any task that could lead to harm, regardless of likelihood of occurrence, should be listed as a critical task.

User tasks related to successful delivery of therapy are often miscategorized as noncritical tasks. Medical device regulators expect to see evidence that a device can be used for its intended purpose, thus providing benefit to the patient.

How Realistic Does the Simulated-Use Environment Need to Be?

The goal of a simulated-use usability study is to observe users interacting with the device interface independently and naturally. There are three human factors considerations that interact to produce "device use": the user, the device interface, and the use-environment. Ultimately, the simulated-use environment should be realistic enough to understand how the device, user, and use-environment all interact. The suitability of a study environment should be treated on a case-by-case basis. At minimum, it must meet the demands and characteristics of the particular study at hand.

The design of simulated-use usability studies should consider all aspects of the use-environment that may influence the user's interaction with the device. Some of these aspects include:

- Representative lighting and ambient noise levels
- Presence of multiple models of the same devices and labeling materials (forcing users to identify the correct ones)
- Ancillary materials (e.g., isopropyl alcohol, lint-free cloths, PPE)
- Distractions
- Whether the area is busy or cluttered
- How fast or slow-paced the environment is

Can We Make Changes to the Device or Instructions After the Validation Usability Study?

Changes can be made after the validation study. However, it is likely the regulatory agency would ask for a reevaluation of the tasks that were affected by the change. Only the tasks affected by the manufacturer's change would need further evaluation. For example, if the wording of only one step in the instructions were changed, only that step would need to be reevaluated.

Completing a reevaluation of only a few tasks is obviously a simpler and quicker affair, but it is important to note that a full 15 (for FDA) new participants would have to be included. Additionally, in order to reevaluate Step 4 of the instructions, participants would be asked to complete Steps 1–9 to ensure proper context.

What Is the Purpose of Identifying Known Issues and How Do We Identify Them?

Known issues refer to usability issues that have already been documented from previous versions of the device, predicate devices, or devices that are similar. Identifying issues experienced with other devices intended to prevent the same mistake from being made twice. By identifying usability issues with current and previous devices, manufacturers can avoid developing medical devices with the same problems.

There are multiple places to look when searching for known issues related to usability, including but not limited to:

- Internal customer feedback and complaints '
- Sales representatives and clinical support staff
- Observing or interviewing current users
- External databases such as the FDA's Manufacturer and User Facility Device (MAUDE) database, FDA's MedSun Network, FDA's Safety Communications, and FDA's Medical Device Recalls website

What Characteristics Can Be Used to Define a "User Group"?

Many medical devices are used by a diverse group of healthcare providers. For example, an injection device may be used by a nurse, physician, medical assistant, pharmacist, or a layperson with no training. User groups are simply these different roles within healthcare. In the case of an insulin pen, a doctor or nurse might demonstrate proper injection technique to their patient. The patient would then be responsible for performing his or her own injections thereafter. User groups for the insulin pen would be nurses, physicians, and patients (lay users). While performing an injection requires the same technique for each group, nurses, physicians, and lay users vary from one another in that they approach the task with different knowledge, experience, and training.

Can Nurses and Physicians Be Included in One User Group?

The FDA recommends that validation usability studies include 15 participants per user group, but no specific number is provided in IEC 62366-1. In an effort to save time and resources, medical device manufacturers often attempt to combine user groups. While there are situations where this may be pertinent, it is not typically appropriate.

Despite working in similar environments, nurses and physicians are distinct from a HFE perspective. In addition to contrasting training and experience, nurses' and physicians' roles within healthcare vary greatly. Responsibilities tasked to a nurse do not typically align with those of a physician, and are therefore usually treated separately during validation usability testing.

How Do You Recommend That We Incorporate User Research into Our Design Process? How Often and When Should We Conduct User Research? What Are the Best Strategies?

Incorporating HFE into the design process is most effective when done in the early stages of the product life cycle. Having the HFE perspective early on promotes informed decisions when the design is still flexible. Fundamental to early HFE success, is the fact that late changes are difficult and expensive, and early ones aren't. Early human-centered design is the most time, effort, and resource efficient strategy. In 2017, AAMI released a technical information report describing methods for incorporating HFE into design controls (see AAMI TIR59:2017 Integrating human factors into design controls).

The key is iteration. Conduct user research to understand the user's needs. Make design decisions with that information. Implement those designs into something testable. Lastly, have users interact with that prototype and get their thoughts on it. It doesn't matter how crude or developed it is; users will have something to say. Research, design, create, test, repeat.

Many manufacturers find themselves severely confined by years-old engineering or design decisions that, had user research been included early on, wouldn't exist. Investing in the HFE perspective earlier, rather than later, is strongly recommended.

Is There a Fast and Effective Way to Get Feedback on the Usability of My Device Without Having to Do an Actual Study with Users?

Usability studies are a powerful means of understanding the usability of a medical device. Accordingly, they tend to be the most time-consuming and expensive option as well.

There are quicker, more cost-effective methods of evaluating usability. A heuristic analysis completed by a HFE expert compares a medical device to a set of design principles (often called heuristics). A heuristic analysis can be an extremely valuable, cost-effective tool to not only identify usability issues, but begin the process of making design improvements in the device's user interface.

Resources

- Privitera, M. B. (Ed.). (2019). *Applied human factors in Medical Device design*. Cambridge, MA: Academic Press.
- Sethumadhavan, A., & Sasangohar, F. (2020). *Design for health: Applications of human factors*. Amsterdam: Elsevier.
- Speer, J., & Foster, B. (2019). *Integrating human factors into design controls to improve patient outcomes* [Audio podcast]. https://soundcloud.com/medical-device-podcast/integrating-human-factors-into-design-controls-to-improve-patient-outcomes.
- Wiklund, M., Birmingham, L., & Larsen, S. A. (2018). *Writing human factors plans & reports for medical technology development: Association for the Advancement of Medical Instrumentation*.

References

Branaghan, R. J., Hildebrand, E. A., & Foster, L. B. (2020). Designing for medical device safety. In *Design for Health* (pp. 3–29). Cambridge, MA: Academic Press.

Faulkner, L. (2003). Beyond the five-user assumption: Benefits of increased sample sizes in usability testing. *Behavior Research Methods, Instruments, and Computers, 35*(3), 379–383.

Holtzblatt, K., & Beyer, H. (1997). *Contextual design: Defining customer-centered systems*. Amsterdam: Elsevier.

Hutt, P. B. (2018). The evolution of federal regulation of human drugs in the United States: An historical essay. *American Journal of Law & Medicine, 44*(2–3), 403–451.

International Electrotechnical Commission. (2015). IEC 62366-1:2015 Medical devices—Part 1: Application of usability engineering to medical devices.

International Electrotechnical Commission. (2016). IEC TR 62366-2:2016. Medical devices—Part 2: Guidance on the application of usability engineering to medical devices.

Lee, J. D., Wickens, C. D., Liu, Y., & Boyle, L. N. (2017). *Designing for people: An introduction to human factors engineering*. Scotts Valley, CA: CreateSpace.

Medicines & Healthcare Products Regulatory Agency. (2017). *Human factors and usability engineering for medical devices including drug-device combination products*.

Privitera, M. B. (Ed.). (2019). *Applied human factors in Medical Device design*. Cambridge, MA: Academic Press.

Semler, E. (2019). *The history of the FDA's fight for consumer protection and public health*. Silver Spring, MD: US Food and Drug Administration.

Sethumadhavan, A., & Sasangohar, F. (2020). *Design for health: Applications of human factors*. Amsterdam: Elsevier.

U.S. Food and Drug Administration. (2016). *Applying human factors and usability engineering to medical devices*. Guidance for Industry and Food and Drug Administration Staff.

Wiklund, M., Birmingham, L., & Larsen, S. A. (2018). *Writing human factors plans & reports for medical technology development: Association for the Advancement of Medical Instrumentation*.

Wiklund, M. E., Kendler, J., & Strochlic, A. Y. (2011). *Usability testing of medical devices*. Boca Raton, FL: Taylor & Francis/CRC Press.

Chapter 10
Controls: Designing Physical and Digital Controls

10.1 Introduction

Controls enable users to change how one or more aspects of a device works. As such, it can have a profound impact on the user experience (UX), occurrence of use-error, task completion time, and even attitudes toward the medical device. Controls come in a variety of shapes, sizes, colors, functionality, and modality (e.g., digital, analog). These need to be carefully considered with regard to who will use the control, the tasks they perform, their environment, and situational constraints. For instance, a finger operated control such as an on-off switch will be challenging to use if the user is wearing gloves or other restrictive personal protective equipment (PPE). The goal is to choose control types that minimize risk and maximize effectiveness.

This chapter begins by introducing a variety of control "coding" guidelines. These are design heuristics to follow for any type of control—physical or digital, large or small. Afterward, we discuss control shape, size, and activation forces. This chapter excludes specialty control types, such as voice recognition, eye tracking, and gesture-based controls. While these topics represent the "bleeding edge" of technology, their application and merit is beyond the scope of this book.

10.2 Control Coding Guidelines

Due to the prevalence of controls, people have developed expectations about how unfamiliar controls should work. Controls often communicate their operation through codes like color, size, location, shape, labeling, and mode of operation (Sanders & McCormick, 1993). Of course, there are strengths and weaknesses associated with each coding type. A valuable heuristic is to use more than one coding type in a given control to ensure that potential use-errors are sufficiently mitigated

© Springer Nature Switzerland AG 2021   227
R. J. Branaghan et al., *Humanizing Healthcare – Human Factors for Medical Device Design*, https://doi.org/10.1007/978-3-030-64433-8_10

(Van Orden, Divita, & Shim, 1993). For example, a STOP button might be shaped like a stop sign (i.e., shape coding), as well as colored red (i.e., color coding) to signify danger. No matter what control coding scheme(s) a design employs, the goal should always be to identify a coding method(s) that is highly discriminable, quick to identify, and accurately identified (Sanders & McCormick, 1998). The following subsections look at these control coding guidelines in detail.

Color Coding

Color coding frequently communicates what a control does when activated. This is often achieved by leveraging stereotypes and associations that people have developed toward colors. For instance, in many regions of the world the color red has a consistent association with danger and its related terms (e.g., fear, threat, stop; Courtney, 1986). Likewise, the color green has a strong association with action-relevant terms (e.g., go, start). You have probably seen these color associations in traffic signals or "accept" and "reject" buttons on smartphones.

Manufacturers should aim to use colors that have consistent meaning among a user population. Research suggests that colors such as red, green, and yellow have the highest consistency across cultures (though, not universal). By contrast, colors such as blue, purple, and white differ considerably from one population to the next (Chan & Courtney, 2001). Colors that have established associations should not be used in controls that have contradictory actions (e.g., green light turns on when a "stop" button is pressed).

Gao et al. (2007) advise that the relationship between color and meaning is multifaceted, and is not always consistent across age groups, gender, nationality, and cultural backgrounds (Elliot & Maier, 2012; Gao et al., 2007). Some Chinese populations, for example, associate red with the term, "caution," whereas most American populations associate this term with yellow or orange (Chan & Courtney, 2001). Figure 10.1 illustrates findings from a few studies (i.e., Courtney, 1986; Wang & Or, 2015) looking at these cross-cultural comparisons of color associations. These results are a testament to the fact that seemingly small differences like color choice can introduce risk into a control's design, especially when that control will be used in international markets without modification. In these situations, it is valuable to test these designs across different countries to determine if region-specific color associations are violated.

While only two colors associations differ, there is far greater consistency in color association among Americans. Additionally, the concepts of "off" and "on" have low agreement in both countries. A similar study by Bergum & Bergum (1981) echoes this finding.

Deviations from color associations can be seen in a number of medical devices on the market. For example, the air and water valves on the handle of the flexible endoscope in Fig. 10.2 are color coded as red and blue, respectively. While the association between "blue" and "water" is salient to a trained user and lay person alike,

Chinese		CONCEPT	Americans	
Color	%		Color	%
Green	60.1%	Safe	Green	61.4%
Green	74.6%	Go	Green	99.2%
Green	51.8%	On	Red	50.4%
Red	50.4%	Hot	Red	94.5%
Red	60.4%	Danger	Red	89.8%
Red	59.0%	Stop	Red	100%
Red	46.8%	Off	Blue	31.5%
Blue	39.6%	Cold	Blue	96.1%
Yellow	51.1%	Caution	Yellow	81.1%

Fig. 10.1 The differences in color associations between Chinese and Americans (based on data from Wang and Or (2015) and Courtney (1986), respectively)

Fig. 10.2 Flexible endoscope showing the air/water valves. (Image by Beloborod/shutterstock. com)

the relationship between "red" and "air" is far less clear. Keep in mind that devices such as endoscopes require substantial training before a user uses the device on an actual patient. As a result, users rely on far more than color coding to draw distinctions between these two valves. However, not all devices incorporate such extensive training. In some cases, the same types of color-coding issues are present, and users are left on their own to figure it out.

Color coding can also be used to create new conventions that are device specific. That is, colors with no pre-existing associations might be used to differentiate controls that are close in proximity to one another, but serve different functions (Helander, 2006). This is especially important in control design, when neighboring controls are similarly sized, shaped, or operated in a similar manner. For example, the membrane buttons on the front of the user interface in Fig. 10.3 are identical in size and shape, but differ in a few other aspects. The most distinguishing and identifiable difference is their color. Proctor and Van Zandt (2018) recommend that when controls are spaced far apart, no more than five unique colors should be used; users have to hold in their memory the meaning of each color, as well as where it's located. However, if the controls are collocated (i.e., side-by-side), and the use-environment is well illuminated, more than five unique color categories can be acceptable (Proctor & Van Zandt, 2018).

Size Coding

Size coding refers to the association people make between the size of a control and its influence on the medical device. Simply put, the larger the control, the larger the movement or response from the medical device users expect. Small controls, by comparison, are anticipated to have small effects (see Fig. 10.4). Of course, in the real-world these expectations are often violated as part of other (useful) design trade-offs.

There are a variety of examples in healthcare where size coding is violated. For instance, many flexible endoscopes have a set of concentric knobs located along the control section (i.e., handle area). These knobs affect the "up/down" and "left/right" angulation movements at the distal tip of the device, which change the device's position and view in a patient during a procedure. By virtue of the "concentric knob" design, the top lobed knob is smaller than the bottom lobed knob (see

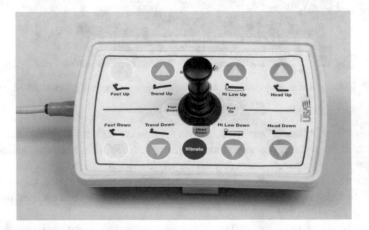

Fig. 10.3 Multiple color controls when controls are co-located. (Image by Aaron Goldsmith/shutterstock.com)

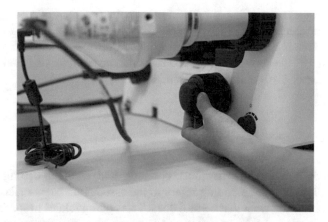

Fig. 10.4 A microscope is a good example of size coding at work—the "coarse adjustment knob" is larger, and the "fine adjustment knob" is smaller. (Image by Charoenrak Sonthirat/shutterstock. com)

Fig. 10.5 Concentric knobs on an endoscope. (Image by Robert Przybysz/shutterstock.com)

Fig. 10.5). This design violates the size coding construct, since each knob's size has no relationship to the magnitude of its impact on the device; they simply control different aspects of the device's articulation. It's likely that this design ultimately won out due to other important design principles (e.g., co-locating controls with similar functions).

Several alternative endoscope designs have been proposed over recent years, some of which have taken aim at the concentric knob design ubiquitous to flexible endoscopes. A good example of this innovation comes from Invendo Medical's Invendoscope. The concentric knob concept has been replaced with a digital control pad reminiscent of a modern video game controller directional pad (i.e., d-pad). For example, pressing up or down on the Invendo directional pad causes the endoscope's distal tip to articulate upward or downward, respectively. This design likely offers a more natural and intuitive mapping between control and device movements (see Fig. 10.6).

Remember that size coding is just a starting point in a control's design. In some cases, a larger control (e.g., finger-operated knob) should be used to control fine movements. This is especially important in digital devices where fine adjustments are made through the software itself. You must find an appropriate balance between a control's ease of use, comfort, and accuracy while it's in use.

Location Coding

Location coding is among the most common and resilient coding principles (Helander, 2006). Location coding refers to the association users make between a control's position and what that control does. This coding relates to spatial information—that is, *where* objects are expected to occur in our environment relative to other things (Reed, 2012). For example, there is a strong location coding of a car's brake pedal relative to the accelerator pedal; no matter what make or model car a person drives, the brake pedal is expected to be to the left of the accelerator. This expectation remains consistent even if other aspects of the pedal design differ between vehicles, such as their size, shape, or spacing (see Fig. 10.7).

Location coding is often learned through training and hands-on experience. For instance, a driver must learn early that the brake pedal is to the left of the accelerator; this is not universally understood among nondrivers. What is more, this knowledge often becomes proceduralized over time. That is, an experienced driver reacts instinctively when faced with a stressful situation where they must brake quickly. The foot moves without conscious awareness or planning.

This procedural knowledge can be both a blessing and curse for medical devices. On the one hand, experienced users may learn to use parts of a medical device's interface effortlessly, thereby making them faster and promoting satisfaction during use. Conversely, however, when the control layout of that device changes (i.e., new model, different manufacturer), use-errors may be introduced due to negative transfer of training—interference of previous knowledge with new learning. This is a perennial problem for medical devices, as patent rights may require novel devices to be designed in less familiar ways to established users. For these reasons, location coding is not a sufficient way to design controls on its own (Proctor & Van Zandt,

Fig. 10.6 This single-use flexible endoscope from Invendo Medical uses a directional pad to control the distal tip's angulation. (Image courtesy of Ambu Medical with permission)

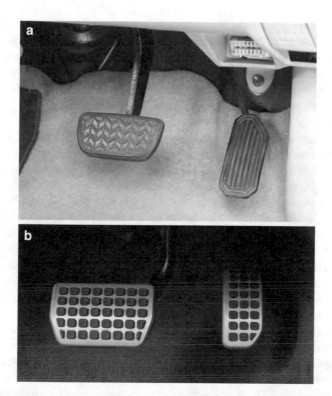

Fig. 10.7 Brake pedal control comparisons—across models and manufacturers, the brake pedal is always located to the left of the accelerator pedal (**a** by Winai Tepsuttinun/shutterstock.com; **b** by Gargantiopa/shutterstock.com)

2018). At minimum, location coding should be coupled with at least one other coding type to minimize or prevent use-error.

Shape Coding

Shape coding uses the shape of a control to inform what the control will do when the user interacts with it. Shape coding can manifest in a few ways. One way is to associate an outcome with pre-existing shape associations from our daily lives. For example, a button intended to stop or abort an action could be designed to mimic a standard octagon-shaped stop sign. Other shapes such as arrows and triangles have similar associations with movement and directional cues, such as "up/down," "left/right," "forward/backward."

The second use of shape coding is to create shape associations that are unique to specific tasks or outcomes. For example, the Bayer MEDRAD® Centargo CT Injection System offers Radiological Technologists a uniquely shaped "abort" button on the CT injector unit that champions this principle (see Fig. 10.8). This button is effective because it's uniquely paired to this one function. In other words, the same button shape is not used elsewhere on the device for a different purpose. This design's effectiveness is further increased through the use of color coding (i.e., association between red and stop). As discussed above, it's a good practice to use multiple coding constructs whenever possible to get the control's design intention across to the user.

Some studies suggest that arbitrary shape coding is less effective when designers try to incorporate this strategy on multiple controls on the same interface (Roscoe, 1980; Slocum, Williges, & Roscoe, 1971). In other words, each control gets its own arbitrary shape. Users simply cannot remember each control shape's purpose or function when too many are added into the interface at once.

Shape coding is used most often on finger-operated push buttons and levers, and less commonly among hand controls such as knobs and wheels. One reason for this is that our fingers are better at discriminating shapes than our hands or feet due to the greater concentration of *mechanoreceptors* in our fingertips—sensory cells that help us detect small changes in mechanical forces. And, to state the obvious, we have more fingers than palms or feet—each finger works with the others to approximate distance and feel for useful landmarks (e.g., points, edges).

Even with minimal training, users quickly learn to associate many control shapes to functions. Under high stress, however, the upper limits for control shape coding drop to less than one dozen, even among highly trained users (Woodson & Conover, 1964). Simple shapes such as circles, squares, and triangles take the least time to discriminate and identify, whereas shapes with several points such as octagons and multi-point stars take substantially more time (Ng & Chan, 2014). What's more, the closer in shape two controls are to each other, the more likely users are to mix them

Fig. 10.8 Bayer MEDRAD® Centargo CT Injection System. (Image courtesy of Bayer AG with permission)

up unless other mitigation efforts are implemented. Some control designs—such as the knobs shown in Fig. 10.9—have a complex shape that makes them easier to tell apart from standard, cylindrical knobs. However, this may come at the cost of slightly extra time required during detection through touch alone.

The challenge with incorporating shape coding into device controls is that few shapes are universally understood without additional training. One exception to this is the strong relationship between arrow shapes and directionality. The use of arrow-style or triangle shaped buttons is common to navigating through a linear workflow (i.e., "next," "back"), or mapping to up, down, left, and right (see Fig. 10.10). While this is an acceptable approach, designers must be cognizant of potential tradeoffs, such as creating a smaller surface area on the control itself by changing its shape. If the control's surface area is too small, the user may not be successful at pressing the button.

Shape coding should be used in situations where the control always supports the same behavior or output. For example, a button shaped like a camera should always be used to take photos or video; it shouldn't stand in as an "ENTER" button in some use-cases. Steiner and Burgess-Limerick (2013) advises that when shape coding is used across a suite of devices, the relationship between shape and function remains consistent to prevent negative transfer of training. They also suggest that this relationship should not be capable of being altered in error during use or ongoing maintenance (Steiner & Burgess-Limerick, 2013).

Graphical user interfaces (GUIs) provide a seemingly limitless set of options with respect to control design. The mock GUI in Fig. 10.11 incorporates a variety of conventional control designs found in digital products. Draw your attention to the

Fig. 10.9 Lobed knobs designed with finger-operated controls afford finger placement around each of the lobe. (Image by dny3d/shutterstock.com)

Fig. 10.10 Directional
arrows on a glucometer.
(Image by
nechaevkon/shutterstock.
com)

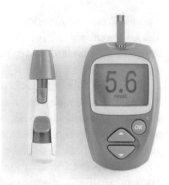

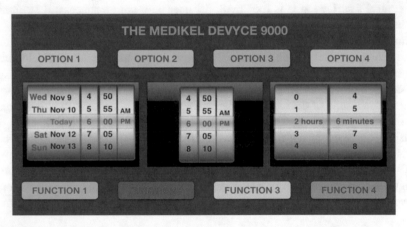

Fig. 10.11 A mock graphical user interface (GUI) with standard design assets
four buttons at the top (i.e., option 1–4), as well as the four buttons at the bottom
(i.e., function 1–4). These buttons use color and label coding effectively, but miss an
opportunity to differentiate the buttons through shape coding. Depending on the
intention of the button, shape coding would be a valuable way to help users quickly
discern between the options available on this screen.

Label Coding

Provide labels for your controls. They should be concise, meaningful, and specific
to the lexicon of actual end-users.

Labels should be placed at or next to its corresponding position along the control.
For example, a vertically aligned on-off switch should have the "on" label at the top
of the switch, whereas the "off" label should be at the bottom. The shared label, "on/
off," should be avoided when possible, since there is no association with which
direction the control needs to be moved to cause it to turn something "on" or "off."
Additionally, whenever possible, the control's label should not be blocked from
view by the user's body, hands, or fingers during intended use. The power switch

control in Fig. 10.12 demonstrates a violation of this concept. Sometimes this cannot be avoided due to limited space on a bezel or user interface. In these cases, the label may need to be placed on the surface of the control with which the user directly interacts. This is often the case with touchscreen controls, where screen real estate is at a premium.

Control labels (in English) should read left to right. Avoid vertical and rotated labels, as they take longer to read and understand (Helander, 2006). If numeric values need to be labeled next to a control, they should be spaced as closely as possible to help strengthen their association with the control. Figure 10.13 shows a few examples of "good" and "bad" examples of these recommendations. The top settings bar is the "good" example—the label is horizontal, and the value label (i.e., 40%) is placed right next to the control's point of interaction. The middle settings bar has two minor issues. The label is vertical, but still maintains the expected left-to-right letter orientation. The value label is close to the point of interaction, but not in an ideal spot. The bottom settings bar is the "bad" example. The label is vertical, and the value label is not only far away from the point of interaction, but also on the far-right side. This means the user would have to look back and forth across the screen to make sense of this one control.

Mode of Operation

Mode of operation coding refers to two aspects of interaction with a control. First, it relates to the control's feel or response during use. Second, it refers to the way a control is operated. Both help distinguish between controls when you can't see them. For example, however brake and accelerator pedals provide their own distinct

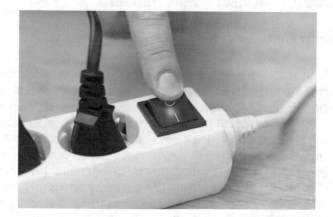

Fig. 10.12 Surge protector with an on-off switch has labels in a less than ideal location; when the user presses either end of the control, the label is blocked from view. (Image by Bacho/shutterstock. com)

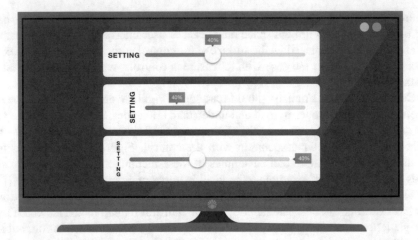

Fig. 10.13 Example of label designs with different text label orientations

feel when pressed. The brake pedal usually feels "squishier" as additional force is applied, while the accelerator pedal has a consistent resistance through its travel. This is helpful since drivers' shoes make it difficult to detect the shape of each pedal.

In other cases, a control may have a unique activation method to differentiate it from nearby controls. For example, instead of using an "on-off" button or switch, you might use a key switch in some situations, such as for tasks that involve high risks or security clearances. Not only does this prevent accidental activation, but it requires the user to physically manipulate the control in a unique way (i.e., turning a key) that no other control offers on the interface.

10.3   Control Movement Considerations

Directionality Considerations

Users often have expectations about how a control's movements relate to changes or directionality in a medical device. These are reinforced through controls that we use on a daily basis. For example, the relationship between a counterclockwise rotation and "turning left" is reinforced through the steering wheel in an automobile. A counterclockwise movement also has a strong association with "decreasing," as one would do to turn down the volume on a volume control.

These expectations for control movements are called control movement compatibility (CMC). CMC is a user's expectation for similarity between a control's movements and the corresponding action (i.e., movement) by the device or system. Studies demonstrate that improving CMC can decrease reaction time, time spent in training, error rates, and can even bolster satisfaction during use (Hoffmann & Chan, 2018). By contrast, errors increase when control and response movements are in opposite or perpendicular directions from one another (Müsseler & Skottke, 2011; Steiner, Burgess-Limerick, & Porter, 2014). That is, pressing an "up" arrow should never make the device respond by lowering a value or setting.

Table 10.1 outlines several of the most common control movements and what (most) users expect the device to do.

Bear in mind that while many users will have mental models of how a control's movement will affect a medical device, there is no guarantee these patterns are universally understood. Even the movement terms themselves should be considered against the region and context of use. For example, the term "counterclockwise" is common to North American countries, but is used less frequently in England. Instead, the term "anti-clockwise" is preferred in most regions. These differences can be especially important to identify among control labels and instructions for use.

Control Travel Considerations

A control's *travel* is the distance that a control is able to move before it's stopped. Controls such as keyboard keys have relatively short travel (i.e., just a few millimeters). Automobile brake pedals have longer travel (i.e., several centimeters). Generally, a control's travel should match the physiological capabilities of the body part interacting with that control. For example, finger operated controls such as switches and buttons should have shorter travel, whereas hand or foot operated controls should have longer travel.

Travel has been researched most with respect to finger operated controls. For computer keyboard controls, the Association for the Advancement of Medical Instrumentation (2009) recommends a key travel of 3 mm. In practice, however, most commercial keyboards produced today have travel specifications closer to the lower end of these ranges (Asundi & Odell, 2011). Insufficient travel can make a control feel as if it's not working appropriately, or that the user did not press hard enough on the control to activate it. This can lead to double-pressing a control and delivering unwanted therapy or changing a setting past an intended threshold.

Table 10.1 Recommended control movement device outcome relationship

Control movement	Device outcome
Up, forward, pull	On
Down, rearward, push	Off
Up, rearward	Up
Down, forward	Down
Up, right, forward, clockwise	Increase
Down, left, rearward, counterclockwise	Decrease
Clockwise, right	Right
Counterclockwise, left	Left

A control's influence on a medical device or system should remain predictable and consistent over the length of its travel. For limited, discrete state controls such as an on-off switch or a keyswitch on a keyboard, this recommendation does not apply, since the control only has two states (e.g., "on" and "off"). However, for controls that offer multiple discrete states, or controls that provide continuous scaling (e.g., volume control knob), the change from one setting to the next should be proportional. For example, changing a volume control from "1" to "3" should have the same relative effect on the system as changing the volume from "8" to "10."

Control Gain

All controls require time to interact with in order to produce changes. With finger operated controls (e.g., on-off switch, keyswitch), the time is milliseconds. For larger controls that might have several discrete states, or involve longer travel distances, the time required to make adjustments can be several seconds. This time loss quickly adds up in some situations, such as a surgical procedure where operating room costs can average $35 per minute (Childers & Maggard-Gibbons, 2018).

One reaction to this dilemma is to make control movements as fast as possible for users. For example, a rotary knob control might be mapped to software in a way that a mere quarter turn of the control causes the system to jump from setting "1" to setting "100." Although this makes the control faster to use, it's likely difficult for the user to dial in a specific value. This relationship—amount of control movement vs. amount of system response or movement—is referred to as the *control's gain*.

When a computer mouse or trackpad is set to a 1:1 gain ratio, the cursor moves about the same physical distance that the user moves the mouse across the table. A higher gain ratio results in the cursor moving *less* than the mouse, while a lower ratio makes the cursor move *more* than the mouse. Although this seems backward at first, you can simplify things by adding values and measurements to examples like these:

- 3:1 gain ratio (high ratio): The (physical) mouse moves 3 in. for every 1 in. the cursor moves on the screen. Thus, the cursor would be perceived as moving slowly.
- 1:5 gain ratio (low ratio): The (physical) mouse moves 1 in. for every 5 in. the cursor moves on the screen. Thus, the cursor would be perceived as moving quickly.

A low control gain ratio one reason why players at control-operated arcade prize games rarely win. In the "claw game," for example, the claw moves a lot, even when the player barely touches the joystick (see Fig. 10.14). Small movements of the joystick cause large and unpredictable amounts of movement in the claw. There is also a slight delay (e.g., 250–500 ms) between control movement and system response making it difficult to make small adjustments, as movement feedback is delayed by just a few hundred milliseconds. Humans simply aren't equipped to react fast enough to such sudden movements without substantial training and experience.

In some cases, the control's gain should be adjustable. For example, experienced computer users may prefer to set their cursor movement settings higher to allow them to move their cursor quickly. Errors such as undershooting or overshooting a target on the screen are minimized due to their experience, knowledge of the system, and their ability to recover from an error quickly and with minimal effort. However, a less experienced user who may not understand how the system works, may select a slower cursor setting to minimize error. The goal is to identify the best tradeoff between speed and accuracy.

Keep in mind that a 1:1 gain ratio is not always ideal for all types of medical device controls. In some cases, it should be purposely reduced to augment the user's capabilities or promote selection accuracy. Many robotic surgery system controls on the market use this approach to provide greater precision and control during minimally invasive surgery (MIS). For instance, a broad wrist movement by the surgeon using da Vinci® surgical system may map to just a few millimeters or centimeters of instrument movement inside the patient (see Fig. 10.15). This facilitates surgical accuracy while operating within a small space inside the patient.

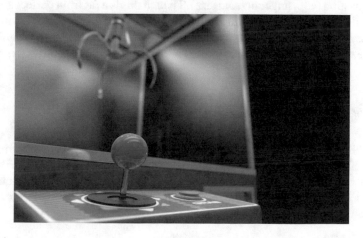

Fig. 10.14 A claw machine control. (Image by Inked Pixels/shutterstock.com)

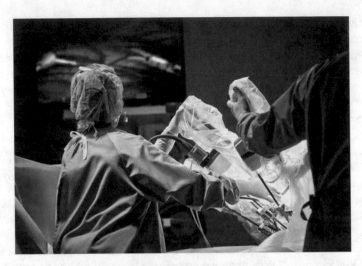

Fig. 10.15 Bed-side during a Robotic Assisted Surgical (RAS) procedure. (Image courtesy of Intuitive Surgical, Inc. with permission. ©2020 Intuitive Surgical, Inc.)

10.4   Control Size and Shape Considerations

The Size of a Control(s) Should Be Comfortable, Accurate, and Consistent Use

This can be challenging when the medical device will be used by a variety of users with different anthropometric characteristics such as finger, hand, and foot size. As a general guideline, controls should be designed to support safe and effective use among users with the largest physiological characteristics. Often, the upper limit of use is designed for users in the 95th percentile in respect to the anthropometric feature in question (e.g., finger or hand size). Though this is a far from perfect approach, it's usually easier for users with smaller physical features to interact with larger controls, than vice versa. Your specific user populations should be examined up front in early research efforts and factored into the design of the controls themselves.

Surface Area Is King

One of the most important factors to consider when designing a control is the amount of surface area the user will make contact with. Designers must strike a balance between how much space allocated to a control and how large (or small) that control needs to be given, how often it's used, it's criticality, placement, etc. Several factors affect surface area:

• Overall size

- Shape
- Surface textures
- Edge details (e.g., chamfer, radius)
- Whether it has a flat, convex, or concave profile

For example, a simple change from a square to circular finger-operated button shape will result in about 25% less surface area. Similarly, a triangular shape has only half as much surface area as a similarly sized square shape. As shown in Fig. 10.16, a 3D rendering of a (fake) hand-held device prototype has three similarly sized buttons in square, circular, and triangular shapes. The square offers the greatest amount of surface area of the three buttons. Designers must be cognizant of buttons being large enough to be accessible, but small enough to avoid accidental activation.

Think of the surface area of the control as the "target" with which the user interacts. A control with too little surface area creates a small target. On the other hand, too much surface area can increase the rate of accidental activation (i.e., "fat-finger" accidents). Oversized controls may also crowd the user interface, making neighboring controls harder to find, access, or use.

Square finger-operated controls such as buttons, switches, and levers, should have surfaces of at least 20 mm (Association for the Advancement of Medical Instrumentation, 2009). This provides a good balance between interaction accuracy, speed, user comfort, and space conservation on the user interface. Controls can be smaller than 20 mm if the manufacturer is comfortable with accepting the risks of greater inaccuracy from that control. For example, most keys on a computer keyboard are about 16–18 mm wide and tall. It's acceptable for these keys to be smaller than the recommended 20 mm size because errors in typing can be deleted without harm. Smartphone GUI "keys" on a virtual keyboard are even smaller. As a result, most people experience more spelling mistakes when texting than compared to typing on a physical keyboard (at least when auto-correct is disabled).

Smaller control surfaces work best for medical devices where the user's hands remain in a fixed position (e.g., typing position). If the user has to move away from this "ideal" position frequently to use other aspects of the medical device, then the control's surface area should be increased. This is particularly important when movements between control areas must be completed quickly (Fitts & Seeger, 1953).

Certain cylindrical-style knob designs offer more surface area than others. For example, lobed and fluted knobs are designed with distinct cutaways to help cradle

Fig. 10.16 3D rendering of a prototype with three button shapes

the user's fingers, thereby increasing the surface area in contact with them. The skirted knob shown in Fig. 10.17 is another good way to increase surface area. Skirted knobs lend themselves to multifinger grips. The term, "skirted," refers to the bottom portion of the knob that flares outward. Incidentally, these designs also decrease the need for the user to apply as much grasping pressure to any one finger, as it spreads out the load over more touch points and surface area.

When Possible, Reduce, or Eliminate the Need for Fine Motor Control

Fine motor control requires concentration to identify the correct target or stopping point. Many users will not have the cognitive bandwidth needed to use these controls effectively during stressful situations, or when they are fatigued. If fine motor control is required, the control should support a variety of grips that benefit from larger muscle groups. In some cases, this may be as simple as enabling multiple fingers to provide leverage or stability during the control's movement, or help support the weight of the user's hand while they remain idle in between interactions (e.g., waiting to press the "photo" key on an ultrasound device).

Finger-Operated Controls Should Support Multifinger Use

Rotary controls that require fine motor control should permit the use of at least three fingers to promote comfort and accuracy during use. A three-finger grip—known as a *three-jaw chuck grasp*—provides better control over small components like knobs, levers, switches, and sliders. By comparison, a two-finger grip with the thumb and index finger—known as a *pincer grasp*—offers less control, especially during rotational movements. Furthermore, a pincer grasp can be challenging or painful for older adults as well as users with some types of peripheral neuropathies.

Fig. 10.17 Skirted volume knobs

Of course, there are some trade-offs with multifinger use on a control. The obvious one is that there are fewer fingers left to interact with adjacent controls. In some cases, this could lead to (slightly) slower task completion rates when the user is expected to input a series of commands or adjust multiple controls back-to-back. Relatedly, control movements in general tend to be slower when multiple fingers are used on a control. This issue is most pronounced when all the fingers must be used in tandem with a hand control. The human wrist can only rotate so far, so the user may need to readjust their grip part way through the control's movement if the travel is too long.

Textures Help Improve Suboptimal Control Shapes

Rotary knob designs can be improved by adding knurling or a rubberized texture to the surface. This decreases the likelihood of the user's fingers slipping off during use. These designs can be an excellent supplement to the other strategies discussed above. Unfortunately, textures like the "diamond" knurling on a set of stacked knobs shown in Fig. 10.18, can be difficult to clean, posing an infection risk due to the potential of harboring dirt and debris. The term "knurled" dates back to Middle English where a "knur" referred to a knot in wood. Shakespeare used the term, "knurled" as a variant for "gnarl." Knurling comes in a variety of forms, and "rough" textures are a hallmark of a knurled surface.

Size and Shape Should Be Scaled to Match Effort, Duration of Use, and Accuracy Requirements

Larger controls encourage users to use larger body parts (e.g., hands instead of fingers) during interaction. Larger muscles hold up better to physically demanding or long-term tasks. Furthermore, larger body parts can work in concert with smaller

Fig. 10.18 Example of
knurled knob

body parts during tasks that require accuracy. For example, your hands and wrists tend to do most of the work when rotating a large rotary control (i.e., clockwise, counterclockwise), and your fingers help provide specificity at the tail end of the movement. Each body part helps provide small movements and adjustments until the target setting is reached. Table 10.2 provides a few guidelines to follow when determining if a control should be large or small.

Be Mindful of Control Resolution in Multistate Controls

Control resolution refers to the "distance" between discrete settings along a control's movement path. Low control resolution means that there is a large area or zone between settings. High control resolution results in jumping from one setting to the next much faster, but comes at the risk of overshooting your target.

"Accuracy" and "resolution" tend to follow a negative relationship—as the need for accuracy increases, control resolution decreases. Generally speaking, you cannot have the best of both worlds in one control, unless there are secondary design mechanisms in place to help you out. One example of this is a "soft close" feature on a kitchen drawer or door. The user can use a fair amount of force with these things because the secondary mechanism (i.e., gas cylinder) will "catch" them just before slamming into the cabinet. Similar concepts can be applied to medical device controls as well, especially when dealing with digital controls where control movement parameters can be strictly controlled through different types of coding and conditional variable states. Without these secondary mechanisms, a high-resolution control that *also* requires high accuracy and consistency during use will force users to slow down, redirect attention and effort, and focus on using their fine motor skills.

For rotary and thumbwheel-style controls, the Association for the Advancement of Medical Instrumentation (2009) recommends a minimum of 30° separation between discrete control states. Although, further separation is preferable when the user is expected to work in stressful or distraction-prone environments or may experience fatigue. This separation is important as it broadens the "target" for each setting, and minimizes the need for fine motor control. Cook and Polgar (2014) add that control resolution should take into account each user group's physical and cognitive capabilities during the design process, since users tend to differ in respect to their skills and abilities.

Table 10.2 Recommendations for the use of large and small controls

The control should be *large* if it…	The control should be *small* if it…
● Requires high activation force(s) ● Demands high levels of accuracy ● It will be held in a fixed position during sustained use ● Will be used frequently	● Requires low actuation force(s) ● Permits some inaccuracies ● Will not be held in a fixed position during sustained use ● Used infrequently

Switch controls with three or more discrete settings (e.g., toggle switches) are also affected by control resolution factors as well. Surprisingly, however, many off-the-shelf switches are designed with less than ideal control resolution specifications. For example, a standard three-way switch may only have about 15° of separation between each of the three control states (see Fig. 10.19). While these controls are functional and widely available, they may require too much fine motor control for users to use them safely and effectively in some types of medical devices. The middle setting is especially affected by this limitation. By comparison, the two outermost switch positions are less problematic since they allow the user to exert force without issue; the switch cannot pass these outside positions.

Many toggle switches (such as the three-way switch shown here on the left) have a limited amount of space between each switch position. This can make it difficult to visually tell which position the switch is in, especially when the switch is small. In total, these types of switches have about 30° of travel split across all three positions. The revised three-way switch on the right-hand side follows the guidelines described above regarding the use of 30° of travel between each control position.

Avoid Sharp Edges Along Control Surfaces

Sharp edges can make a control uncomfortable or difficult to use. This is especially true among finger operated controls where people are sensitive to discomfort due to the increased concentration of nerve endings in the fingertips (Mancini et al., 2014). Sharp edges are also prone to damage (e.g., chipping) from bumping into other equipment during assembly, disassembly, or maintenance. Depending on the material, this can create small burrs that snag and tear open personal protective equipment (e.g., gloves), or even cut the user's skin. Simple fixes such as incorporating a radius or chamfer to the outside edge of the control will help mitigate this issue. In Fig. 10.20 the red block simply refers to a rounded edge, whereas the green block is a 45° inward plane. The Association for the Advancement of Medical Instrumentation (2009) recommends a minimum radius edge of 3 mm to reduce discomfort. These "eased edges" reduce the discomfort caused by sharp, perpendicular edges on controls.

Fig. 10.19 Example of two 3-way toggle switches

Fig. 10.20 A comparison
of radiused (red) and
chamfered edges (green)

10.5   Control Feedback Considerations

Feedback is integral to make sense of the relationships between controls and devices. Well-designed controls incorporate one or more types of feedback. Often, this comes in the form of confirming a new event, reminding the user about his or her current status, or simply informing the user about which choices are and are not available. Regardless of how feedback is presented (e.g., visually, auditorily), here are a few useful guidelines (Association for the Advancement of Medical Instrumentation, 2009). The acronym, "RIMS" is a useful way to remember these tips:

- *(R) Redundancy*—Users benefit from two or more forms of feedback co-occur, especially during critical events or tasks. Redundant feedback should be presented multimodally (e.g., visually and auditorily). Redundancies can also be used in circumstances where the user should reconfirm a critical selection, such as with a drug dose confirmation message presented on a GUI.
- *(I) Immediacy*—Feedback should be provided immediately or as soon as possible after manipulating a control. For example, the user should feel the button click when pressing it. Likewise, the system should indicate that a program is loading after the user initiates the command(s) for it to begin. Delays or lags in feedback can result in double-pressing a control or moving past a desired setting.
- *(M) Modality*—Feedback should suit the existing scenario's demands and use-environment. For example, an alarm that registers at the same frequency bands as human speech may be ill-fitting for a medical device used in places where many people are talking at once. As a rule, only so much information can be passed along each sensory channel (e.g., audition, vision) before the user can no longer process it effectively. Feedback should be supplied through modalities that aren't overwhelmed with competing stimuli (Burke et al., 2006), yet still provide sufficient "signal strength" to cue the user.
- *(S) Specificity*—Feedback should be clear and understandable to as wide of a user population as possible. For instance, a tactile "click" while pressing a button

is not limited by geographic or language constraints—users from all walks of life understand that the click indicates their press was registered by the control (or they can quickly learn it).

Visual Feedback

Of all of our sensory modalities, vision is most capable of handling complexities (Gallace & Spence, 2009). In fact, some argue that for adults in particular, vision is the main channel through which humans make sense and recall the world around us (Hirst, Cragg, & Allen, 2018). This, visual feedback is common in medical device design.

Visual feedback comes in several mediums. Common forms include the use of lights/colors, text, and icons. Of course, these mediums are not exclusive; combinations and variations of each are commonly implemented into controls. The following information outlines several recommendations and guidelines that designers and manufactures should follow in respect to visual feedback.

Account for Environmental Luminance and Color Spectrum Factors

Many controls use lights to signal that a control has been activated (or deactivated). In some cases, lights may be used in a pattern to indicate some type of change or "status" of the medical device. These lights should be designed to take into account the predominant color spectrum and light intensity (i.e., luminance) present in the use-environment. This is especially important when the user's vision may already be dark-adapted (i.e., scotopic vision). For example, a continuous positive airway pressure (CPAP) device used while the user sleeps may have control lights that are too bright, or use flashing lights to signal the device's status. These lights can be distracting and disruptive to users in a dark bedroom.

Ambient light can also affect how lights are perceived. For instance, some operating rooms will use green spectrum lighting during procedures in which the OR team relies on video displays. The (slight) green hue to the ambient light helps make content on visual displays easier to see.

Use Icons When Possible

Icons are an excellent way to forego the difficulties of translating text for international devices. They also save space on a user interface, compared to text labels. However, icons are not always understood. Indeed, icon interpretation is influenced by factors such as the user's technological background and familiarity, experience working with similar products or systems, and even their cultural background.

Use Appropriately Colored Text

Labels should be presented in fonts that are easy to read, taking into account the control panel or GUI's size and brightness. Verify that the contrast between the label (i.e., foreground) and the surrounding control panel (i.e., background) makes the label easy and fast to read. A contrast ratio of at least 3 to 1 should be used for large text labels, and increased to at least 4.5 to 1 for smaller text. Bear in mind that label color can significantly affect contrast. As WCAG accessibility guidelines (2018) points out, the same label presented in pure red, pure green, and pure blue differ considerably in terms of contrast.

Select Appropriate Font Types

Fonts generally fall into two categories: serif and sans serif. The term, "serif" refers to the addition of small strokes at the ends of characters and letters, while the term, "sans" simply means the absence of those details. Times New Roman is a standard "serif" font, whereas Arial is a prototypical "sans serif" font. In respect to control labels, san serif fonts are usually preferred as they are slightly easier and faster to read on signage and system GUIs (Dogusoy, Cicek, & Cagiltay, 2016; Moret-Tatay & Perea, 2011).

Choose Appropriate Text Size

Font size affects a user's ability to discern text. For instance, Bernard, Liao, and Mills (2001) reported that older adults (avg. 70 years old) found 14-point fonts more legible than 12-point fonts while reading a digital display. Younger adults, on the other hand, generally experience better legibility and reading speed when text is presented in 12-point font. For device controls that may be operated by younger children (i.e., 6–7 years old), text should be presented in larger fonts, line length should be minimized, and simple words should be selected (Katzir, Hershko, & Halamish, 2013). However, for slightly older children approaching adolescence (i.e., 10–11 years old), reading comprehension, accuracy, and speed is improved by using smaller fonts (e.g., 12-point font). The main reason for this shift is that younger children tend to be "glued to the print." Meaning, reading tends to be letter-dependent and requires substantial effort to decode unfamiliar text (Katzir et al., 2013).

Make Labels Durable

Physical control labels are susceptible to wear and tear following repeated use. Devices that are commonly used outdoors (e.g., wearable devices) may experience photofading due to exposure to light and ozone-related chemicals (Allen, 1994).

Generally speaking, violet and blue dyes experience less pigment breakdown, while reds, yellows, and greens fade faster.

Labels on physical controls should be *embossed* or *debossed* when possible. Embossing refers to raising the control label characters above a common surface, whereas debossing refers to lowering it below that surface (see Fig. 10.21). Both options reduce the chances of the control label's ink being rubbed off.

Generally speaking, embossed characters are easier to read. The edges of alphanumeric characters tend to appear sharper and crisp compared to debossed edges.

One limitation with embossed and debossed labels is that the small, recessed areas of the label can harbor unwanted debris and bacteria. In some cases, this introduces a potential risk of infection. These advantages and disadvantages should be taken into account given the expected use-environments, as well as how and when the device will be serviced or replaced.

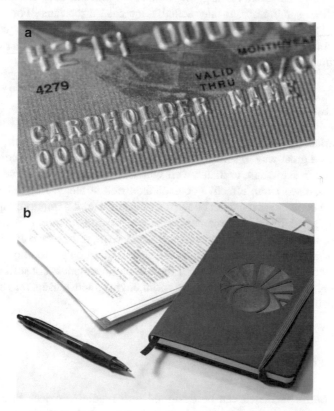

Fig. 10.21 (**a**) Credit card numbers with embossed numbers. (Image by Andrew Berezovsky/shutterstock.com.) (**b**) Debossed leather notebook

Present Visual Feedback in Close Proximity to the Control

Confusion may result from control feedback that is presented too far away from the control itself, for example a light turning on in response to pressing a button on the opposite end of the medical device. Sometimes this issue cannot be helped, such as when the medical device requires a secondary display separate from a set of physical controls. In these situations, issues can be mitigated by ensuring the feedback is promptly timed upon activation of the control, or by taking a multimodal approach with supplemental forms of feedback (e.g., audible chime).

Use Backlighting on Nonalphanumeric Keys Only

Backlighting on membrane buttons help users locate critical buttons in low lighting (e.g., perimeter of a darkened operating room). Many manufacturers leverage backlighting further as a feedback mechanism—when the user presses the button, the key lights up or turns off. However, keep in mind that fast interactions with these types of controls can make backlighting feedback seem out of sync with the user's interactions. In some cases, it may actually increase error rates. For this reason, backlighting feedback should only be used on controls where slower or single interactions occur. Its use among alpha-numeric buttons is not advisable, due to the speed at which many users type.

Use Color Sparingly

Many manufacturers are quick to use their company's color palette on their controls. After all, it's a great way to solidify branding efforts. However, most company colors are not diverse enough, or match well with existing color stereotypes and associations to be used as an effective control feedback strategy. Regardless of what color-coding convention a medical device's controls use, it's important that aspects about the user and the use-environment be taken into consideration. For example, if a potential user may have a color vision deficiency (e.g., red-green color deficiency), additional measures should be taken in the design of a control to allow the user to use the device successfully. Likewise, if the use-environment is not sufficiently illuminated (e.g., periphery of an operating room environment), users may not be able to distinguish between two colors.

Auditory Feedback

Auditory control feedback can be designed into medical devices in a variety of ways, as long as it suits the acoustic variability and characteristics of the use-environment, as well as the user's hearing capabilities. In some use-environments such as a hospital ICU, it is common for nursing staff to work with a variety of medical devices that beep and buzz and click at different frequencies, rates, and volumes throughout their shift. Medical device manufacturers rarely think about how their device's control sounds blend—or fail to blend—into this environment. Instead, their attention and concern are directed toward how their alarms fit in the soundscape.

In some cases, medical devices serving completely different functions must co-occur in the same environment. Each device may occupy the same frequency ranges, making it difficult for users to hear and figure out important details coming from the devices. This can put users who are responsible for managing multiple devices at once (e.g., nurses) in a confusing and mentally fatiguing position, which in turn, contributes to unwanted behaviors such as "alarm fatigue" or muting auditory feedback altogether (Sendelbach & Funk, 2013). Substantial efforts must be directed toward striking a balance between capturing the user's attention when needed, and retreating into the background during nonessential tasks (Weinger, Wiklund, & Gardner-Bonneau, 2011).

Every use-environment will have a constant churn of ambient noises competing for the user's attention. Many of these sounds are so enmeshed in the sonic fabric of our environments that we forget they are there (e.g., HVAC system, outside street traffic). These sounds occupy every stretch of a user's auditory range; that is, from the lowest to highest discernable frequencies.

Thankfully, however, much like our color perception's favoritism toward green-yellow hues (Gegenfurtner & Sharpe, 1999; Mollon, 1982), our auditory systems are evolutionarily optimized to hear frequencies within the same range we use in speech. Details of speech in the 2–4 kHz range contribute significantly to our perception of "intelligibility" of a voice. That is, the voice's crispness, understandability, and for lack of a better description, the quality that makes your voice sound like "you." This speech frequency range also happens to be the part of our auditory range where humans are capable of picking out distinct details without substantial increases in volume (i.e., intensity) from the sound source.

For these reasons, auditory control feedback should be designed to cover at least some of this frequency range. Indeed, the Association for the Advancement of Medical Instrumentation (2009) recommends that control feedback is most effective when presented between 400 and 1500 Hz. One caveat to these recommendations is that users with the most common types of hearing impairments (e.g., tinnitus, noise induced hearing loss) often benefit from auditory control feedback below 3 kHz, since their detection of high frequency sounds is compromised.

Pair Auditory and Visual Feedback Together on Tasks with Low Cognitive Workload Demands (Avoid It on High-Demand Tasks)

Multimodal feedback (e.g., audition and vision) can be a useful way to ensure that the user receives sufficient feedback while using a control. However, presenting this feedback redundantly over too many sensory channels, or presenting stimuli over the "wrong" sensory channels can impede user performance. According to Burke et al. (2006), the multimodal pairing of auditory and visual feedback is best used in situations where the task demands on cognitive resources are low. This may include nonemergency or noncritical tasks.

Minimize Auditory Feedback Duration and Intensity for Frequently Used Controls

Frequently used controls (e.g., keyboards, mice, numeric keypads, directional controls) can be fatiguing when they are paired with drawn out or loud auditory feedback. What's more, they can even be distracting, taking away from other critical elements of the user's experience with the medical device or the tasks they are performing. For example, an infusion pump may be designed to beep in response to the user pressing a button on the interface. While this is a helpful cue that the button has been successfully pressed, it can be annoying when the user is trying to remain discrete in a public setting while administering treatment.

Auditory control feedback such as a simple "beep" may be easier to program into a medical device's controls, but that doesn't mean it's the most helpful form of feedback to the user. Tactile or visual feedback may be more beneficial in some situations, even if it's more difficult to incorporate from an engineering standpoint.

At minimum, designers should consider implementing ways for the user to adjust their control's auditory feedback volume. This may come in the form of preselected settings (e.g., low, medium, high), or in the form of a continuous volume adjustment knob. Be sure to think through all the "strengths" and "weaknesses" of giving users control over these sound parameters.

Avoid "Pure Tones" When Presenting Auditory Control Feedback

Pure tone sounds are single tone frequencies, similar to the sound a tuning fork produces when struck. While pure tones are generally easier to create and implement into medical device controls, they can be disorienting and unsettling to users. They can also be difficult to use as an aid in locating the source of the sound (i.e., sound localization). Additional details and tips on this topic can be found in the "Auditory Displays and Alarms" section of Chap. 11.

Avoid Using Sound as the Only Mechanism for Control Location Feedback

Humans hear via a binaural auditory system; meaning, we have two points (i.e., ears) of receiving sound. This arrangement makes us better at detecting the location of a sound source along horizontal planes (i.e., left to right), rather than vertical planes (i.e., up and down). We are also better at sound localization when the sound source is in front of us compared to when it's behind us (Middlebrooks & Green,

1991). For these reasons, sound should not be used as the only way to signal the user about the location of a control, or where their attention should be directed. A more effective strategy involves using multimodal feedback, such as pairing auditory and visual feedback together Black et al. (2017).

10.6   Activation Force Considerations

Activation force is the amount of force required to interact with a control. A general guideline is to identify a force threshold that is easy to achieve for the majority of users, but not so easy that accidental activation becomes an issue. Finding this balance may take several iterations of designing and researching a control. Each iteration should be tested with a mix of representative users from each user group to determine the threshold. For some medical devices, such as insulin pens or autoinjectors, user groups may differ considerably from one another in terms of physical capabilities. Each groups' capabilities should be accounted for in the final design of the device. In some cases, this might mean creating separate lines of products with different force thresholds.

Factors such as the control's expected duration of use, as well as the user's body position (i.e., posture, hand positions) have significant impact on activation forces. Generally speaking, users can sustain longer durations or higher frequency of use when their body position is kept in a neutral state. For example, hand controls that require frequent flexing at the wrist can place pressure and strain on the median nerve and flexor tendons. Over time, this compression can lead to a common Repetitive Strain Injury (RSI) referred to as Carpal Tunnel Syndrome. Controls that require nonneutral body positions during operation, high activation forces, and prolonged or repetitive interactions are most likely to lead to the occurrence of RSIs (Plerhoples, Hernandez-Boussard, & Wren, 2012).

However, keep in mind that users can be surprisingly resourceful when faced with activation forces that exceed their capabilities or comfort. They may resort to secondary tools, equipment, alternative grips, or even different parts of their body to activate a control if needed. For example, a user with a peripheral neuropathy may instinctively press a control button with their proximal interphalangeal joint (PIP), rather than their fingertip to minimize finger discomfort. Users with chronic pain or lack of sensitivity in their fingertips may adapt by using the proximal interphalangeal joint (PIP) to interact with hard control surfaces (Fig. 10.22). These types of use behaviors should be taken into account when designing the shape, size, and textures of controls.

While this is an impressive display of adaptability, this behavior changes some of the underlying assumptions the manufacturer must make about how users interact with their control. As such, these unexpected behaviors should be evaluated to determine if additional risk is introduced, and whether the control itself needs to be redesigned. Such design efforts may look at reducing activation forces, as well as a combination of other aspects (e.g., control surface textures, location, surface area).

Fig. 10.22 Control
interaction with proximal
interphalangeal joint (PIP)

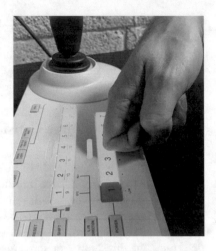

The following table presents some guidelines adapted from the Association for the Advancement of Medical Instrumentation (2009). The guidelines relate to activation force considerations for various control types as well as the part of the user's

Table 10.3 Activation force considerations by body part and control type

Body part	Type of control	Activation force recommendation (range)
Finger	Button (e.g., on-off switch)	0.2–1.5 N
Finger	Key switch on a keyboard	0.25–0.8 N
Finger	(Resistive) touchscreens	0.25–1.5 N
Finger	Membrane buttons	0.7 N (minimum)
Finger	Thumbwheel	1.7–5.6 N
Finger	Toggle switch	3.1 N (minimum)
Finger	Levers, sliders	3.1 N (minimum)
Finger	Rocker switch	3.1 N (minimum)
Hand	Track ball	0.25–1.5 N
Hand	Joystick	3.4–8.9 N
Hand	Hand-sized push button	58 N
Hand	(One-handed) Handwheel	127 N (minimum)
Hand	(Two-handed) Handwheel	245 N (minimum)
Hand	(One-handed) Large lever; forward-backward movement	8.9–113 N
Hand	(One-handed) Large lever; left-right movement	8.9–89 N
Hand	(Two-handed) Large lever; forward-backward movement	8.9–222 N
Hand	(Two-handed) Large lever; left-right movement	8.9–113 N
Foot	Foot pedal (without foot resting on pedal while idle)	17.8–88.9 N
Foot	Foot pedal (including foot resting on pedal while idle)	44.5–88.9 N

body that is intended to operate them. See Table 10.3. Note that the International System of Units (SI) discusses "force" in terms of newtons (N), rather than pounds (lbs) or kilograms (kg). To convert newtons to pounds, multiply the newton value (N) by 0.22481.

10.7   Control Placement Considerations

There are a variety of ways that controls can be laid out on a medical device. As a starting point, importance, frequency of use, and sequence of use should lead early design efforts (Karwowski, 2005). Specifically, controls that are both important and used frequently should be placed front and center of the user interface. This makes it easier to identify the control, and in some cases, can promote faster and more accurate interactions (Bullinger, Kern, & Braun, 1997).

Controls with similar functions should be placed near one another. For example, the angulation knobs on a flexible endoscope are placed concentrically on top of each other, since both controls affect the movements in the distal tip of the endo-scope. Flexible endoscope users (e.g., colorectal surgeons) are accustomed to inter-acting with both controls in a fluid, singular movement.

In other cases, controls with similar functions are used as part of a sequence or procedure. Additional efforts should be directed in the design process to understand what these sequences are, and how the user interface(s) might be improved to sup-port these behaviors and expectations. Importantly, this does not mean that controls must always be placed in a sequence in a single row or column arrangement. In some cases, research will reveal different "clustering" behaviors for different tasks. That is, each set of tasks may require a unique sequence of control interactions. Designing control placement to support only one of these tasks may come at the detriment of other tasks completed with the same controls. Research will help reveal these "clustering" patterns, which can then be used to help optimize where controls are placed.

Mind the User's Reach Envelope

The term, "reach envelope" refers to all the possible positions that a user can com-fortably and consistently reach. A traditional reach envelope starts at the user's shoulder and extends to their fingertips. However, depending on the task and use-environment, a reach envelope can be defined for each arm independently, from the forearm downward, including bending at the waist. In all cases, the medical device's reach envelope specifications should be based on users whose physical size is at the fifth percentile of the population. This ensures that small users can still safely and effectively interact with controls. Larger users can reach past these envelope thresh-olds, so they are accounted for as well.

Reach envelopes can be divided into zones to help determine where controls should be placed. Usually, two or three zones are defined (Yang & Abdel-Malek, 2009).

– Primary zone: used for controls that are frequently used. The user's reach should be comfortable or "neutral," covering about 1/3rd of the user's entire range of motion (ROM). All controls in this zone should be reachable through movement in the elbow joint (and below). For tasks completed while sitting at a console or table, this distance is around 9–12 in.
– Secondary zone: used for controls that are occasionally used. The user's reach should be "moderate," covering about 2/3rd of the user's entire range of motion (ROM). All controls in this zone should be reachable through movement in the shoulder joint (and below). For tasks completed while sitting at a console or table, this distance is around 22–24 in.
– Tertiary zone: used for controls that are rarely used. The user's reach should be their "maximum" capacity, covering the user's entire range of motion (ROM). For tasks completed while sitting at a console or table, this distance is around 24–26 in.

Keep in mind that reach envelopes are different based on gender and physical capabilities. The average reach length for females is approximately 13.5% smaller than for males (Sengupta & Das, 2000). Similarly, people often experience reductions in their range of motion (ROM) as they age, resulting in smaller reach envelopes among older adults (Molenbroek, 1998).

Reach envelopes also change as a result of the user's stance and direction of reach. For example, standing reach envelopes are somewhat larger than when the same user is sitting down. This is due in part to the fact that the user has added range of motion from bending at the hips and waist. Reach also tends to be slightly longer for each arm (about 10%) when the user reaches out to his or her sides, rather than straight in front of them.

Dead Space Between Neighboring Controls Limit Accidental Activation

Dead Space refers to the gaps between adjacent controls or settings within a control. These range from gaps between keys on a keyboard, to the button placement on medical device's GUI (see Fig. 10.23). Dead space allows successful control interactions with less demand for accurate finger placement. In other words, the user doesn't have to hit dead center on the control's surface each time they interact with it. Without sufficient dead space, the user may hit neighboring buttons, or two buttons simultaneously (Saunders & Novak, 2012).

Dead space requirements differ between control types. For example, finger operated button controls can get away with a relatively small amount of dead space

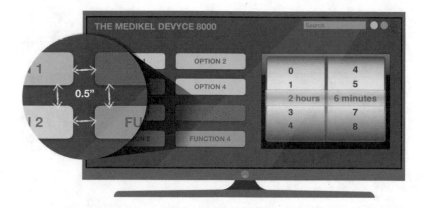

Fig. 10.23   GUI with a dead space of 0.5″ between adjacent buttons

because of the top-down activation movement by the user. However, dead space requirements for rotary knobs are significantly larger, since these controls involve multiple fingers during use. For these types of controls, Proctor and Van Zandt (2018) recommend a dead space of at least 1 in. for rotary controls.

Controls Placed Together Naturally Suggest a "Familial" Relationship

The Gestalt Psychology movement during the turn of the twentieth century introduced the world to several foundational design "laws" based on how people perceive and think about objects (i.e., stimuli) in their environment. The unifying theme across these Gestalt laws is that people are inherently pattern seeking beings—we aim to make sense of the world around us, even if that means filling in the perceptual gaps between objects.

The Gestalt law of "proximity" suggests that when objects are placed near one another, people perceive these objects as part of a unified group (Wagemans et al., 2012). Figure 10.24 demonstrates this phenomenon well—though there are the same number of individual circles in the left and right matrices (i.e., 36 circles), the left matrix appears to be a single "group," while the right matrix appears to be broken up into three "groups." With respect to control placement, this means that when naïve users encounter a new device, they approach this device with expectations based purely on whether controls are positioned close to one another. Once they learn the function of one control in this grouping, some may develop expectations about what its neighboring controls affect. It's important that designers and manufacturers embrace these expectancies.

In Fig. 10.24, the left and right sides are both made up of 36 circles. However, the left side appears as one grouping, while the right side appears as three separate

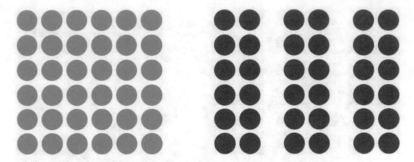

Fig. 10.24 Example of Gestalt law of proximity

groupings. Interestingly, even dividing these matrices into a "left" and "right" is an example of the law of proximity.

10.8 Touchscreen Considerations

A touchscreen interface is a combination of a display and input device (Orphanides & Nam, 2017). Over the last two decades, touchscreen interfaces have become ubiquitous. Touchscreens create a direct, physical pathway between the user and a GUI. By comparison, a desktop computer relies on a mouse and keyboard that are operated on a different visual plane than the display. The eyes and hands work in two different worlds.

Touchscreens afford new releases and bug fixes. Things like the control's gain, button size, contrast. can all be adjusted to a high degree of precision to promote safe, effective, and satisfying interactions. Touchscreens also provide designers and engineers with a more compact form factor (Orphanides & Nam, 2017).

Importantly however, while touchscreens are different in a number of ways from devices with physical controls, many of the guidelines and recommendations for physical controls also apply to touchscreen controls. For example, recommendations specific to the control's color, shape, etc. will almost always be identical regardless of whether it's a physical or digital GUI control.

Of course, there are a few logistical differences that make touchscreen controls different from physical controls. For example, due to the limited space on handheld touchscreens (e.g., smartphones, tablets), not every control on the GUI can be presented at an optimal size. Whereas a key on a standard QWERTY keyboard is generally recommended to be between 16 and 18 mm, this size wouldn't be feasible on a 4.3″ smartphone touchscreen. There simply isn't enough space on the screen. Designers and manufactures must acknowledge these types of tradeoffs, and determine what concessions must be made.

Types of Touchscreens

There is no such thing as a single type of "touchscreen." Multiple touchscreen technologies have been developed over the years, some as early as the 1970s (e.g., Accutouch by Elo TouchSystems) when computer technology was still in its infancy (Saffer, 2008). For the sake of simplicity, this book will address a few of the more common—and commercially successful—touchscreen technologies. However, please note that additional technologies do exist, and warrant further consideration for their application in medical device design.

Capacitive Touchscreens

To say that capacitive touchscreen technology is pervasive would be an understatement. As of the publication of this book, capacitive touchscreens are the cornerstone of almost every single new smartphone. Not surprisingly, capacitive touchscreens also happen to be the most common touchscreen technology used in medical devices to-date.

This technology works by sensing the difference in electrical charge between the object touching it and the charge sent across the surface of the screen itself. The screen is coated with very thin layers of a transparent, conductive material (e.g., copper). A small electrical current is passed uniformly through these conductive layers whenever the device (or screen) is powered on. When a user touches the screen's surface—or uses a conductive object to touch the screen—it changes the electrical current at that specific x, y coordinates on the screen. Capacitive touchscreens produced in the last few years have begun to support multitouch inputs as well, enabling the screen to read x, y coordinates for several fingers touching the screen simultaneously.

Compared to other types of touchscreens (see below), capacitive touchscreens have excellent clarity, and can be sealed to prevent liquids, debris, and other contaminants from entering behind the screen and damaging components (Bhalla & Bhalla, 2010). One downside, however, is that some types of personal protective equipment (PPE) will block the electrical current of the user's skin. Usually, these types of PPE are made from nonconductive materials (e.g., rubber), or are simply too thick. In other cases, users may have selected a glove that is too loose, leaving an air gap at the fingertip that interferes with their touch. All things being equal, however, standard hospital gloves made from nitrile—though, technically a synthetic "rubber"—usually work with capacitive touch screens without degrading performance.

Burnett, Large, Lawson, De-Kremer, and Skrypchuk (2013) observed that successful performance was 27% higher when participants completed tasks with a capacitive, rather than resistive touchscreens (Fig. 10.25).

TOUCHSCREEN

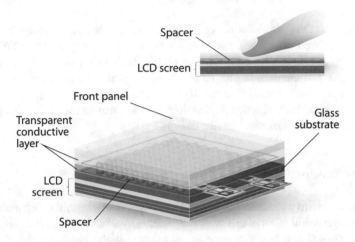

Fig. 10.25 Capacitive touchscreen technology. (Image by Designua/shutterstock.com)

Resistive Touchscreens

Although a somewhat older technology, resistive technologies are used in many products. Resistive touchscreens are made from two conductive screen layers separated by a thin air gap less than the width of a human hair. When the user presses the screen, the two conductive layers touch and pass on voltage to the system. The system detects where contact was made, and translates this location into an x, y coordinate (Fig. 10.26).

Fig. 10.26 Resistive touchscreen that consists of: (1) a glass panel, (2) resistive coating, (3) micro-insulators, (4) a conductive membrane

While relatively inexpensive to manufacture, resistive touchscreen has pros and cons that can make them ideal for some use contexts, and a nonstarter for others. Table 10.4 lists out the strengths and weaknesses of resistive touchscreens.

Surface Acoustic Wave (SAW) Touchscreens

A surface acoustic wave (SAW) touchscreen relies on sound waves to detect finger placement instead of an electrical current used by capacitive and resistive screens. When a user's finger(s) touches the glass' surface, some of these sound waves are absorbed at the point of contact. The system translates this information into a set of x, y coordinates. Some technologies can also transmit the amount of sound wave dampening (i.e., hard presses) as a z-axis factor. Meaning, it can use touch force as an extra touchscreen dimension. Table 10.5 lists out the strengths and weaknesses of SAW touchscreens.

Size Considerations

Touchscreens come in a variety of sizes, ranging from smaller, handheld smartphones and tablets to large computer monitors or wall panel displays. Each plays an important part in how information is presented, as well as the type, size, and placement of controls. For taller (e.g., wall display) touchscreens, make sure that the GUI controls are placed at an acceptable height and reach envelope given the device's user population criteria. Users should never have to stand on their toes or crouch down to reach a control, even if it's technically feasible to design the interface this way.

Place the controls at a height that shorter users can safely and consistently reach the highest touchpoint on the screen. Designers and engineers will often use the user population's fifth percentile height to define this value. In general, however, this maximum height should be approximately 74 in. tall (Chengalur, Bernard, & Rodgers, 2004). Outside of these parameters, however, wall display touchscreen controls maintain the same size considerations as hard button controls.

Table 10.4 Pros and cons of resistive touchscreens

"Pros" of resistive touchscreens	"Cons" of resistive touchscreens
● Not affected by PPE ● Can use a (nonconductive) pointing device (e.g., eraser tip) ● Provides a (slight) tactile feedback when depressed ● User can rest their hand or fingers on the screen for support without making a selection on the GUI	● Many do not permit multitouch GUI entries ● Multitouch gestures are not possible (e.g., pinch to zoom) ● Negative transfer of training from (capacitive) touchscreens in users' daily lives ● Must physically depress the screen

Table 10.5 Pros and cons of SAW touchscreens

"Pros" of SAW touchscreens	"Cons" of SAW touchscreens
● Screen size is technically feasible at large scales ● Screens could be transparent (e.g., glass window pane) ● Dual-sized touchscreens are feasible ● Could theoretically be applied to any hard, flat surface (e.g., very thick glass, metal) ● Image clarity is great as a result of not needing embedded wire mesh ● Not affected by PPE ● Can use a (nonconductive) pointing device (e.g., eraser tip) ● Supports multitouch inputs	● The object that touches the screen must be a soft material (e.g., skin, rubber stylus tip) to absorb/disrupt the sound waves for the technology to work ● Water droplets and debris can negatively impact screen accuracy ● Current technology requires the use of larger bezels to protect and hide the sensors

For handheld devices (e.g., smartphones and tablets), some studies have revealed that icon and button placement on the screen can affect accuracy and user preferences. Park and Han (2010) found that (right-handed) users were most accurate at touching GUI targets displayed on the left side of a handheld touchscreen device with their thumb. By comparison, they performed worse when attempting to touch the same sized targets positioned in the bottom-right corner of the screen. The reasoning behind this is simple: the far left side of the screen allows right-handed users full range of motion in their thumb; the bottom right corner is the most restrictive area, requiring the user to bend their thumb and potentially change to a less effective grip of the device overall. This study also reported that users reached targets in the middle of the screen faster (Orphanides & Nam, 2017), especially when these targets were at least 7 mm wide.

A related study by Hwangbo, Yoon, Jin, Han, and Ji (2013) found that larger targets should be used for elderly users. They recommended a maximum target size of 12 mm. This study also found that a larger dead space (e.g., 3 mm) between GUI controls provided greater success and lowered task completion time (Orphanides & Nam, 2017). For GUI controls that are not involved in typing tasks (e.g., on-off button), the Association for the Advancement of Medical Instrumentation (2009) recommends a dead space of about 6 mm.

Giving users the option of interacting with a handheld touchscreen in "portrait" or "landscape" orientation is a valuable way to change the size of the GUI. Since "landscape" layouts afford larger left-to-right screen space, buttons and labels can often be scaled slightly larger. Alternatively, users may have preferences in respect to using two hands or one hand on a small touchscreen device. Providing the GUI in both formats allows them to pick and choose how they want to use it.

Embrace Existing Gestures and Shortcuts

Several multitouch gestures have become standardized across various consumer industries. These include pinch to zoom, double-tapping, dragging, and rotating. Controls designed for touchscreen use should factor in these expectations, and adjust the design accordingly.

Keep in mind, however, that older adults may struggle with some gesture finger movements. A study by Chang, Tsai, Chang, and Chang (2014) reported that older adults struggled more often than younger adults with touchscreen gestures like dragging and zooming. These tasks were even more difficult when they were attempted on smaller touchscreens ranging from 4.3″—10″. Older adults tend to have worse fine motor movement control and visual capabilities than younger adults (Chang et al., 2014), making it difficult to interact with small controls in a relatively narrow space (e.g., 4.3″ touchscreen).

Touchscreens tend to be better for discrete selection tasks, rather than tasks that involve fine motor movements and precision. Users tend to perform well on touchscreen tasks where they can pick one target (e.g., on-off button) in a list or matrix arrangement. Some studies have reported that users actually perform *better* at these types of discrete selection tasks on a GUI when compared to using physical buttons (Burnett et al., 2012; Orphanides & Nam, 2017). However, performance often deteriorates when GUI controls require high degrees of movement precision, such as with a virtual rotary controller knob or slider control like the one presented in Fig. 10.13, earlier in this chapter.

There are several reasons why users do worse on high precision tasks with touchscreens. One issue is that touchscreens offer little in the way of contact surface with the user's finger. Often, it's just the tip of the user's finger that facilitates contact with the screen. Additionally, it can also be taxing for the user to hold their arm in a fixed position while interacting with the touchscreen over long periods of time (Baldus & Patterson, 2008), leading to fatigue and dissatisfaction over long, continuous periods of use. Lastly, people have a tendency to use more force than necessary on touchscreens, especially for complex gesture movements such pinching or dragging (Asakawa, Crocker, Schmaltz, & Jindrich, 2017). As a result, they tend to be slower in data input tasks, such as an open-ended text response in a patient's medical chart.

Touchscreens may not be a good option for high movement tasks or environments prone to vibration. Multiple studies have demonstrated that touchscreens can be difficult to use successfully during tasks with unexpected movements (Goode, Lenne, & Salmon, 2012; Lin, Liu, Chao, & Chen, 2010; Orphanides & Nam, 2017). What's more, this issue generally cannot be improved through training and continued experience (Goode et al., 2012). For these types of tasks and environments, consider using a physical control such as a mouse or trackball for pointing tasks.

Activate Controls on "Up" Triggers

When we adjust a physical control, the change is (usually) instantaneous on the medical device. However, a touchscreen control can be programmed to implement a change to the system based on specific events or circumstances. In some cases, it may be better to design system responses based on when the user lifts their finger off the GUI button, rather than when they first make contact with it (Association for the Advancement of Medical Instrumentation, 2009). This gives the user opportunity to drag their finger off an unintended selection (for example) and toward the desired control, or into a neutral space with no effect on the system.

Resources

- Association for the Advancement of Medical Instrumentation. (2009). *ANSI/ AAMI HE75-2009: Human factors engineering—Design of medical devices.* Arlington, VA: Association for the Advancement of Medical Instrumentation.
- Helander, M. (2006). *A guide to human factors and ergonomics.* Boca Raton, FL: CRC Press.
- WCAG 2 accessibility guidelines. (2018). Contrast and color accessibility understanding WCAG 2 contrast and color requirements. (n.d.). Retrieved August 27, 2020, from https://webaim.org/articles/contrast/.

References

Allen, N. S. (1994). Photofading and light stability of dyed and pigmented polymers. *Polymer Degradation and Stability, 44*(3), 357–374.

Asakawa, D. S., Crocker, G. H., Schmaltz, A., & Jindrich, D. L. (2017). Fingertip forces and completion time for index finger and thumb touchscreen gestures. *Journal of Electromyography and Kinesiology, 34*, 6–13.

Association for the Advancement of Medical Instrumentation. (2009). *ANSI/AAMI HE75–2009: Human factors engineering—Design of medical devices.* Arlington, VA: Association for the Advancement of Medical Instrumentation.

Asundi, K., & Odell, D. (2011). Effects of keyboard keyswitch design: A review of the current literature. *Work, 39*(2), 151–159.

Baldus, T., & Patterson, P. (2008). Usability of pointing devices for office applications in a moving off-road environment. *Applied Ergonomics, 39*(6), 671–677.

Bernard, M., Liao, C. H., & Mills, M. (2001, March). The effects of font type and size on the legibility and reading time of online text by older adults. In *CHI'01 extended abstracts on human factors in computing systems* (pp. 175–176).

Bhalla, M. R., & Bhalla, A. V. (2010). Comparative study of various touchscreen technologies. *International Journal of Computer Applications, 6*(8), 12–18.

Black, D., Hettig, J., Luz, M., Hansen, C., Kikinis, R., & Hahn, H. (2017). Auditory feedback to support image-guided medical needle placement. *International Journal of Computer Assisted Radiology and Surgery, 12*(9), 1655–1663.

Bullinger, H. J., Kern, P., & Braun, M. (1997). Controls. In G. Salvendy (Ed.), *Handbook of human factors and ergonomics* (pp. 697–728). Hoboken, NJ: Wiley.

Burke, J. L., Prewett, M. S., Gray, A. A., Yang, L., Stilson, F. R., Coovert, M. D., et al. (2006, November). Comparing the effects of visual-auditory and visual-tactile feedback on user performance: A meta-analysis. In *Proceedings of the 8th international conference on multimodal interfaces* (pp. 108–117).

Burnett, G. E., Large, D. R., Lawson, G., De-Kremer, S., & Skrypchuk, L. (2013). A comparison of resistive and capacitive touchscreens for use within vehicles. *Advances in Transportation Studies, 31*, 5–16.

Chan, A. H., & Courtney, A. J. (2001). Color associations for Hong Kong Chinese. *International Journal of Industrial Ergonomics, 28*(3–4), 165–170.

Chengalur, S. N., Bernard, T. E., & Rodgers, S. H. (2004). *Kodak's ergonomic design for people at work*. Hoboken, NJ: Wiley.

Childers, C. P., & Maggard-Gibbons, M. (2018). Understanding costs of care in the operating room. *JAMA Surgery, 153*(4), e176233–e176233.

Cook, A. M., & Polgar, J. M. (2014). *Assistive technologies E-book: Principles and practice*. Amsterdam: Elsevier Health Sciences.

Courtney, A. J. (1986). Chinese population stereotypes: Color associations. *Human Factors, 28*(1), 97–99.

Dogusoy, B., Cicek, F., & Cagiltay, K. (2016, July). How serif and sans serif typefaces influence reading on screen: An eye tracking study. In *International conference of design, user experience, and usability* (pp. 578–586). Cham: Springer.

Elliot, A. J., & Maier, M. A. (2012). Color-in-context theory. In *Advances in experimental social psychology* (Vol. 45, pp. 61–125). Cambridge, MA: Academic Press.

Fitts, P. M., & Seeger, C. M. (1953). SR compatibility: Spatial characteristics of stimulus and response codes. *Journal of Experimental Psychology, 46*(3), 199.

Gallace, A., & Spence, C. (2009). The cognitive and neural correlates of tactile memory. *Psychological Bulletin, 135*, 380–406. https://doi.org/10.1037/a001532.

Gao, X. P., Xin, J. H., Sato, T., Hansuebsai, A., Scalzo, M., Kajiwara, K., et al. (2007). Analysis of cross-cultural color emotion. *Color Research & Application: Endorsed by Inter-Society Color Council, The Colour Group (Great Britain), Canadian Society for Color, Color Science Association of Japan, Dutch Society for the Study of Color, The Swedish Colour Centre Foundation, Colour Society of Australia, Centre Français de la Couleur, 32*(3), 223–229.

Gegenfurtner, K. R., & Sharpe, L. T. (Eds.). (1999). *Color vision*. Cambridge: Cambridge University Press.

Goode, N., Lenne, M. G., & Salmon, P. (2012). The impact of on-road motion on BMS touch screen device operation. *Ergonomics, 55*(9), 986–996.

Helander, M. (2006). *A guide to human factors and ergonomics*. Boca Raton, FL: CRC Press.

Hirst, R. J., Cragg, L., & Allen, H. A. (2018). Vision dominates audition in adults but not children: A meta-analysis of the Colavita effect. *Neuroscience & Biobehavioral Reviews, 94*, 286–301.

Hoffmann, E. R., & Chan, A. H. (2018). Review of compatibility and selection of multiple lever controls used in heavy machinery. *International Journal of Industrial Ergonomics, 65*, 93–102.

Hwangbo, H., Yoon, S. H., Jin, B. S., Han, Y. S., & Ji, Y. G. (2013). A study of pointing performance of elderly users on smartphones. *International Journal of Human-Computer Interaction, 29*(9), 604–618.

Jenkins, W. L., & Connor, M. B. (1949). Some design factors in making settings on a linear scale. *Journal of Applied Psychology, 33*(4), 395–409.

Katzir, T., Hershko, S., & Halamish, V. (2013). The effect of font size on reading comprehension on second and fifth grade children: Bigger is not always better. *PLoS One, 8*(9).

Lin, C. J., Liu, C. N., Chao, C. J., & Chen, H. J. (2010). The performance of computer input devices in a vibration environment. *Ergonomics, 53*(4), 478–490.

Mancini, F., Bauleo, A., Cole, J., Lui, F., Porro, C. A., Haggard, P., et al. (2014). Whole-body mapping of spatial acuity for pain and touch. *Annals of Neurology, 75*(6), 917–924.

Molenbroek, J. F. (1998, October). Reach envelopes of older adults. In *Proceedings of the human factors and ergonomics society annual meeting* (Vol. 42, No. 2, pp. 166–170). Sage, CA/Los Angeles, CA: SAGE Publications.

Mollon, J. D. (1982). Color vision. *Annual Review of Psychology, 33*(1), 41–85.

Moret-Tatay, C., & Perea, M. (2011). Do serifs provide an advantage in the recognition of written words? *Journal of Cognitive Psychology, 23*(5), 619–624.

Müsseler, J., & Skottke, E. (2011). Compatibility relationships with simple lever tools. *Human Factors, 53*, 383–390.

Ng, A. W., & Chan, A. H. (2014). Tactile symbol matching of different shape patterns: Implications for shape coding of control devices. In *Proceedings of the international multiconference of engineers and computer scientists* (Vol. 2).

Orphanides, A. K., & Nam, C. S. (2017). Touchscreen interfaces in context: A systematic review of research into touchscreens across settings, populations, and implementations. *Applied Ergonomics, 61*, 116–143.

Park, Y. S., & Han, S. H. (2010). Touch key design for one-handed thumb interaction with a mobile phone: Effects of touch key size and touch key location. *International Journal of Industrial Ergonomics, 40*(1), 68–76.

Plerhoples, T. A., Hernandez-Boussard, T., & Wren, S. M. (2012). The aching surgeon: A survey of physical discomfort and symptoms following open, laparoscopic, and robotic surgery. *Journal of Robotic Surgery, 6*(1), 65–72.

Proctor, R. W., & Van Zandt, T. (2018). *Human factors in simple and complex systems*. Boca Raton, FL: CRC Press.

Reed, S. K. (2012). *Cognition: Theories and applications*. Boston, MA: Cengage Learning.

Roscoe, S. N. (1980). *Aviation psychology*. Ames: Iowa State University Press.

Saffer, D. (2008). *Designing gestural interfaces: Touchscreens and interactive devices*. Sebastopol, CA: O'Reilly Media, Inc..

Sanders, M. S., & McCormick, E. J. (1998). Human factors in engineering and design. *Industrial Robot: An International Journal*.

Saunders, K., & Novak, J. (2012). *Game development essentials: Game interface design*. Cengage Learning, Boston, MA.

Sendelbach, S., & Funk, M. (2013). Alarm fatigue: A patient safety concern. *AACN Advanced Critical Care, 24*(4), 378–386.

Sengupta, A. K., & Das, B. (2000). Maximum reach envelope for the seated and standing male and female for industrial workstation design. *Ergonomics, 43*(9), 1390–1404.

Slocum, G. K., Williges, B. H., & Roscoe, S. N. (1971). Meaningful shape coding for aircraft switch knobs. *Aviation Research Monographs, 1*(3), 27–40.

Steiner, L., Burgess-Limerick, R., & Porter, W. (2014). Directional control-response compatibility relationships assessed by physical simulation of an underground bolting machine. *Human Factors, 56*(2), 384–391.

Steiner, L. J., & Burgess-Limerick, R. (2013). Shape-coding and length-coding as a measure to reduce the probability of selection errors during the control of industrial equipment. *IIE Transactions on Occupational Ergonomics and Human Factors, 1*(4), 224–234.

Van Orden, K. F., Divita, J., & Shim, M. J. (1993). Redundant use of luminance and flashing with shape and color as highlighting codes in symbolic displays. *Human Factors, 35*(2), 195–204.

Wagemans, J., Elder, J. H., Kubovy, M., Palmer, S. E., Peterson, M. A., Singh, M., et al. (2012). A century of Gestalt psychology in visual perception: I. Perceptual grouping and figure–ground organization. *Psychological Bulletin, 138*(6), 1172.

Wang, H., & Or, C. K. (2015). A study of the relationship between color-concept association and occupational background for Chinese. *Displays, 38*, 50–54.

WCAG 2 Accessibility Guidelines. (2018). Contrast and color accessibility understanding WCAG 2 contrast and color requirements. (n.d.). Retrieved August 27, 2020, from https://webaim.org/articles/contrast/

Weinger, M. B., Wiklund, M., & Gardner-Bonneau, D. (2011). *Handbook of human factors in medical device design*. Boca Raton, FL: CRC Press.

Woodson, W. E., & Conover, D. W. (1964). *Human engineering guide for equipment designers*. Berkeley: University of California Press.

Yang, J., & Abdel-Malek, K. (2009). Human reach envelope and zone differentiation for ergonomic design. *Human Factors and Ergonomics in Manufacturing & Service Industries, 19*(1), 15–34.

Chapter 11
Displays

11.1 Introduction to Displays

A display is anything that presents dynamic information to the user through a sensory modality. There are two critical components in this definition that are worth dissecting. Dynamic information is periodically updated as new data becomes available. The other component—sensory modality—refers to this data being presented to any of our main sensory systems (e.g., vision, hearing, touch).

A visual display is the most common and familiar type of display—and for good reason. Visual displays communicate rich information to users quickly, conveniently, and consistently. We make sense of this information because our visual perception is the most complex and developed sensory modality we have (Gerrig, Zimbardo, Campbell, Cumming, & Wilkes, 2015).

Other sensory modalities, such as audition and touch, also have their own types of displays. Despite the strangeness of using the term "auditory" and "display" in the same breath, auditory displays are common among medical devices. For example, an auditory warning system is a type of auditory display. It checks both boxes of our criteria above. It's based on a primary sensory modality (i.e., audition), and it presents dynamic information to the user.

Tactile displays are less common, but they exist. Generally speaking, there are two types of tactile displays:

- Those that use tactile feedback or sensations for information exchange
- Those that create a physical representation of a 3D object

The latter might be used to allow the user to feel the shape, surface texture, or tactile parameters of the object itself (Chouvardas, Miliou, & Hatalis, 2008). The former could use a variety of tactile sensations, such as vibration, temperature, pressure, or texture to communicate multiple states or options. For example, temperature could be used on a two-dimensional trackpad to communicate how close the user is to a specific target. Likewise, a trackpad capable of rapidly changing surface

© Springer Nature Switzerland AG 2021   271
R. J. Branaghan et al., *Humanizing Healthcare – Human Factors for Medical Device Design*, https://doi.org/10.1007/978-3-030-64433-8_11

texture could be used to communicate 3D depth. Practically speaking, however, tactile displays in healthcare remain limited. What's more, when they are used, they are almost always used in parallel with visual or auditory displays as the primary source.

Although haptic (i.e., vibration) feedback is a part of a tactile display, it is merely one method by which dynamic tactile information is communicated to the user. Tactile displays can be indispensable for some users. For example, blind and visually impaired users—as well as some who might be deafblind—might use a refreshable braille display to read an alpha-numeric text (see Fig. 11.1). These devices often work by electro-mechanical means; rounded pins representing the six co-located dots in a braille cell will raise and lower based on the line of text the user has accessed. The pins refresh their position to create new characters when the user switches to a different text segment.

This chapter will discuss the two most common types of displays—visual and auditory displays—as they pertain to medical devices and human factors engineering (HFE). This is not to suggest that tactile displays are not important. However, given the limited scope of medical devices that use tactile displays beyond control-generated feedback (i.e., clicks, vibration), this topic is better addressed elsewhere in the literature. For an introduction to tactile displays, see Chouvardas et al. (2008). This chapter will also forgo a discussion on the emerging topic of augmented reality displays. This is a growing area of interest in the medical community but beyond the scope of this book. Interested readers are encouraged to review the resources (listed at the end of this chapter) for a cursory introduction to this topic spanning the past two decades.

Fig. 11.1 Refreshable braille displays are often used with computers in lieu of a visual display. (Image by zlikovec/shutterstock.com)

11.2   Visual Displays

Common Types of Visual Display Technologies

Medical device manufacturers, at some point, will need to choose the *type* of display their product will use. In this decision, there are a number of factors to be considered, such as display size, power consumption, viewing angle, resolution, and luminance. In-depth discussion of these display types and their properties is beyond the scope of this book. However, Table 11.1 presents a few of the most common visual display technologies in use, as well as their strengths and weaknesses.

Luminance Considerations

Luminance is the technical term for what we refer to as "brightness." The distinction between luminance and brightness is that the former is an objective (quantifiable) measure, whereas the latter is our perception of the light itself. For simplicity, this chapter will use the terms luminance and brightness interchangeably, unless there is a reason to point out specific distinctions.

The internationally recognized measure of display luminance is "candelas per square meter," usually written in abbreviated form as, "cd/m^2" One important distinction about the measure of display luminance is that we are referring to the display's luminance over a specific, two-dimensional surface area. That is, the surface area of the display's screen. This is different from other measures of luminance, such as the "lumen," since a lumen refers to a single point or source (e.g., lightbulb). Devices measured in lumens are usually expected to display a source of light that will disperse over an area. Displays, on the other hand, should have the same (or as close as possible) luminance regardless of whether the user is looking at the center of the display or its edges. In fact, ANSI/HFES 100 recommends that adjacent areas of a display should maintain a luminance ratio difference of less than 1.7:1 to prevent use-related issues.

Contrast

Luminance contrast is important for discovering and discerning text, graphics, and media on a display. This refers to the contrast between the darkest values on the screen and the brightest ones. Generally, the higher this ratio, the more realistic, crisp, and better-defined content will appear. When contrast is too low, it forces users to dedicate attentional resources for simple tasks, such as reading text or making sense of icons on a washed out, dull screen. This can lead to eye fatigue, slowed performance, and use-errors.

Table. 11.1 A list of common visual display technologies

Technology	Size capabilities	Power consumption	Luminance capabilities	Resolution	Contrast potential	Compatible lighting levels	Viewing angles
Cathode ray tube (CRT)	Small to medium	High	Low to medium	Low to medium	Medium	Low to medium	Wide
Electro-luminescent (ELD)	Small to medium	Low	High	Low to medium	Medium to high	Low to medium	Wide
Vacuum fluorescent (VLD)	Small	Medium	High	Low to medium	Medium to high	Low to medium	Wide
Liquid crystal, super twisted nematic (STN)	Small to medium	Low	Low to medium	Medium	Low to medium	Low to medium	Low
In-plane switching (IPS)	Small to large	Medium	Low to high	High	High	Low to medium	Wide
Electrophoretic (e-paper)	Small	Very low	Low	Low	Low to medium	Medium to high	Very wide
Organic light emitting diode (OLED)	Small to large	Low	High	Medium to high	High	All levels	Wide
Analog displays	Small	N/A	N/A	N/A	High	Medium to high	High

The specific contrast requirements of a display will depend on the nature and complexity of the tasks. For example, pulmonologists using a display to discover cancerous nodules from lung CT scans may need a display with excellent contrast, whereas a blood pressure monitor with large characters may suffice with the lower contrast ratios.

At minimum, the display's contrast ratio should be 3:1 (ANSI/HFES100). Figure 11.2 shows a few examples of what these contrast ratios look like. When you are dealing with RGB or HEX values (or similar), the maximum contrast ratio is 21:1. This corresponds with a true black (HEX code: #000000, RGB (0, 0, 0)) and a true white (HEX code: #FFFFFF, RGB (255, 255, 255)). Higher contrast ratios are achieved by increasing the luminance of the "white" to appear brighter, while keeping the "black" as dark as possible. Most LCD displays have no issue achieving the minimum contrast ratio recommended by ANSI/HFES100 (2007). In fact, almost all new-to-market smartphones, laptops, tablets, and televisions today have contrast ratios that are several times beyond this threshold.

For simpler displays, such as non-back-lit, segmented displays (Fig. 11.3), contrast ratios can be problematic in low-light use-environments. One reason for this is that the background of these displays is never truly "white," but rather light gray. Likewise, since there are no backlighting capabilities, the luminance can never be increased. These work well in indoor environments with predictable, ambient lighting conditions, but will not be the right match for bright, outdoor environments, nor in dark environments (e.g., CPAP machines used in a dark bedroom).

Use-Environment

Displays intended for indoor use should maintain a minimum luminance limit of at least 35 cd/m^2 (Association for the Advancement of Medical Instrumentation, 2009). This includes use-environments where overhead lighting or ambient lighting is present, but not overwhelming. However, if the display will be used in exceptionally bright locations (e.g., outdoors, mid-day sun exposure), this minimum will not

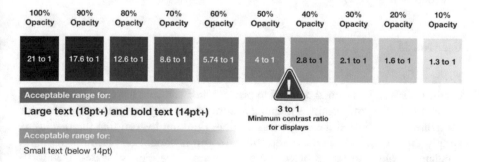

Fig. 11.2 Contrast ratios comparisons—WCAG guidelines for accessible text contrast are shown by the green bar and the ANSI/HFES100 (2007) recommendation for the minimum display contrast ratio is indicated by the red caution symbol

Fig. 11.3 Segmented,
reflective LCD display

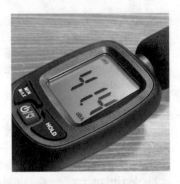

be sufficient. For example, many modern laptops and smartphones will have an
upper luminance level that is over five times higher than the minimum recommendation above (e.g., 300–500 cd/m^2). Display manufacturers have come to understand that users will use these types of displays in a wide range of environments.

Keep in mind that in exceptionally bright environments, the display must over-
come several constraints for the user to see content on the screen successfully. For
example, a highly reflective glass surface will reflect substantial light from the environment—referred to as optical glare or specular glare (Association for the
Advancement of Medical Instrumentation, 2009). It will also highlight smudges and
oil left on the glass' surface, thereby making it more difficult to see the screen.
Indeed, Liu, Zafar, and Badano (2014) reported that in a study of handheld devices
with glass displays, performance on a series of medical detection tasks was worse
when performed in "bright" and "very bright" conditions similar to outside light.

In addition to affecting the display itself, bright outdoor environments may also
bring about changes in user behaviors and PPE. For example, the user might choose
to wear polarized sunglasses. It is not realistic to expect users to remove their glasses
every time they need to interact with a medical device's display, especially if that
device is intended for mobile use. Similar to other situations discussed above, it's
important that the device manufacturer understand these potential constraints at the
outset of design, so that they select a display that is suitable for these use behaviors
and conditions.

User Needs

Users should generally be allowed to adjust the brightness to their preferences or
needs. However, it is a good practice to prevent the user from lowering the display's
brightness to the "off" position manually. Some users may not realize that they (or
someone else) adjusted the brightness setting, thinking instead that the display is
simply turned "off." Brightness adjustment controls should be placed in familiar,
predictable, quick to access places on the user interface. For software-based technologies, this might mean putting the control only one layer "deep" in the information architecture, such as in a drop-down or swipeable menu.

Some modern display technologies offer adaptive brightness capabilities. These adjust their brightness automatically based on how much light they receive through an ambient light sensor. These technologies can be useful for large, singular changes in environmental brightness (e.g., going from indoors to outdoors), but they aren't always accurate or reliable for smaller or fluctuating changes in ambient light (Ma, Lin, Hsu, Hu, & Hou, 2012).

Even if your display offers adaptive brightness, be sure to offer the user a way to easily and quickly override these settings. No user should be reliant on the technology to select an optimal brightness for their display, since everyone's eyes operate differently. For example, adults over the age of 60 years are more sensitive to glare and bright lights. They may need displays with enhanced brightness adjustment capabilities to accommodate lower lows and higher highs.

Keep in mind that our ability to discern subtle differences in display contrast is also affected by our environment's lighting conditions. Specifically, humans are better at detecting subtle contrast differences when lighting conditions are similar to daylight, and our visual system is operating in a photopic state. As ambient lighting decreases, our vision changes from mesopic (i.e., twilight), and eventually to scotopic (i.e., nighttime) states. As this happens, we need larger jumps in contrast between objects to detect differences or smaller details.

Viewing Angle

Although displays have improved over the decades, the brightness (and clarity) of a display will decrease as the user goes from an "ideal" viewing position to a more oblique (i.e., off-axis) one. An ideal viewing position is when the user is directly in front of the display, with their eyes positioned at the same height or slightly above the screen's midpoint. As the user moves away from this position (i.e., left, right, up, or down), the display will appear darker. The extent of this difference has a lot to do with the viewing environment, as well as the display technology itself.

It is important to evaluate the effects of off-axis viewing. Some devices will be viewed from various angles. For example, a patient using a self-infusion pump mounted to an IV stand may not take time to reposition the pump each time they decide to sit or stand during infusion. They may even choose to leave the IV stand in the far reaches of their visual periphery, while attending to other things. Similar inconsistencies are common in the hospital environment. For instance, ICU nurses completing rounds may monitor and record vitals at several positions around the room or from a fixed position.

In some cases, limiting screen viewing angles can be an advantage. With increased restrictions and security around personally identifying information (PIP), it's important to protect sensitive information. Displays with restricted viewing angles can make sure that only the person in front of the display can view this information.

It's important to understand how users anticipate interacting with the display. Testing the usability of a display in a simulated environment is valuable, but may not tell the whole story. After all, your research participant (i.e., user) is solely focused on the task at hand. They do not have as many extraneous stimuli competing for their attention, nor are they as potentially fatigued as they might be at the end of a long shift. Alternative research methods—for example, contextual inquiry—can help researchers gauge how competitive products are used (or misused). This can inform the viewing angle requirements of a new medical device's display.

In addition to testing viewing angles with users, collect actual photometric measurements of the display from common viewing angles and distances from the display (Association for the Advancement of Medical Instrumentation, 2009). These provide objective data regarding whether a display is suitable for these viewing conditions given factors such as luminance, color reproduction capabilities, and contrast ratios. Measurements taken at the maximum viewing angle(s) should be used to weed out displays early on that do not meet your usage requirements (Association for the Advancement of Medical Instrumentation, 2009).

Hardware Considerations

Brighter displays consume more power than dimmer ones. For instance, an Amazon Kindle with a low-luminance, "e-Ink" display can function for weeks or months on a single charge. By comparison, a similarly sized LCD tablet display might only last hours under the same use conditions. Display-related power consumption usually isn't a main consideration for stationary devices; however, it can be important for devices that operate on batteries.

LCD displays have improved in recent years, especially in respect to power efficiency. Generally, any LCD display that incorporates backlighting will consume more battery than those lit through natural light. For example, reflective LCD displays use ambient light or a light facing the screen to "reflect" off the back part of the screen to illuminate the screen itself. A similar type of LCD display—referred to as a transflective LCD—also leverages ambient light, but is equipped with a backlight for low-light conditions.

In some cases, displays remain active for hours or days before being switched to a bypass mode or turned off. Because of these lengthy use cycles, manufacturers should evaluate the display's susceptibility to screen burn-in or ghosting—a discoloration of areas of the screen that constantly displays "bright" content, such as the white pointer of a computer mouse. A display with a burned-in area will give the subtle appearance that the content is still shown, even when that content is removed from view. Although this issue is more common among older display technologies (e.g., CRT displays, plasma displays), it can still be a problem with some types of back-lit displays. A similar issue, referred to as image persistence, is more likely to occur with modern LCD displays, wherein the habitually displayed content will persist for a brief period of time after it is removed from view. Usually, the effect will slowly dissipate after a few seconds, and the display will return to normal.

One way to prevent screen burn-in and image persistence is to allow the display to enter into a sleep mode, periods where no content is displayed following a period of inactivity by the user. Although this option consumes little power, it's not always clear whether the device is turned off or nonfunctional without further actions. An alternative option is to use a sleep screen in which an icon or text moves randomly around the screen. This communicates that the device is operational, while eliminating burn-in and image persistence.

Color Considerations

Our visual systems rely on millions of rods and cones to inform us about the colors and shades of the world around us (see Chap. 5). Cones are most sensitive to three regions in the visual light spectrum: red, green, and blue. Electronic displays exploit this physiology by relying on (closely packed) red, green, and blue filtered light, referred to as RGB.

RGB color is the backbone of color displays. Although the exact mechanism by which these colors are illuminated is different depending on the technology (e.g., VLDs vs. LCDs), almost every display relies on these three colors to generate the colors found in medical technology. Each color contributes a portion to the color shown in a network of pixels (or similar) on the display. For example, when R is 100%, B is 0%, and G is 0%, that portion of the display shows a pure red. When each RGB value is equal to the others, the resulting "color" is gray. Secondary and tertiary colors are created by further combinations of colors in specific proportions.

Importantly, a display's color gamut will never match exactly what human vision is capable of seeing; it cannot display the same depth of color you see in the real world. This is especially true in the "green" region of vision. This is unlikely, however, to diminish performance or even personal preference.

Keep in mind that low and mid-market displays do not always cover the full Standard Red, Green, Blue (sRGB) gamut. This may matter more for some medical device applications than others. For example, accurate color representation of patient anatomy during surgery warrants the use of fuller gamut displays. On the other hand, it's likely overkill in other cases where a display is used only to present a simple, 2D GUI.

Resolution and Clarity Considerations

Regardless of the technology, every display has a definable resolution; that is, the visual dimensions of the display. This is different from the physical dimensions of the display—in other words, its length and height as measured in inches or centimeters. Think of visual dimensions as columns and rows. A display with a resolution of 1920 × 1080 means that it has 1920 columns (i.e., spanning across the display's width) and 1080 rows (i.e., spanning the height of the display). Importantly, a resolution of 1920 × 1080 looks very different on a 10″ tablet compared to a 100″ television display; 1080 rows spanning the height of 10″ are 10 times smaller than those spanning the height of 100″ (see Fig. 11.4).

Low resolution in itself is not necessarily problematic for medical devices. After all, if the only information the user needs to know from the display is whether it lights up solid green vs. solid red at different points in time, then a low-resolution display would be sufficient. Low resolution only becomes a problem when it is used on tasks that need higher resolution. Recommendations from AAMI's HE75 (, 2009) suggest that for text-based tasks, a resolution of 72 pixels per inch (PPI) is adequate. This means that on a 10″ wide display, the resolution should be a minimum of 720p. Other types of media such as pictures will require greater detail. In these cases, consider using a display with a resolution ranging between 150 and 300 PPI. On a 10″ display, that would require a resolution of greater than 1080p.

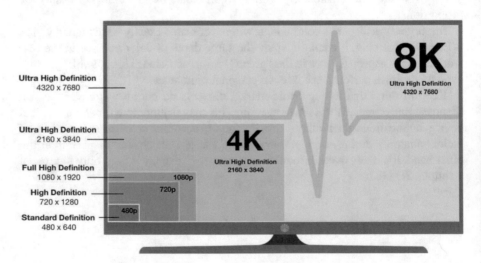

Fig. 11.4 Size comparison of various display resolutions reflecting the screen size at various resolutions if pixel size remained consistent across each display shown

Task Demands

A task that requires extreme precision such as identifying the boundaries of delicate patient anatomy during surgery will warrant displays with high resolution, whereas tasks requiring less precision may permit lower resolution displays. Higher resolution is needed to quickly make sense of qualitative data. During a robot assisted surgery (RAS), for example, the surgeon controlling the robot relies on visual display information to make judgments, decisions, and actions. Indeed, for many RAS systems, this visual information is the only (direct) feedback the surgeon gets; haptic and auditory feedback may be missing altogether. For some RAS systems, the capabilities of the display are bolstered by presenting the surgical site in high definition, as well as in 3D.

Displaying solely quantitative information usually requires only low resolution. For example, a multiparameter patient monitor presents a wealth of alpha-numeric data about the patient's heart rate, blood oxygen saturation, respiration rate, temperature, and so forth. These data (e.g., "BP 120/80") remain the same whether it's presented on an older CRT computer monitor, or a best-in-class 4k LCD display. On the other hand, you could argue that the former display may result in more misreads of information (e.g., was that a "3" or an "8"?), especially when viewed from across the room, or by users with poor visual acuity.

Tasks that require sensemaking from a mix of qualitative and quantitative data should err on the side of higher resolution. For example, sensemaking of an EKG waveform relies on both numeric data as well as the (qualitative) shape of the wave itself. In these cases, displays must meet the resolution requirements for the qualitative aspects of the sensemaking tasks.

Fig. 11.5 Small display on a fitness band that tracks SpO2 levels from a pulse oximeter

Shape and Size

Displays range from tiny, wrist-sized LCDs (see Fig. 11.5) to massive networks of LEDs chained together, spanning the length of large walls.

Most displays are rectangular. Within the last decade, however, manufacturers have explored nonrectangular displays—referred to as "free form" displays—to fit niche use-cases and applications. This allows you to design things using a shape that optimizes information presentation.

These displays work by incorporating the driving technology that drives the screen into the pixels themselves instead of in its perimeter like traditional rectangular displays. Although a free form display can theoretically take on the form of any shape, it's most common application is in circular shapes, such as the Nest thermostat (Fig. 11.6).

The appropriate size of a display depends on the content displayed on the screen, the display's native resolution, the user's range of viewing distances, and the need to keep the display in close proximity to the controls that regulate it. From an engineering standpoint, display size is also influenced by space constraints on (or around) the medical device, as well as the amount of power consumed during use. The following subsections present a few considerations related to users and the display's potential use-environments.

User Considerations

A chief consideration when choosing display size is the range of distances from the user. Some products are more predictable in this respect than others. For example, smartphone apps will usually be viewed at arm's distance (i.e., less than 2 feet). Likewise, a medical device with controls at a fixed location typically has one or more displays in close proximity as well.

For other devices, "routine use" may include a wide range of viewing behaviors. Unless there are specific conditions in which the device must be viewed closer, the minimum viewing distance should be no less than approximately 15–16 in.

Fig. 11.6 Nest thermostat with a circular display. (Image by Joni Hanebutt/shutterstock. com)

(Association for the Advancement of Medical Instrumentation, 2009). On the other hand, the maximum viewing distance is contingent on the information presented as well as tasks the users must conduct with this information. For example, reading text will often require a larger display than simply detecting if an icon has turned red vs. green.

Identifying this upper limit of viewing distance requires studies with your users. A good practice is to identify displays sizes that are clearly too small (or too large), as well as offering several variations between these two upper and lower bounds. Identifying these two boundaries is important later on, since there will be no such thing as the "perfect" screen for all users. Interviews with Subject Matter Experts (SMEs) can be a quick and easy way to rule out display sizes that will not work. The sizes that remain in the running can be vetted through usability testing.

Furthermore, even devices with predictable display viewing behaviors will not always be used as expected. For example, a smartphone application intended to help OR teams communicate with experts outside the OR environment might require atypical behaviors, such as a nurse holding up the phone to show specific areas of the OR, or even serve as a "human tripod" to allow the outside person to speak with someone in the sterile field. Depending on the prevalence and risks associated with these behaviors, it will be important for manufacturers to incorporate these considerations early on in the development process.

Especially for smartphone and tablet-based products, part of these early display size decisions will be constrained by the types of equipment your user can realistically carry with them. For example, nurses in a hospital environment may or may not be able to carry a smartphone with them depending on their hospital's policies. Some hospitals will issue smartphones or tablets to staff for work-related tasks, though the exact make and model may vary from hospital to hospital, or even from department to department. As a result, your platform will need to be designed with flexibility to account for these variations.

In other cases, your team may have an existing product or prototype developed for use on a tablet-sized display. As you find out that users need this content in a smaller, smartphone sized footprint, you will have to make some difficult decisions about your design. Although it is tempting to simply "scale down" all of the tablet-sized content by a certain percentage, this decision can lead to myriad issues that won't be caught until tested sufficiently with users. After all, scaling all content down uniformly means that touchscreen GUI controls (e.g., buttons, sliders) scale down too. If the original design was already on the lower bounds of acceptable control size, then the scaled down version will not be usable. The better approach long-term is to reconsider the design through the lens of this smaller form factor. This will involve difficult decisions about what information needs to be paired together, and what information can be separated.

On the opposite end of the spectrum, displays can also be too large for some users to interact with. A good rule of thumb is that a user should be able to look at the entire display without having to adjust their head left, right, up, or down. Their eyes should cover all of the movement aspects required by the display.

Placement Considerations

User Considerations

It's a good practice to assess the user's visual capabilities when choosing displays. One factor is accommodation capability. As discussed in Chap. 5, visual accommodation is a person's ability to focus on objects at various distances. There is a lower limit—referred to as the near point—in which users can fixate on objects closer to their eyes and still remain in focus. And, there is an upper limit—referred to as the far point—which describes the point at which objects in the person's distant visual field are no longer in focus. For a young person with 20/20 vision, the near point will be approximately 10 in. from their eyes (Helander, 2006), and the far point will be approximately 20 feet.

The near point increases as people get older and their vision declines; meaning, objects must move farther away to remain in focus. While this seems counterintuitive, it's due to the fact that close-up objects require the lens of our eyes to compress as much as possible. The lens loses elasticity with age, limiting the amount of compression possible. The exact degree to which a display must be moved away from the user depends on the size of the content. Smaller content, like text or important details in an image or video, is more affected by near point restrictions than larger content.

These parameters dictate the bounds of display placement. Under no circumstances, should displays be placed outside of these bounds, or else use-errors are all but guaranteed.

For hand-held medical devices with displays embedded or attached, the near point will often be more difficult to reconcile than the far point. Conversely, displays placed along walls, carts, or desks will be constrained mostly by far points. In both cases, however, suboptimal display placement will lead users to use poor

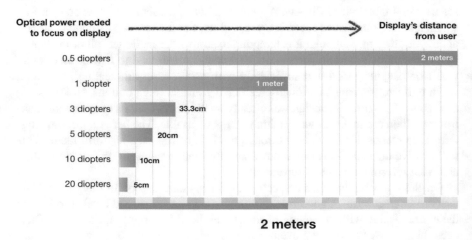

Fig. 11.7 Display distance and diopter chart

postures to make the display more visible (Helander, 2006). Near-sighted users will tend to lean in toward the display; far-sighted users will lean away from it. Both conditions can lead to discomfort and musculoskeletal issues following sustained, habitual use.

The near point of visual accommodation is measured in diopters—a measurement of "optical power." One diopter (D) is equal to the reciprocal of focal length measured in meters. The basic idea is that the more diopters (i.e., optical power) a user has, the closer they can focus on things. Figure 11.7 provides a few examples to help you understand the concept better. Keep in mind that these examples assume the user has 20/20 vision, and no visual impairments. The diopter values listed relate to the optical power needed to bring that object into focus.

Use-Environments

Display location is critical in shared use settings, such as an operating room or a hospital ICU. Complicated tasks such as those completed by surgeons in laparoscopic procedures are made more complicated by the fact that they have to switch between multiple areas of their environment to get the information they need. These visual scanning behaviors reduce performance, lead to fatigue, and increase error rates during surgery (Rogers, Heath, Uy, Suresh, & Kaber, 2012; Matern, Faist, Kehl, Giebmeyer, & Buess, 2005; Wang & MacKenzie, 1999).

Display placement in surgery has been extensively studied but has produced conflicting recommendations, due in part to the technologies studied (Rogers et al., 2012; van Det, Meijerink, Hoff, Totté, & Pierie, 2009). For example, van Veelen, Jakimowicz, Goossens, Meijer, and Bussmann (2002) and van Veelen, Kazemier, Koopman, Goossens, and Meijer (2002) reported that the center of a CRT display should be placed at a height of approximately 5 feet, 5 in., and about 5 feet away from the surgeon viewing it. By comparison, the same researchers found in a self-reported survey that surgeons preferred a slightly lower height, and a maximum distance of about 2 feet (van Veelen, Kazemier, et al., 2002).

A single, stationary display in a busy use-environment where users are constantly in motion will not always provide an optimal viewing distance for users. For example, the display may not afford sufficient viewing angles for all users. Or, one display may simply not be big enough to see important information. In these cases, the next best option is to provide multiple displays strategically positioned around the use-environment. Each should show the exact same content to promote consistent viewing experiences. This also means that displays should possess the same capabilities (e.g., luminance, contrast, resolution, size) to prevent different interpretations of information from team members. In some cases, displays might even be placed directly next to each other, but simply at different angles to account for the left and right sides of the room.

Identifying the ideal placement of these displays requires upfront research. Observational research is preferred over interviews for this purpose. Researchers should catalog user movements, the types of artifacts in the environment, and the interactions between users if multiple users are working together. These factors will help define key behaviors and heuristics that can be applied to the use-environment. It's important that manufacturers, hospitals, and designers understand that the placement of displays in shared environments is not a static, one-time decision. Instead, they are part and parcel of a dynamic, ever-changing environment, and may need to be reevaluated over time to remain an effective tool for users.

11.3   Auditory Displays and Alarms

Overview of Auditory Displays and Alarms

An auditory display communicates dynamic, sound-based information. Like visual displays, they update users about circumstances or tasks happening with a device, and in some cases, elicit a behavioral response (i.e., call to action).

Auditory displays are often misunderstood as only alarms, due to two factors that make them effective at signaling emergencies (Stanton & Edworthy, 1999):

- Loud alarms, especially those exceeding 90 dBA, are good at getting *and* holding our attention. You cannot look away from an auditory alarm like you can a warning message on a visual display.
- When an auditory alarm goes off, there is a natural tendency to want to resolve it quickly (i.e., urgency).

Also, auditory alarms are more prevalent in technology compared to nonalarming auditory displays (Walker, Nance, & Lindsay, 2006). However, with the introduction of audio-based interfaces such as Amazon's Echo (Fig. 11.8) and Google Home, the ratio between the two is beginning to shift. Since 2014, Amazon has sold well over 100 million Alexa-enabled devices. Despite the growing popularity of

Fig. 11.8   Amazon Echo Generation 2. (Image courtesy of Zapp2/Shutterstock)

auditory displays in consumer electronics, a similar trend has yet to appear in healthcare. Auditory alarms have often been used without much consideration of user needs (Patterson, 1990). For example, a loud alarm for an infusion pump makes sense in a hospital where a nurse or physician will be in a different room, but that alarm would be too loud at home.

Our auditory system is merely another modality we use to extract information from our environment. As such, designers and manufacturers can leverage hearing to communicate information. Whether to use an auditory display depends on cognitive load and context of use. Multiple authors (see Association for the Advancement of Medical Instrumentation, 2009; Deatherage, 1972; Stanton & Edworthy, 1999) proposed the following conditions in which an auditory display may be a viable solution:

- The information is simple
- The information is brief
- The information is continually updated or changing
- The information comes from one source
- The information will not need to be recalled later (including just a few seconds later)
- The information is time-oriented
- The information should be acted on ASAP
- The user's tasks require unconstrained movement in their environment
- The user's visual system is already occupied with other (necessary) tasks
- The user has difficulty with visual perception (i.e., blind, visually impaired)
- The use-environment is dark and the user's eyes may not be "dark adapted"
- The use-environment is relatively quiet (i.e., about 70 dB or less)
- The use-environment is visually cluttered
- A visual display is not feasible due to space, logistical, or technological constraints

An auditory display may not be best when it's not obvious how the user needs to interact with the device (i.e., which control do I need to use?). In this case, a visual display may make more sense, as visual feedback can be placed near the necessary controls. By contrast, even if a sound speaker is placed next to the intended controls, the relationship can be difficult to detect, especially under stressful circumstances.

Fundamentals of Auditory Displays

As described in Chap. 6, we perceive sound based on frequency, intensity, and timbre. However, another component—let's call it patterns and intervals—communicates information like the urgency of a sound. For the sake of brevity, we will not go into too much depth on this topic. Each factor can be used to communicate to the user, in similar ways that text, graphics, color, size, shape, etc. can be used in visual displays to communicate information. These are the "building blocks" of an auditory display:

- *Frequency (pitch)*: the quality of sound that we use to perceive a sound as "high" or "low" relative to others. Frequency and pitch are similar to one another, but not identical. The former is an objective measure of a sound wave's oscillation frequency. The latter is our subjective experience of that sound.
- *Sound intensity (loudness)*: for lack of a better term, this is the "volume" of a sound. Loudness is the subjective experience of a sound's intensity.
- *Timbre (quality)*: the "quality" of a sound. Timbre (or *timber*) is influenced by the number of overtones (i.e., harmonics) in a sound source, as well as their frequencies and intensities relative to each other. Every naturally occurring sound possesses overtones. It's the aspect of a sound that makes someone's voice sound like them.
- *Patterns and intervals*: the timing, spacing, and ordering of sounds to "build" a sound over time. The great composer, Claude Debussy, once said that, "music is the space between notes" (Koomey & Holdren, 2008). A similar sentiment can be said about auditory patterns and intervals: criticality is the space between sounds. Our perception of urgency generally increases as the time between sound intervals decreases.

Creating Discoverable Sounds

The initial hurdle of any auditory display or alarm is simply to make its sounds discoverable to users. After all, a user cannot act on information if they don't know it's there in the first place. Generally speaking, a discoverable sound is one that:

- Is loud enough—the sound must be at least 15 dB above ambient noise levels (National Research Council, 1997; Patterson, 1989).
- Take into consideration the user's cognitive bandwidth and capabilities—additional sonic cues may be needed to get the user's attention during periods of intense concentration or duress. Similarly, those with a hearing impairment may not be able to hear certain sound frequencies.
- Uses the appropriate frequency ranges given the environmental constraints—takes into account where and how far away a user will be relative to the sound source (Alain, Du, Bernstein, Barten, & Banai, 2018).

A medical device's use-environment will always contain at least some background noise. An average ICU hospital environment, for example, tends to have about 60 dB of background noise happening regardless of the time of day (Qutub & El-Said, 2009). Yet even in the quietest environment, you will still hear noises such as someone breathing (about 10 dB), or the soft whirring of an HVAC system (about 40 dB). This collection of auditory stimuli in a use-environment make up what is called the ambient noise level.

The upper limit of the ambient noise level is referred to as the masking threshold. An auditory display must exceed this level to be discoverable by a user for a given frequency. In practice, however, auditory displays should be at least 15 dB louder than the masking threshold (Patterson, 1989; Robinson & Casali, 2000), but no greater than 25 dB (i.e., maximum) above it.

Keep in mind that each frequency (i.e., low and high pitch sounds) has its own masking threshold. A crowded environment with lots of people will have a much higher masking threshold in the frequency range specific to speech: approximately 300–3000 Hz. In the same environment, higher frequencies (i.e., greater than 3000 Hz) may have a relatively low masking threshold. This means it doesn't take as loud of a sound in this frequency range to be heard.

Other environments will experience different masking thresholds based on their specific acoustic characteristics, distance of the listener, equipment in the room, and user behaviors and tasks. For example, Privopoulos, Howard, and Maddern (2011) recommend that signals below 1000 Hz should be used for outdoor alarms because low frequencies are less affected by atmospheric absorption, and are more effective at getting around barriers with minimal changes to the sound itself. As a result, it's important to design an auditory display with these specific masking thresholds in mind to optimize the types of sounds presented to users.

Different frequencies require different sound levels for them to be perceived with the same loudness. For example, low frequency tones require a higher sound level than higher frequencies to be perceived at the same "loudness." If all frequencies were consistently played at the same decibel level as each other, we would not perceive the low frequencies as well (if at all); the high frequencies would dominate our hearing. Figure 11.9 shows how hearing thresholds are affected by sound frequencies.

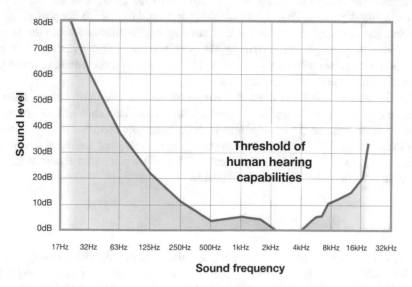

Fig. 11.9 Threshold of human hearing capabilities—sound frequencies compared against decibel levels across the human hearing range (20 Hz to 20 kHz)

Note that your hearing is so refined in the 2–4 kHz range that it takes hardly any sound intensity for you to detect these frequencies.

HF engineers have long understood that effective alarms are ones that gain attention, but avoid startling or scaring the user (Patterson, 1982; Stanton & Edworthy, 1999). Striking this balance between discoverability and distraction can take a considerable research effort, especially when a medical device can be used in a wide array of use-environments with differing ambient noise levels. A good first step is to identify the quietest and loudest use-environments and design to those specifications. Keep in mind however that, if there is a large disparity between these two extremes, an auditory display (at least on its own) may not be a viable option.

Another good practice is to "cover the lows" and "cover the highs" in each sound produced; meaning, always include at least one tone—though, more may be preferable—in the lower frequency range, and a few others in the higher ones. The Association for the Advancement of Medical Instrumentation (2009) recommends that a sound contain at least three frequencies. If you think of each frequency as taking a shot at a target, including a few tones from different ends of the spectrum means that your user will be more likely to hear it and understand where it's coming from. That is, you are more likely to hit the target with at least one of those shots.

There are other reasons to include sounds with both high and low frequencies. Environmental factors such as walls, doors, and even large empty rooms can affect the sound source before it reaches the user. If you have ever wondered why a sound coming from inside a closed room sounds "muffled," it's because specific sound frequencies are being filtered out. Low frequency sounds tend to pass through walls and doors, using those materials almost as a resonating chamber. High frequency sounds tend to reflect or get absorbed by the walls and doors themselves. As such, the higher frequencies are effectively "removed" by the time the sound reaches the user outside the room. This is an especially important detail to keep in mind for wireless medical devices where the user can separate themselves from the part of the device responsible for providing auditory feedback and updates. This also applies to caregivers who may not be in the same room as their patient or loved one while they are connected to the medical device.

Lastly, bear in mind that the technology used to produce these sounds must be capable of meeting the frequency properties. Small speakers such as those found in a smartphone tend to perform poorly at producing low frequency sounds below (approximately) 500 Hz. This is important when evaluating the form factor and technology that goes into a medical device. Make sure to choose components that meet the specifications of the sound's design.

Localization of Sounds

Sound localization helps us direct visual attention (Akeroyd, 2014), telling us generally where a sound is coming from, so our eyes can handle the fine tuning afterward. Ostensibly, the more accurate and faster that first effort is, the more likely that we can avoid a threat and survive. As discussed in Chap. 6, our brains automatically process sound location based on the difference between when and how a sound hits each ear. These differences include things like the sound's intensity, frequency spectrum, and timing details (Blauert, 1997). For example, if the sound is coming directly from your left side, it will take just slightly longer for that sound to continue traveling and reach your right ear. And, since our heads are a relatively solid object, it will also mask parts of the remaining sound passing by, giving it an "acoustic shadow" by the time it hits our other ear (Akeroyd, 2014). However, when the sound source is directly in front of us, the sound will hit both of our ears at the same exact time. The technical term for this concept is the Interaural Time Difference (ITD).

A related concept—interaural level difference (ILD)—is based on the same idea described above, but deals with sound level/intensity instead of frequencies. Generally speaking, ILD and sound localization correlate positively—meaning, as ILD increases, so does our ability to localize a sound source. This means we will experience the greatest reduction in sound level from our left ear to our right ear when we are dealing with higher sound frequencies. As a result, we are able to pick up on this cue with greater accuracy. The reason for this is that our heads physically "filter out" a lot of the high frequencies passing from left to right. Lower frequencies are less affected, and thus, continue to the next ear largely unscathed. This effect is most pronounced when the sound source is coming directly from our left or right side (i.e., 90°). While still present, the relationship between frequency and ILD is less significant as the sound source moves in front of us (i.e., 0°) or behind us (i.e., 180°). Figure 11.10 shows how ILD changes in respect to a sound source revolving from in front of you to behind you on the left side. Imagine drawing a semi-circle

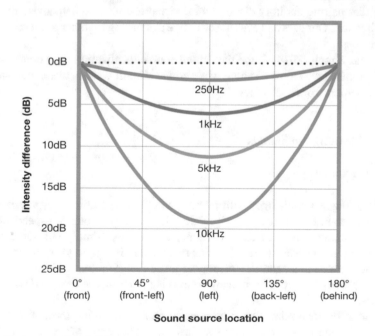

Fig. 11.10 Interaural level difference

around your head, starting at the center of your forehead (0°) and moving counter-clockwise (left) toward the center of the back of your head (180°). This figure shows how certain sound frequencies change in intensity as they come in from different positions in this semi-circle. High frequencies have the biggest drop in sound inten-sities as they approach from your left side.

Pitch is also a vital aspect of identifying the location of a sound. As a rule, people are better at localizing a sound source when that sound covers a broader range or "band" of frequencies, rather than a narrow band or a single sound frequency. This is partially the reason why a "low battery" smoke alarm warning ("chirp") can be difficult to tell where the sound is coming from. The single, +3000 Hz simple tone does not have enough tonal complexity to afford easy sound localization. A similar issue exists among ambulance sirens. Drivers in the United States, for example, often have difficulty localizing the sirens of an ambulance because these sirens use a relatively narrow frequency band ranging from 500 to 1.8 kHz (Withington & Chapman, 1996).

Here are a few other important considerations about how humans localize sound:

- People tend to perform poorly at localizing sounds coming from behind or above them.
- Sound localization is most accurate and fastest for high frequency sounds (>1 kHz), rather than low frequencies (<1 kHz) (Sanders & McCormick, 1993).
- Sound discrimination and localization is best for sounds closest to the frequency of the human voice, with a slight skew toward higher frequencies (i.e., above about 2 kHz).
- Consider pairing auditory displays with visual elements to help speed up sound localization. Our ears and eyes are effective at working together to identify sound sources.
- Yost and Zhong (2014) found that sound source localization (in a controlled lab environment) was most accurate when sound stimuli were between one and two octaves apart from each other.

Giving "Meaning" to Sounds

Nonspeech Sounds

One of the biggest challenges with making an effective auditory display is giving meaning to sounds. This can be especially challenging among nonspeech sounds such as computer-generated chimes, beeps, and buzzes. Simply put, these sounds lack implicit and universal meaning. For this reason, it's good to limit the number of abstract sounds in an auditory display to just a few (Stanton & Edworthy, 1999); having to process and recall too many sounds is taxing on the user, and will lead to problems.

Recently, designers have used earcons to convey meaning through sound. An earcon is like an icon—but for your ears instead of your eyes. Many computer users recognize the sound of the "recycling bin" when it's (virtually) emptied. Similarly, Mac computer users will recognize the characteristic "Apple startup" earcon upon powering up their computer. Earcons come in two varieties (Blattner, Sumikawa, & Greenberg, 1989; Stanton & Edworthy, 1999): representational and abstract earcons. The specifics of each type are discussed below.

Representational earcons—sounds that mimic or "represent" a known phenomenon or concept. For example, a sound with a downward inflection can be used to indicate values going down on an injector system. On the other hand, a sound with a sharp upward inflection at the tail end may be a useful way to represent a question or system confusion. This is due to the fact that spoken (English) questions naturally end on the same upward inflection. Representational earcons are easy to learn with moderate practice and experience as the association between sound and phenomenon is already primed.

Abstract earcons—sounds that have no ties to existing standards or expectations. An abstract earcon is a lot like any new (visual) icon that users must learn through hard work, time, and experience. It does not aim to build off existing associations in the world. This takes more attentional resources and training, but it is not likely to get mixed up with an existing phenomenon that means something else to the user.

Each type of earcon is generally built around short sequences of notes (Sanderson, 2006). An earcon will often rely on changes in timing and frequency to communicate its intention. It's less common to vary in sound intensity. For example, an earcon demonstrating the action of "going down" (e.g., lowering values) may play three progressively lower notes in succession. The opposite pattern can be used to give the illusion of "going up" (e.g., raising values) as well. They can also imply complex things like "uncertainty" by ending on an upward inflection, similar to how English speakers raise the pitch of their voice at the end of a question.

Earcons benefit from following melodic normalcy; meaning, the jumps between frequencies used to build the sound remain in the same (western) musical key. Additionally, these frequencies shouldn't rapidly jump between octaves (i.e., a doubling or halving of a frequency) unless there is a good reason to do so. A good earcon is a lot like a good musical jingle: it's mentally sticky, and uniquely its own. The former aids in speeding up learning as well as recall time. The latter reduces the risk of misidentifying the earcon to mean something else.

Savvy designers will build "families" of earcons, in the same way that they might build "families" of visual icons. For example, an earcon representing "down" and another indicating "up" might both start their melody on the same frequency (and musical key), then head in opposite sonic directions. Or, they may speed up a motif altogether to signal increasing urgency.

Earcon families are also effective when a manufacturer has a line of related products which incorporate similar pairings between auditory messaging and device actions. This can be especially useful when a user is likely to use multiple products from the same manufacturer, or when they eventually switch to the latest generation of a model or device. The user can apply what they've learned from one device and speed up familiarization and learning on the new one. On the flip side, though, this can work against the manufacturer and user if the same set of sound "rules" are not consistently followed across products.

Speech-Based Sounds

Speech is complex and informationally rich, so it makes sense to use it in auditory displays. Speech-based auditory displays have several benefits and a few drawbacks. Here are a few of both:

Benefits of Speech-Based Sounds

- *Minimal learning* Native speakers of a language can be expected to have a rich and thorough understanding of the basic words used in the commands and prompts of an auditory display. However, designers must remain aware of the limitations of each user group's language comprehension and vocabulary.
- *Ability to communicate (somewhat) complex commands and actions* Generally speaking, speech-based displays can supply the user with slightly more information than nonspeech sounds without compromising performance. This being said, speech-based commands should still aim to be concise and simple.
- *(Synthetic) speech can be easily edited and integrated* Computer-generated speech (i.e., synthetic speech) has come a long way in recent years. Revising synthetic speech is simply a matter of rewriting content for the computer to produce.

Drawbacks of Speech-Based Sounds

- *Words with multiple meanings* Some words will have multiple meanings, which can lead to confusion. Manufacturers must be careful about which words they choose in a speech-based auditory display.
- *Human-generated speech can be expensive* Using a human-recorded voice for a wide-range of customized applications can be challenging and costly to achieve (Chiou, Schroeder, & Craig, 2020). What's more, when edits to instructions or prompts are invariably required, the manufacturer must decide between trying to rehire the same voice actor, accept having two different voices, or invest in re-recording the entire auditory command list.
- *Bound by linguistic and dialectical limitations* Efforts need to be duplicated for each language used in the auditory display (e.g., English, Spanish). It's not simply a matter of hiring a translator to "copy over" validated content into a new language. For example, there are many language-specific factors and shortcuts that work in English, but do not translate effectively to Japanese, and vice versa (Alami, 2017).
- *Sounds blending into the environment* In environments with a lot of talking, auditory displays must compete with the environment, and may result in delayed user

Fig. 11.11 The Auvi-Q®
auto-injector uses a
combination of both
speech and text
instructions. (Image
courtesy of Auvi-Q® with
permission)

responses or completely missed prompts altogether (Mattys, Davis, Bradlow, & Scott, 2012).

One example of a good use of a speech display is provided by the The Auvi-Q® auto-injector, which uses a combination of written instructions, graphics, and speech to guide users through delivering an emergency dose of epinephrine. The auditory display helps with this type of device because a user's hand may block parts of the written instructions during use, or the lighting conditions may be too dim to read the instructions. Furthermore, the speech instructions count aloud for the duration of the injection (i.e., 1… 2… 3… 4… 5…). During emergency situations, people have a tendency to count much faster than directed, leading to partial dosing. This feature in the auditory display helps mitigate this issue (Fig. 11.11).

Human-Generated Speech vs. Computer-Generated Speech

Until recently, users benefited most from auditory displays with human-generated speech, rather than computer-generated speech (i.e., synthetic speech). More recent synthetic speech algorithms have undergone vast improvements, with some varieties having little negative impact (if any) compared to human-generated speech. A few recent studies have demonstrated that training outcomes were similar, regardless of whether the user was supplied with instructions from a human or synthetic speech from a modern voice engine (Chiou et al., 2020; Craig & Schroeder, 2017; Wester, Watts, & Henter, 2016).

Ultimately, synthetic speech engines are far from perfect for all applications. For example, Hui, Jain, and Watson (2019) reported that synthetic speech can present comprehension issues when commands are made up of unusual or overly complex sentence structures. Human-generated speech commands with the same constraints also lead to lowered comprehension, just not as much as synthetic speech. The authors also found that nonnative speakers had more comprehension issues with synthetic speech engines, when compared to native speakers (Hui et al., 2019). Synthetic speech also tends to be more difficult for those with hearing impairments, even when hearing is corrected through devices such as cochlear implants (Ji, Galvin III, Xu, & Fu, 2013).

Rate of Speech

Generally, speech comprehension improves when the rate of speech is kept at a moderate to moderately low level. Older adults with declining hearing capabilities tend to benefit from a slightly slower than average rate of speech (Schneider, Daneman, & Murphy, 2005). Additionally, comprehension also improves with brief but purposeful pauses between longer runs of commands or instructions (Jacobs, 1988). One explanation for this is that these slight pauses give users a moment to process the run of speech they just heard, before the next part begins. Another reason is that it allows for better articulation of individual phonemes. As discussed in Chap. 6, older adults with hearing loss tend to have a difficult time hearing certain consonant-based phonemes, as shown in Fig. 11.12.

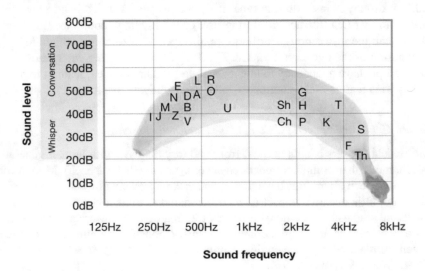

Fig. 11.12 The "banana" shape of speech

Vowel Spacing

Auditory display speech comprehension can be affected by the speaker's vowel annunciation, referred to as, "vowel spacing." Generally, wider vowel spaces promote better understandability and comprehension. One possible reason for this is that the English language comprehension tends to be consonant driven, meaning, we extract much of a word's important phonetic cues and meaning from nonvowel sounds (Owren & Cardillo, 2006). As such, a wider vowel spacing affords extra time to process consonant details.

Spearcons

Beyond basic, conversational speech, HF engineers have studied a unique way to "chunk" auditory information called spearcons (spear-based earcons). Spearcons are similar to earcons in the sense that they communicate a single concept, event, or phenomenon to the user. The two main differences with a spearcon are:

- Sound content is speech-based
- The rate of speech is two to three times faster than natural speech

Spearcons have been widely studied in respect to patient monitoring in the hospital environment. For example, Li et al. (2017) found that users of a patient monitoring system were able to learn faster and were more accurate at identifying patient SpO2 vital signs when using spearcons instead of earcons. A more recent follow-up study offered additional support for these findings, adding that spearcons were also effective in a multipatient monitoring task where both normal and abnormal SpO2 and heart rate data were presented.

Interestingly, the general idea behind spearcons is often implemented by blind and visually impaired users who use "screen reader" technologies and applications. If you're not familiar with this technology, the purpose of a screener reader is to help users essentially translate visual information (e.g., buttons, pictures, menu structures) to words. Usually, a screener reader will describe buttons, states, or events in just a couple brief words (e.g., "On/off button… disabled")—similar to a spearcon. On a smartphone or tablet, you can navigate through these items by swiping left and right, or tapping different parts of the screen. If everything is designed appropriately on the back end, the screen reader will follow the information architecture outlined in the design of the website or app. As a result, as a user swipes right, the screener reader will go through the screen's content from top to bottom, left to right—just like how you might scan and read a website or app.

As you might imagine though, there can be a lot of information to translate from visual details into words. And, at a normal speaking rate, it can take a considerable amount of time to get through an entire webpage or app screen. Many screen reader users will get around this issue by increasing the speaking rate of the screener reader up to about 300 words per minute. To an outside listener, the screen reader speech sounds garbled and incomprehensible. But it makes a lot more sense to an experienced screen reader user who has worked with the technology for years.

Creating a Sense of Urgency

It's important that a medical device has the capacity to communicate a sense of urgency and criticality to adverse events while the device is in use. For a visual display, urgent details can be communicated through icons or text labels (e.g., warning vs. caution), colors, size, shapes, animations, etc. For auditory displays; however, there are fewer tools at the designer's disposal. Nevertheless, the tools that it does have can be used quite effectively if the designer understands a few basic principles about how people perceive urgency through sounds.

Reaction times (RT) are based on two factors: signal detection and response initiation (Green & Luce, 1971). In other words, if you want someone to respond quickly to an auditory alarm, the alarm must be clear and easy to detect, as well as have a strong association with the perceived need to respond. Here are a few of the tell-tale characteristics of an "urgent" alarm through an auditory display (Hellier & Edworthy, 1999; Suied, Susini, & McAdams, 2008).

- Sound intervals repeat quickly (i.e., fast rate)
- Loud volume (relative to ambient noise level)
- Higher pitched
- Random and dissonant harmonic frequency pairings

It's important to remember that people generally do not perform well when exposed to an urgent sound (e.g., alarms) for too long. Indeed, if the alarm is too aversive, users may divert their attention from the initial task (e.g., help the patient), and focus instead on ways to stop or turn off the sound itself (Bellettiere et al., 2014). The take-home message here is that while it is easy to create auditory cues that evoke a sense of urgency, this is only half the battle. An effective sound in an emergency situation draws the user's attention, instills a sense of urgency, and then gets out of their way once they begin to act.

Alarm-Specific Considerations

Alarms adhere to the same principles described above for auditory displays. After all, an alarm is simply one part of the total hierarchy of sounds capable by an auditory display—it just happens to sit at the top of the hierarchy in terms of criticality. The ISO/IEC 60601-1-8 guidance on alarm design provides a thorough description of what manufacturers need to and should do to make safe and effective alarms. Here are a few of the required ones (ISO/IEC 60601-1-8):

- Alarms must have at least three prioritization levels (e.g., low, medium, high)

- Higher priority alarms must be louder than lower priority ones (e.g., mediums must be louder than lows)
- Different alarm priority levels must have different sound "burst" patterns and rates (e.g., high priority has ten fast pulses, medium priority has three moderate pulses, low priority has one to two slow pulses)
- High and medium priority alarm pulse patterns must repeat following a rest period known as an "interbeat interval"
- The fundamental frequency of an alarm must be between 150 and 1 kHz
- There must be at least four unique, harmonic frequencies within the alarm between 300 and 4 kHz
- Each of the four harmonic frequencies must be within 15 dB above or below the fundamental frequency
- "Technical alarms" (alarms not directly related to patient harm) must be different than the three priority alarms used to indicate patient-related harm
- Alarms must be validated through usability testing (see Chap. 4: "Usability Inspection Methods")

As it stands, medical device manufacturers often make their own proprietary alarm sounds. In other words, there is no shared, universal alarm type to indicate specific issues. This is unfortunate since each manufacturer must essentially "rein vent the wheel." A shared framework and hierarchy for alarm design would likely do a lot of good for patients and users alike (Wiklund & Smith, 2001).

The remainder of this chapter will discuss several additional alarm-specific considerations that you should keep in mind that relate to HFE.

Alarm Fatigue

Alarm fatigue is a habituation or desensitization to frequent alarms, especially when the alarms are nonactionable. It's akin to the response in the fable about the boy who cried wolf. When the wolf finally came, nobody believed the boy.

If a user is presented with enough false alarms, they eventually assume all alarms are false. Furthermore, each alarm can potentially distract from your other work, creating a recipe for annoyance, resentment, slowed response rates, and distrust among users (Ketko, Martin, Nemshak, Niedner, & Vartanian, 2015).

Some areas of healthcare are more affected by alarm fatigue than others. For example, certain hospital environments (e.g., ICU, PACU, NICU) meet all the conditions to elicit alarm fatigue:

- Many patients
- Many alarm-capable devices
- Many competing responsibilities from which to be distracted
- False alarms (i.e., false negatives)
- Nonactionable alarm events/information (i.e., okay, but what do you want me to do about it?)
- Perceived low risk if response is delayed or alarm is ignored

Indeed, one study on alarms in the operating room found that, on average, clinicians were presented with a whopping 359 alarms during a single procedure. Yet less than 20% of these alarms had a true therapeutic consequence (Schmid et al., 2011). A more recent review provides evidence that the rate of alarms that do not warrant clinical intervention may even be as high as 74–99% (Paine et al., 2016). The Joint Commission (2016) specifically calls attention to this issue for medical devices such as cardiac monitors, IV machines, and ventilators. Though, these devices are merely the tip of the proverbial iceberg.

The knee-jerk "answer" to solving alarm fatigue is to build smarter alarms. For example, rather than an IV machine alarming at the slightest hint of air in a patient line, the parameters and thresholds for air could be fine-tuned for real world use. Of course, this is easier said than done. In some cases, the underlying technology is just not sensitive enough to make this type of "smart" system. In other cases, the technology might be possible, but would be cost-prohibitive. This does not mean that manufacturers should not strive for improvement in these areas. Rather, it's a reminder that our expectations must be realistic; not all HFE-related problems can be "designed out" through technology alone.

When "designing out" the issue is not feasible, effective alarm management should be viewed as critical to eliminating or reducing the prevalence rate of alarm fatigue (Cvach, 2012). Here are a few promising strategies identified in literature:

- General modification to alarm parameters and thresholds
- Alarm customization based on individual patient needs
- Better display systems for nurses
- Hospital policy changes
- Hospital training on alarms (i.e., importance, what to do)
- Integrating alarm(s) into other systems
- Permitting delays to alarms or notifications
- Earlier maintenance or replacement of patient-interfacing equipment (e.g., electrodes)
- Adjusting placement of patient-interfacing equipment (e.g., forehead instead of finger probes)

Ruppel et al. (2018) also point out that alarm fatigue-prone environments such as hospitals should examine which devices are responsible for most alarms, and to evaluate whether an intervention is possible on these problematic devices.

Category-Specific Alarm Sounds

Several efforts have been made to standardize alarm sounds and earcons within device-type categories. In other words, the category of "ventilators" would have its own set of unique earcons and sounds, while "heart rate monitors" would have its own set. ISO/IEC 60601-1-8 champions this exact idea. It recommends that a ventilator alarm would have a chime similar to the "NBC" motif, while the heart rate

Fig. 11.13 Melodic alarm earcons where each one matches a note to a specific syllable in a related word (e.g., VENT-I-LATE, CAR-DI-AC)

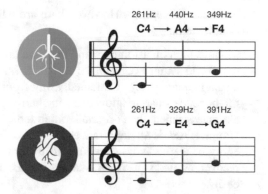

monitor alarm would match the three syllables of "CAR-DI-AC" (Thompson, 2010) (Fig. 11.13).

This is an excellent sentiment but not as effective as hoped. Busy hospitals have lots of sounds, including melodic sounds from music or television playing in the background. Melodic alarms can easily get lost in the mix, leading to slower response times. Second, most medical professionals don't have a refined enough ear to pick up on differences of just a few hundred Hertz, especially during stressful situations. This is perhaps unsurprising given that the "breathing" and "cardiovascular" categories (described above) both begin on the same starting frequency. Indeed, even with significant training, medical professionals such as anesthesiologists and nurses frequently misidentify and miscategorize these types of melodic alarms (Meredith & Edworthy, 1995; Sanderson, 2006; Sanderson, Wee, & Lacherez, 2006).

As it stands, alarm earcons by device category type are largely ineffective. These results may change if researchers and designers made better use of the sonic "tool kit" at their disposal. For example, things may improve if designers incorporated different timbres, intensity levels, and temporal aspects (e.g., pulsing, tempo).

Dismissing and Delaying Alarms

One of the big questions that any medical device manufacturer will face with their alarm's design is how much control they should allow the user over the alarm. On one hand, giving users control over how loud, how long, or even when an alarm is triggered introduces risk. Depending on the device and circumstances, this could result in patient injury or death.

On the other hand, alarms command attention when activated. Thus, if an alarm cannot be dismissed, turned down, or muted, then the user has less cognitive bandwidth to use, which impacts identifying the problem and how to fix it. This can lead to delays in finding a solution, or potentially finding the wrong solution. More frequently, however, this alarm will simply be a disruption (i.e., annoying, inconvenience). Unfortunately, there is no clear-cut solution. The honest answer is one we hate to hear—it depends. However, here are a few guidelines to help you navigate this murky topic:

- Manufacturers must understand the time and attentional constraints on users in real world settings (Grissinger, 2016). If it's unrealistic for a user to respond to an alarm quickly (e.g., distracted, away from the device for long periods), then it might not be smart to allow users to change alarm settings in ways that make the alarm less salient than originally intended.
- "Forgiveness" is an important aspect of a user-friendly medical device. The category of alarms is no exception to this rule. A device should enable users to recover or undo previous alarm settings, and return to the default settings easily.
- Permanent changes to alarm settings should be left to qualified administrators (or similar). This is especially important in a hospital setting where multiple users will often use the same device. You don't want one user's alarm preferences to negatively affect another user's performance.
- Be sure to independently test all alarm options that you plan to offer to the user for its safety and effectiveness. Some options may elicit faster reaction times than others, or may be perceived as more or less "pleasing" to the user. Although these factors are important, no option should compromise safety and effectiveness.
- The user should be reminded periodically when the alarm-inducing condition has not been resolved. This is especially important if the user is allowed to temporarily mute or reduce the volume of an alarm (e.g., 30 s delay, 50% reduction in sound intensity). It's usually not advisable to offer a manual volume control for users to lower an alarm's volume at their discretion. There is simply too much risk that the user will leave the control in a lowered (or off) state in the future (Grissinger, 2016).

Resources

- ANSI/HFES100. (2007). *National standards for human factors engineering of computer workstations*. HFES (pp. 1–98).
- Chen, J., Cranton, W., & Fihn, M. (Eds.). (2016). *Handbook of visual display technology*. Berlin, Germany: Springer.
- Eckert, M., Volmerg, J. S., & Friedrich, C. M. (2019). Augmented reality in medicine: Systematic and bibliographic review. *JMIR mHealth and uHealth, 7*(4), e10967.
- Nicolau, S., Soler, L., Mutter, D., & Marescaux, J. (2011). Augmented reality in laparoscopic surgical oncology. *Surgical Oncology, 20*(3), 189–201.
- Sielhorst, T., Feuerstein, M., & Navab, N. (2008). Advanced medical displays: A literature review of augmented reality. *Journal of Display Technology, 4*(4), 451–467.

- Tang, S. L., Kwoh, C. K., Teo, M. Y., Sing, N. W., & Ling, K. V. (1998). Augmented reality systems for medical applications. *IEEE Engineering in Medicine and Biology Magazine, 17*(3), 49–58.

References

Akeroyd, M. A. (2014). An overview of the major phenomena of the localization of sound sources by normal-hearing, hearing-impaired, and aided listeners. *Trends in Hearing, 18,* 233121651456044.

Alain, C., Du, Y., Bernstein, L. J., Barten, T., & Banai, K. (2018). Listening under difficult conditions: An activation likelihood estimation meta-analysis. *Human Brain Mapping, 39*(7), 2695–2709.

Alami, A. (2017). To what extent does the medicalisation of the English language complicate the teaching of medical ESP to Japanese medical students learning English as a foreign language. *International Journal of Learning Teaching and Educational Research, 16,* 102–110.

ANSI/HFES100. (2007). *National standards for human factors engineering of computer workstations.* HFES (p. 67).

Association for the Advancement of Medical Instrumentation. (2009). *ANSI/AAMI HE75–2009: Human factors engineering—Design of medical devices.* Arlington, VA: Association for the Advancement of Medical Instrumentation.

Association for the Advancement of Medical Instrumentation. (2011). *A siren call to action: Priority issues from the medical device alarms summit.* Arlington, VA: Association for the Advancement of Medical Instrumentation.

Baber, C., Stanton, N. A., & Stockley, A. (1992). Can speech be used for alarm displays in 'process control type tasks? *Behaviour & Information Technology, 11*(4), 216–226.

Bellettiere, J., Hughes, S. C., Liles, S., Boman-Davis, M., Klepeis, N., Blumberg, E., et al. (2014). Developing and selecting auditory warnings for a real-time behavioral intervention. *American Journal of Public Health Research, 2*(6), 232.

Blattner, M. M., Sumikawa, D. A., & Greenberg, R. M. (1989). Earcons and icons: Their structure and common design principles. *Human–Computer Interaction, 4*(1), 11–44.

Blauert, J. (1997). *Spatial hearing: The psychophysics of human sound localization.* Cambridge: MIT Press.

Carey, C. W. (2014). *American scientists.* New York: Infobase Publishing.

Chang, H. T., Tsai, T. H., Chang, Y. C., & Chang, Y. M. (2014). Touch panel usability of elderly and children. *Computers in Human Behavior, 37,* 258–269.

Chen, J., Cranton, W., & Fihn, M. (Eds.). (2016). *Handbook of visual display technology.* Berlin, Germany: Springer.

Chiou, E. K., Schroeder, N. L., & Craig, S. D. (2020). How we trust, perceive, and learn from virtual humans: The influence of voice quality. *Computers & Education, 146,* 103756.

Chouvardas, V. G., Miliou, A. N., & Hatalis, M. K. (2008). Tactile displays: Overview and recent advances. *Displays, 29*(3), 185–194.

Clinton, V. (2019). Reading from paper compared to screens: A systematic review and meta-analysis. *Journal of Research in Reading, 42*(2), 288–325.

Crabtree, M., Mirenda, P., & Beukelman, D. (1990). Age and gender preferences for synthetic and natural speech. *Augmentative and Alternative Communication, 6*(4), 256–261.

Craig, S. D., & Schroeder, N. L. (2017). Reconsidering the voice effect when learning from a virtual human. *Computers & Education, 114,* 193–205.

Cvach, M. (2012). Monitor alarm fatigue: An integrative review. *Biomedical Instrumentation & Technology, 46,* 268–277.

Deatherage, B. H. (1972). Auditory and other sensory forms of information processing. In H. P. Van Cott & R. G. Kincade (Eds.), *Human engineering guide to equipment design* (pp. 123–160). Washington, DC: American Institutes for Research.

Dingfelder, S. (2008). Too discordant for the masses? *Monitor on Psychology, 4*, 28.

Durlach, N. I., Thompson, C. L., & Colburn, H. S. (1981). Binaural interaction in impaired listeners: A review of past research. *Audiology, 20*(3), 181–211.

Eckert, M., Volmerg, J. S., & Friedrich, C. M. (2019). Augmented reality in medicine: Systematic and bibliographic review. *JMIR mHealth and uHealth, 7*(4), e10967.

Gerrig, R. J., Zimbardo, P. G., Campbell, A. J., Cumming, S. R., & Wilkes, F. J. (2015). *Psychology and life*. London: Pearson Higher Education AU.

Green, D. M., & Luce, R. D. (1971). Detection of auditory signals presented at random times: III. *Perception & Psychophysics, 9*(3), 257–268.

Grissinger, M. (2016). Drawn curtains, muted alarms, and diverted attention lead to tragedy in the postanesthesia care unit. *Pharmacy and Therapeutics, 41*(6), 344.

Hart, J. A., Lenway, S. A., & Murtha, T. (1999). *A history of electroluminescent displays* (pp. 1–18). Bloomington: Indiana University.

Hellier, E., & Edworthy, J. (1999). On using psychophysical techniques to achieve urgency mapping in auditory warnings. *Applied Ergonomics, 30*(2), 167–171.

Himmelstein, A., & Scheiner, M. (1952). The Cardiotachoscope. *Anesthesiology, 13*(1), 62–64.

Holzinger, A., Baernthaler, M., Pammer, W., Katz, H., Bjelic-Radisic, V., & Ziefle, M. (2011). Investigating paper vs. screen in real-life hospital workflows: Performance contradicts perceived superiority of paper in the user experience. *International Journal of Human-Computer Studies, 69*(9), 563–570.

Hui, C. J., Jain, S., & Watson, C. I. (2019). Effects of sentence structure and word complexity on intelligibility in machine-to-human communications. *Computer Speech & Language, 58*, 203–215.

Jacobs, G. (1988). *The effect of pausing on listening comprehension*. ERIC.

Ji, C., Galvin, J. J., III, Xu, A., & Fu, Q. J. (2013). Effect of speaking rate on recognition of synthetic and natural speech by normal-hearing and cochlear implant listeners. *Ear and Hearing, 34*(3), 313.

Joint Commission. (2016). R3 report: Alarm system safety.

Ketko, A. K., Martin, C. M., Nemshak, M. A., Niedner, M., & Vartanian, R. J. (2015). Balancing the tension between hyperoxia prevention and alarm fatigue in the NICU. *Pediatrics, 136*(2), e496–e504.

Koomey, J., & Holdren, J. P. (2008). *Turning numbers into knowledge: Mastering the art of problem solving*. El Dorado Hills, CA: Analytics Press.

Lai, J., Wood, D., & Considine, M. (2000). The effect of task conditions on the comprehensibility of synthetic speech. In *Proceedings of the SIGCHI conference on human factors in computing systems* (pp. 321–328).

Li, S. Y., Tang, T. L., Hickling, A., Yau, S., Brecknell, B., & Sanderson, P. M. (2017). Spearcons for patient monitoring: Laboratory investigation comparing earcons and spearcons. *Human Factors, 59*(5), 765–781.

Liu, P., Zafar, F., & Badano, A. (2014). The effect of ambient illumination on handheld display image quality. *Journal of Digital Imaging, 27*(1), 12–18.

Ma, T. Y., Lin, C. Y., Hsu, S. W., Hu, C. W., & Hou, T. W. (2012, May). Automatic brightness control of the handheld device display with low illumination. In *2012 IEEE International Conference on Computer Science and Automation Engineering (CSAE)* (Vol. 2, pp. 382–385). Washington, DC: IEEE.

Mattys, S. L., Davis, M. H., Bradlow, A. R., & Scott, S. K. (2012). Speech recognition in adverse conditions: A review. *Language and Cognitive Processes, 27*(7–8), 953–978.

Meredith, C., & Edworthy, J. (1995). Are there too many alarms in the intensive care unit? An overview of the problems. *Journal of Advanced Nursing, 21*(1), 15–20.

National Research Council. (1997). *Tactical display for soldiers: Human factors considerations*. Washington, DC: National Academies Press.

Nicolau, S., Soler, L., Mutter, D., & Marescaux, J. (2011). Augmented reality in laparoscopic surgical oncology. *Surgical Oncology, 20*(3), 189–201.

Orphanides, A. K., & Nam, C. S. (2017). Touchscreen interfaces in context: A systematic review of research into touchscreens across settings, populations, and implementations. *Applied Ergonomics, 61*, 116–143.

Owren, M. J., & Cardillo, G. C. (2006). The relative roles of vowels and consonants in discriminating talker identity versus word meaning. *The Journal of the Acoustical Society of America, 119*(3), 1727–1739.

Paine, C. W., Goel, V. V., Ely, E., Stave, C. D., Stemler, S., Zander, M., et al. (2016). Systematic review of physiologic monitor alarm characteristics and pragmatic interventions to reduce alarm frequency. *Journal of Hospital Medicine, 11*(2), 136–144.

Patterson, R. D. (1989). Guidelines for the design of auditory warning sounds. *Proceedings of the Institute of Acoustics, 11*(5), 17–25.

Patterson, R. D. (1990). Auditory warning sounds in the work environment. *Philosophical Transactions of the Royal Society of London. B, Biological Sciences, 327*(1241), 485–492.

Phatak, R. (n.d.). *Dependence of dark spot growth on cathode/organic interfacial adhesion in Organic Light Emitting Devices (PDF)*. *UWSpace* (p. 21). Waterloo: University of Waterloo.

Plomp, R., & Levelt, W. J. M. (1965). Tonal consonance and critical bandwidth. *The Journal of the Acoustical Society of America, 38*(4), 548–560.

Privopoulos, E. P., Howard, C. Q., & Maddern, A. J. (2011). Acoustic characteristics for effective ambulance sirens. *Acoustics Australia, 39*(2), 43.

Qutub, H. O., & El-Said, K. F. (2009). Assessment of ambient noise levels in the intensive care unit of a university hospital. *Journal of Family & Community Medicine, 16*(2), 53.

Robinson, G. S., & Casali, J. G. (2000). Speech communications and signal detection in noise. *The Noise Manual, 5*, 567–600.

Rogers, M. L., Heath, W. B., Uy, C. C., Suresh, S., & Kaber, D. B. (2012). Effect of visual displays and locations on laparoscopic surgical training tasks. *Applied Ergonomics, 43*(4), 762–767.

Ruppel, H., De Vaux, L., Cooper, D., Kunz, S., Duller, B., & Funk, M. (2018). Testing physiologic monitor alarm customization software to reduce alarm rates and improve nurses' experience of alarms in a medical intensive care unit. *PLoS One, 13*(10), e0205901.

Sanders, M. S., & McCormick, E. J. (1993). *Human factors in engineering and design* (6th ed.). New York: McGraw-Hill.

Sanderson, P. (2006). Auditory displays in healthcare. *User Experience Magazine, 5*(3) Retrieved from https://uxpamagazine.org/auditory_displays_healthcare/.

Sanderson, P. M., Brecknell, B., Leong, S., Klueber, S., Wolf, E., Hickling, A., et al. (2019). Monitoring vital signs with time-compressed speech. *Journal of Experimental Psychology: Applied, 25*, 647.

Sanderson, P. M., Wee, A., & Lacherez, P. (2006). Learnability and discriminability of melodic medical equipment alarms. *Anaesthesia, 61*(2), 142–147.

Schmid, F., Goepfert, M. S., Kuhnt, D., Eichhorn, V., Diedrichs, S., Reichenspurner, H., et al. (2011). The wolf is crying in the operating room: Patient monitor and anesthesia workstation alarming patterns during cardiac surgery. *Anesthesia & Analgesia, 112*(1), 78–83.

Schneider, B. A., Daneman, M., & Murphy, D. R. (2005). Speech comprehension difficulties in older adults: Cognitive slowing or age-related changes in hearing? *Psychology and Aging, 20*(2), 261.

Sielhorst, T., Feuerstein, M., & Navab, N. (2008). Advanced medical displays: A literature review of augmented reality. *Journal of Display Technology, 4*(4), 451–467.

Stanton, N. A., & Edworthy, J. (1999). Auditory warning affordances. In N. Stanton & J. Edworthy (Eds.), *Human factors in auditory warnings* (pp. 113–127). Surrey, UK: Ashgate.

Suied, C., Susini, P., & McAdams, S. (2008). Evaluating warning sound urgency with reaction times. *Journal of Experimental Psychology: Applied, 14*(3), 201.

Tang, S. L., Kwoh, C. K., Teo, M. Y., Sing, N. W., & Ling, K. V. (1998). Augmented reality systems for medical applications. *IEEE Engineering in Medicine and Biology Magazine, 17*(3), 49–58.

van Det, M. J., Meijerink, W. J. H. J., Hoff, C., Totté, E. R., & Pierie, J. P. E. N. (2009). Optimal ergonomics for laparoscopic surgery in minimally invasive surgery suites: A review and guidelines. *Surgical Endoscopy, 23*(6), 1279–1285.

van Veelen, M. A., Jakimowicz, J. J., Goossens, R. H. M., Meijer, D. W., & Bussmann, J. B. J. (2002). Evaluation of the usability of two types of image display systems, during laparoscopy. *Surgical Endoscopy, 16*(4), 674e678.

van Veelen, M. A., Kazemier, G., Koopman, J., Goossens, R. H. M., & Meijer, D. W. (2002). Assessment of the ergonomically optimal operating surface height for laparoscopic surgery. *Journal of Laparoendoscopic & Advanced Surgical Techniques, 12*(1), 47–52.

Walker, B. N., Nance, A., & Lindsay, J. (2006). *Spearcons: Speech-based earcons improve navigation performance in auditory menus*. Georgia Institute of Technology. Atlanta.

Wester, M., Watts, O., & Henter, G. E. (2016). Evaluating comprehension of natural and synthetic conversational speech. In Proc. speech prosody (Vol. 8, pp. 736–740).

Wiklund, M. E., & Smith, E. A. (2001). Answering the call for harmonization of medical device alarms. *Medical Device and Diagnostic Industry, 23*(10), 118–125.

Winters, B. D., Cvach, M. M., Bonafide, C. P., Hu, X., Konkani, A., O'Connor, M. F., et al. (2018). Technological distractions (part 2): A summary of approaches to manage clinical alarms with intent to reduce alarm fatigue. *Critical Care Medicine, 46*(1), 130–137.

Withington, D. J., & Chapman, A. C. (1996). Where's that siren? *Science and Public Affairs, 2*, 59–61.

Yost, W. A. (2016). Sound source localization identification accuracy: Level and duration dependencies. *The Journal of the Acoustical Society of America, 140*(1), EL14–EL19.

Yost, W. A., & Zhong, X. (2014). Sound source localization identification accuracy: Bandwidth dependencies. *The Journal of the Acoustical Society of America, 136*(5), 2737–2746.

Ziefle, M. (1998). Effects of display resolution on visual performance. *Human Factors, 40*(4), 554–568.

Chapter 12
Human–Computer Interaction

12.1 Introduction

Computing is continuously changing. Just a short time ago we thought of a computer as a box sitting on a desktop with a monitor and printer attached. Prior to that, computers occupied giant climate-controlled rooms. In healthcare, before patient charts were available digitally, everything was done manually on paper. Now, patient charts, prescription orders, imaging orders, lab results, and scans can be updated or shared between healthcare professionals instantly.

Computers have become so ubiquitous that it is difficult to imagine life without them. In everyday life, we wear them on our wrists, embed them in our running shoes, and carry them in our pockets. These diverse computers and devices usually incorporate some type of user interface (UI), including displays and controls.

As discussed in Chap. 1, the user interface (UI) refers to all of the means in which a user interacts with a device. For example, a laptop UI includes a screen, keyboard, trackpad, buttons, and ports, as well as the operating system. However, the graphical user interface (GUI), pronounced "gooey," is a subset of the overall UI and refers to the graphics, menus, icons, and digital buttons on a display. In our laptop example, the GUI would include the operating system and all of the digital touchpoints a user needs to access a website or interact with an application.

Especially in healthcare, GUIs vary in size and complexity, from blood pressure cuffs to CT scanners. For each device, users need to form accurate mental models, navigate information structures, and comprehend nomenclature. The sheer number of devices, and variations among models, can make their operation complicated.

There are two fundamental types of GUIs for medical devices. The first includes devices like infusion pumps and glucose meters, where the GUI resides on the device itself. The second runs on all-purpose computers like desktops, laptops, or tablets. Examples of these include electronic health records (EHR), image viewing software, and appointment scheduling tools. Many of the design considerations for these are the same, but when they differ, they will be highlighted below.

© Springer Nature Switzerland AG 2021
R. J. Branaghan et al., *Humanizing Healthcare – Human Factors for Medical Device Design*, https://doi.org/10.1007/978-3-030-64433-8_12

Designers must accommodate display size, screen resolution, functionality, and the volume of information to devise a simple GUI for addressing user needs. Medical devices are mainly utilitarian: Users need to achieve a goal or outcome, and do not need whiz-bang features. These GUIs need to fit their environment of use and simplify work by fitting with other equipment. Certainly, the device should be attractive, and consistent with the manufacturer's brand image, but that takes a backseat to simple functionality.

Speaking of branding, manufacturers should still design their products with a consistent look and feel. On the other hand, some intrabrand consistency can contradict other manufacturers' GUIs. As a result, it is important to aim for consistency not only within a brand, but also between brands with similar tools, so users can easily move from one device to another.

Users are not trained on every device and model, so UIs should be easy to learn and easy to use immediately. True, a novice user might not need to know how to access all of the advanced functions, but they should be able to accomplish the basic functions right away. This can be accomplished by limiting the number of functions that a device can achieve, providing clear nomenclature such as labels and screen titles, and providing informative feedback.

Users may work with devices rarely or intermittently, and healthcare environments can be distracting with many interruptions. To accommodate these realities, consider providing step-by-step sequences that guide users through tasks. This is often called progressive disclosure (Springer & Whittaker, 2019), and enables users to make small decisions at each step rather than holding information in working memory.

12.2   User Experience (UX)

User experience (UX) refers to all the effects felt by a user when interacting with a device. A UX cannot be designed, but it instead must be felt by the user. The best that designers can do is facilitate positive UX through thoughtful design. Components of a UX include the product's usefulness, usability, and emotional impact (both during the actual interaction, and in memory after the interaction). Usefulness refers to the device's functionality or capabilities. It enables the user to achieve something with the device that they could not achieve without it. For example, a continuous blood glucose monitor enables the user to ascertain their blood sugar levels immediately without needing to prick their finger.

Usability is the practical component of user experience (UX), and includes the device's ease of learning, efficiency of use, and ability to be remembered (International Standards Organization, 2008; Nielsen, 1994). If usefulness provides the device's functionality, usability provides easy access to that functionality.

Emotional impact refers to the affective component of UX that influences peoples' feelings. This includes trust, confidence, pleasure, aesthetics, and appeal. It can even involve deeper factors such as self-identity, and pride of ownership. For example, Hassenzahl (2008) makes a distinction between do goals and be goals. "Do" goals are evaluated by how well they enable us to do things. These are evaluated by usability and usefulness measures—their "pragmatic quality." "Be" goals on the other hand, refer to our emotional needs, including self-identity, relating to others, and satisfaction. These are the goals that prompt people to tattoo Harley Davidson logos on their body. There is nothing functional about it, but it serves an important emotional purpose. Trust and confidence epitomize emotions needed in medical devices. These emotions increase better functionality, better usability, and better emotional impact: In short, they are impacted by the whole UX.

12.3 Design Principles

Based on scientific principles of perception and cognition outlined in other chapters, some important guidelines for good human–computer interaction are outlined in this section. They are not rote recipes or hard and fast rules. In fact, at times you will need to violate one principle just to achieve another. This highlights one of design's key principles: Design is a continuous exercise in tradeoffs. Each design decision carries advantages and disadvantages, so that making one good decision can create another disadvantage.

Support Mental Models

A mental model is an approximation or understanding of a device's purpose, and of how it works. The device's user interface (UI), including information architecture, screen layout, nomenclature, messaging, and interactions can either help people develop an appropriate mental model or hinder them from developing such a model. Mental models can be facilitated by making the organization of the device as visible, simple, and understandable as possible. People tend to understand the mental model better if the device has fewer elements and reflects how tasks actually get done. Norman (2013) points out two important qualities of any user interface: (1) discoverability—can I find what I need? and (2) understandability—do I comprehend how things work? These qualities feature prominently in the development of mental models as well. You can reinforce the development of mental models by making sure users can always answer the following questions, no matter what they are doing

- Where am I? Make sure users know where they are currently in the UI and where they have been in the structure of the system.
- What can I do? Make sure users understand the actions available to them at every point in the interaction. Include enough information, including descriptive words

and labels, for users to choose correct actions. Long labels are not necessarily bad, because adding words can add precision. Just avoid extraneous words.

- What has been done? Keep users aware of task progress, what has been done, and what is left to do. And make sure users know how to return to the home screen.
- What is the meaning of each UI component? Help users understand what system features exist and how they can be used in their work context. Use the language of the user rather than technical jargon.
- How do I do it? Provide informative cues to indicate how each interaction works. Use a positive tone of voice, avoiding violent, negative, or demeaning terms such as "illegal," "invalid," and "abort."

Allocate Tasks Wisely

People and computers each have strengths and weaknesses, and they tend to be complementary. For example, people are great at creative thought, planning, pattern recognition, and supervisory control. But they are terrible at making complex calculations, storing and retrieving information, and doing the exact same thing repeatedly. Computers, conversely, are great at making calculations and remembering information. They do not get tired, nor do they get bored. But their creative and supervisory abilities leave something to be desired.

Choosing which tasks and activities the human does and which the device does is called task allocation. We enable people to do what they do best and devices to do what they do best. Table 12.1 highlights some of these differences.

Table 12.1 Comparison of people and machine capabilities

Humans are better	Machines are better
Understanding based on sensation and perception; hearing, vision, touch, smell, taste and so on	Detecting energy that is outside peoples' ability to sense (e.g., infrared, gamma, ultraviolet, and radio waves)
Adapting to novel and unexpected situations	Routine, repetitive, tedious operations
Choosing alternative plans if the first plan does not work	Fast, consistent response
Planning future actions	Storage and retrieval of detailed information
Supervisory control	Making fast and accurate calculations
Using information from the external environment	Conduct several activities simultaneously
Intuitive thinking	Never gets fatigued
Requesting help from other people	
Flexibility	

Consistency

To reduce confusion and facilitate learning, the UI should be as consistent as possible. This includes:

- Consistent sequences of action for similar types of functions
- Consistent labeling of all items
- Consistent use of colors to indicate meaning
- Consistent use of graphics and icons
- Consistent button shapes
- Consistent placement of major areas for navigations and interaction

Minimize Memory Load

Working memory (WM) is one of two main bottlenecks in cognitive work; the other is attention. As a result, it is important to minimize the user's memory load while using your device. For example, enable users to recognize objects and items rather than needing to recall them. This in fact is the best reason to provide an interface that requires seeing and selecting rather than typing commands in a command line. Other ways to improve memory is by using familiar labels, icons, and interaction styles, as well as grouping-related items.

Provide Informative Feedback

People need to stay informed through consistent, timely, and informative feedback. This feedback should:

- Provide system status (e.g., is the product running? Is it processing my input?). This is especially important during long operations.
- Indicate how the task is progressing. It tells users what has been done and what still needs to be done.
- Tell users where they are in the information hierarchy, and how to find their way home.
- Solicit confirmation before potentially destructive actions. Confirmations should provide information about alternatives, and help users make informed responses to the dialogue box questions.

Feedback should include enough information for users to fully understand the interface output, based on their inputs. This will enable the user to feel confident that their actions worked, and to understand the reason when their actions do not work. Feedback should be easy to see, and placed within the user's focus of attention, where it is likely to be noticed. Status message lines often do not work if they are not overtly conspicuous. Pop-up messages next to the cursor will be far more noticeable.

Make Tasks Efficient

Make tasks as efficient and quick as possible by eliminating unnecessary steps and providing automatic completion where possible. It is also helpful to provide hints for formatting to facilitate data entry and decrease error (Fig. 12.1).

Utilitarian/Minimalist Design

One way to make tasks more efficient is to pursue a minimalist design. Minimalist design can be achieved by:

Fig. 12.1 Hints in the text entry fields facilitate data entry and decrease error

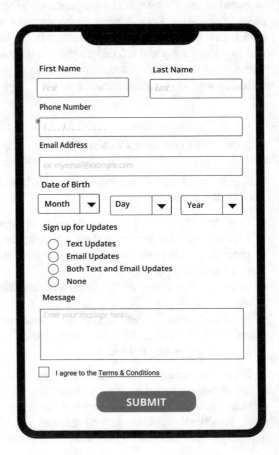

- Providing only the information necessary to help people conduct their tasks.
- Grouping objects according to their relatedness, especially if the objects are used to conduct the same tasks or functions.
- Breaking down complex instructions into simpler steps.

Error Prevention and Error Handling

Usability is influenced by how well a device handles errors, but the first priority is to prevent errors in the first place. This can be done by:

- Graying out inappropriate actions
- Enabling item selection rather than free form typing
- Providing automatic completion during text entry
- Providing intelligent defaults in selections

The second approach is to provide the ability to easily reverse actions. This includes providing the ability to "Go Back" or "Undo" multiple times. This provides several benefits. For example, it makes interaction with the product more resilient, so that a simple error is not catastrophic. Further it enables and even encourages people to explore the user interface, thus enabling them to learn it better. Finally, it provides more trust and confidence in the product. The third approach is to make error messages as helpful as possible. Some tips for helpful error message are:

- Be specific and precise about the nature of the error, what might have caused it, what to do about it, and how to get additional information.
- Make error messages specific, informative, positive in tone, and constructive.
- Offer constructive help for error recovery. Where possible, indicate what users should do to correct the problem.

12.4   Interaction Styles

Interaction styles determine how users work with, or communicate with, the device to input information. Different interaction methods are appropriate for different types of tasks and data, and often reside in the same UI. Some tips for choosing interaction styles are

- Group similar interaction styles together. For example, if there are several form fill-in interactions, consider putting them together so the user does not need to move frequently from keyboard input to mouse input or screen input.
- Aim for fewer interactions to increase productivity and decrease error.

Form Fill-in

Many devices require substantial data entry of patient name, physician notes, diagnostic codes and so on. Form fill-in interfaces, which resemble simple paper forms, are appropriate for this. Advantages include familiarity; that is, little instruction is required to enter data correctly. Additionally, it works well for keyboard input and use of TAB, ENTER and RETURN keys to navigate from one field to another. In Fig. 12.2, the screen on the left shows several form fill-in interactions grouped together so the user does not need to move back and forth between keyboard input and mouse input or screen input. On the other hand, form fill-in is prone to errors. Below are recommendations to reduce the likelihood of those errors.

- Place data fields in related groups.
- Provide meaningful labels.
- Provide visual cues indicating data entry fields versus other types of text on the screen. Provide these cues even if the fields are not editable.
- Make the length of the data entry field roughly the same as the longest expected text. This provides a cue to the user about what type of information gets entered there.

Fig. 12.2 Grouping form fill-in

- Left justify text in data fields.
- Right justify numbers in data field (Fig. 12.3).

- Align numbers by the decimal point in text fields.
- Where possible limit the number of sizes of the data fields. For example, choose two data field lengths; one that accommodates the shortest text and one that accommodates the longest text. This simplifies the appearance of the UI.
- Provide guidance about the format of the input. For example, a telephone number might have separate blocks for the international code, the area code, the exchange and the number. This helps people enter information correctly.
- Left justify labels on data fields, and place them either to the left or directly above the data entry field.
- Provide units of measure (e.g., "psi," "mL,"). Be careful not to mix English and metric units, unless each medical convention dictates that (AAMI:HE75), and avoid using acronyms unless they are well known.
- When possible, provide sensibility checks on the data to make sure the values are correct and within a safe range. Where appropriate, enable users to override the limits as necessary. This could be done with a confirmation dialog box that enables the user to cancel the action.

Menus

In menu systems, users see and choose items from categorized lists, pull down menus, toolbars, menu bars, pop-up menus and so on. These are familiar and are especially appropriate for novice or intermittent users. They are also helpful for users with poor typing skills or with no access to a keyboard. To improve menu system usage:

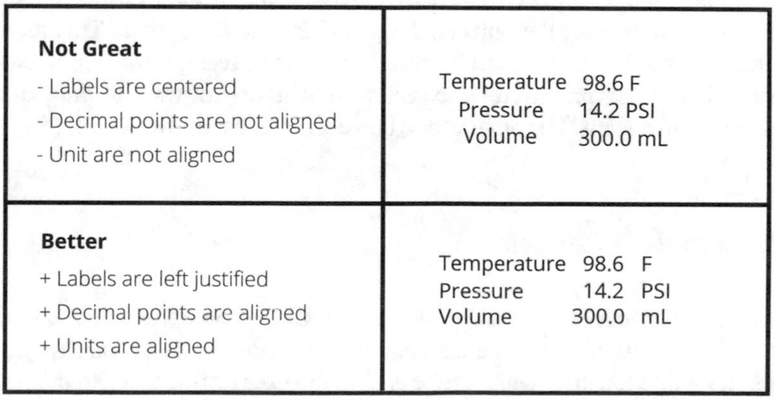

Not Great	
- Labels are centered - Decimal points are not aligned - Unit are not aligned	Temperature 98.6 F Pressure 14.2 PSI Volume 300.0 mL
Better	
+ Labels are left justified + Decimal points are aligned + Units are aligned	Temperature 98.6 F Pressure 14.2 PSI Volume 300.0 mL

Fig. 12.3 Comparison of labels and decimal alignment

- Limit the number of menu items to about six. This reduces search time (Lee, Wickens, Liu, & Boyle, 2017). This number can be increased, however, by separating items with a dividing line. When doing this, users simply need first to identify the appropriate chunk (delineated by a line) and then choose among items in the chunk.
- Gray out inactive menu items.
- Provide concise but complete menu labels.

Direct Manipulation

Direct manipulation provides a natural and realistic experience to device users by enabling them to act directly on visible objects on the screen (Hutchins & Holland, 1986). It uses interactions like drag and drop, as well as metaphors like desktops and folders to represent file systems. Direct manipulation is powerful because it matches the user's mental model of the task itself. As a result, they can be easier to remember than other interaction styles. They are especially good for intermittent users who are not fast typists. To improve direct manipulation:

- Use well-designed, understandable and consistent icons.
- Design familiar icons with concrete (visualizable) meanings.
- Label your icons, when possible.

Command Line

Command line systems require users to type expressions at a prompt. These expressions typically include a command, an object to operate on, and perhaps a set of modifiers (arguments) to the command. This enables extremely experienced users to access functionality quickly. The disadvantage is that it forces users to recall the names of the commands, the objects, the modifiers and the syntax. This requires a great deal of knowledge retrieved by recall rather than recognition. Command line interactions are often preferred by expert users who do not want to drag and drop items for repeated steps. For other users, however, it does not work well (Fig. 12.4).

Gestures and Multitouch

In most direct manipulation and touchscreen UIs, people use only one finger. Multitouch interactions use more than one finger to read gestures, such as pinching and spreading, on touch screens. These enable the user to zoom in and out, to navigate from one screen to another, and to scroll up and down. A similar type of

Fig. 12.4 Command line ("acp command" by xmodulo is licensed under CC BY 2.0)

approach enables gestures to work without even touching the screen. This approach uses motion tracking to capture hand, arm and body gestures designed to provide a more natural and intuitive way of interacting with a device (Shneiderman, 1998). Lee et al. (2017) suggest these tips for good multitouch/gesture design:

- Provide cues to signal which gestures can be used and what the function is
- Make gestures easy to perform
- Minimize fatigue and stress associated with repetitive use
- Ensure gestures are easily differentiated by the computer

Most UIs combine interaction styles to get the job done. The key is to match the interaction style to the task being conducted, the environment of use, and the user characteristics. As a general rule, interaction styles like menus, question-and-answer, and form fill in, are preferred to the command line for most intermittent users.

Dialog Boxes

Dialog boxes mix these interaction styles. For example, the dialog box below is used to alert clinicians prescribing the drug Droperidol (Fig. 12.5).

12.5   Information Architecture

Information architecture focuses on making information findable and understandable (Rosenfeld, Morville, & Arango, 2015). Good information architecture reduces the likelihood of information overload (Toffler, 1970), and increases the ratio between signal and noise in our user interfaces. That is, people can find the information and functionality they need without wading through unnecessary material. This requires structuring, organizing, and labeling information and components of the user interface, so that people can find, understand, and manage this information.

Organizing items into meaningful structures results in faster selection time and higher user satisfaction. Since healthcare providers (HCPs) are often under time pressure, it is important to present critical information quickly and legibly (AAMI HE75:2009/(R)2018). Further, this critical information might include not only

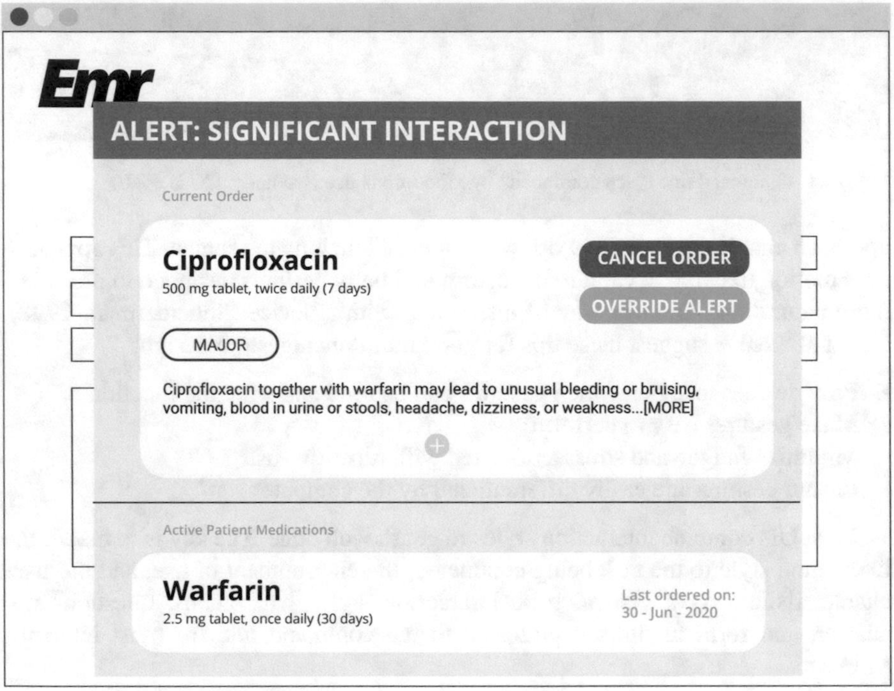

Fig. 12.5 Dialog box indicating significant interaction

current measures but time-based trends. For example, the HCP might need to know how blood pressure has changed over the past 5 min. In addition, the most important information should be the most prominent.

Depth vs. Breadth

There is often a tradeoff between the depth and breadth of a menu system. A deep menu structure might only provide three options at the top layer, but require the user to traverse several layers to complete their task. Shallow menu systems might offer options at the top level, but require traversing fewer levels. Generally, the shallow structure is preferable; it affords a better bird's eye view of the whole system, and improves the user's understanding (mental model) of the system. It also reduces the likelihood of burying functionality that people need. To design a shallow menu structure correctly though, the top level has to be well organized, with options arranged according to urgency, importance, frequency of use and then grouped according to relatedness.

Serial Choice

The organization of screens and information should match the tasks people are conducting. For example, a linear or serial choice (progressive disclosure) UI walks the user through a series of screens in a fixed order. Each screen usually has just a few choices. This reduces the likelihood of skipping steps, especially if the user gets interrupted. This is helpful for intermittent users, new users, and users with little training. It also reduces demands on WM.

Branching

A branching UI (AAMI, 2018) presents a greater variety of options than serial choice. These are more appropriate for experienced people who use that system frequently. An example of this can be seen in many Electronic Medical Record (EMR) systems which allow healthcare professionals to access a patient's medical information digitally. There are many EMR systems out there, and their designs vary. However, in general, EMRs allow HCPs to track various types of patient information over time, including a patient's history, current status, lab work, imaging orders, prescriptions, and previous procedures. EMRs also allow healthcare professionals to add patient data, write notes, and place orders for labs, imaging, and prescriptions. In short, there is a lot one can do with an EMR. Therefore, EMR systems that provide quick access to the most important functions speed up the interaction and make the system feel more intuitive. For example, Fig. 12.6 shows two paths for how a user can access important lab results. The path outlined by the black arrows requires multiple button presses, whereas, the path outlined in red requires only one.

When designing such a structure, it can be helpful to map out and visualize the structure or hierarchy of computer screens. Often this simple task will give you an intuitive feel for whether the system is too complicated.

Networked

Finally, a networked system is like a web, with many screens connected to many others. In fact, the World Wide Web itself is a networked structure rather than linear or branching. It enables people to follow alternative paths to what they need. It feels more like a tool box rather than just a single tool, making it most appropriate for experts using specialized systems who need to perform several tasks at a time to be efficient (Fig. 12.7).

Fig. 12.6 Two paths to
access lab results; the
black arrows require
multiple button presses, the
path in red requires only
one

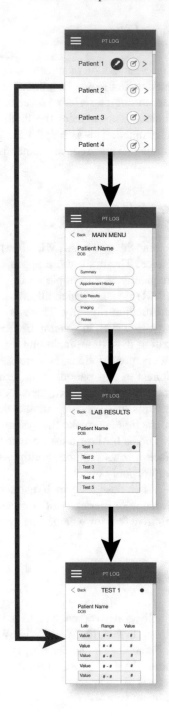

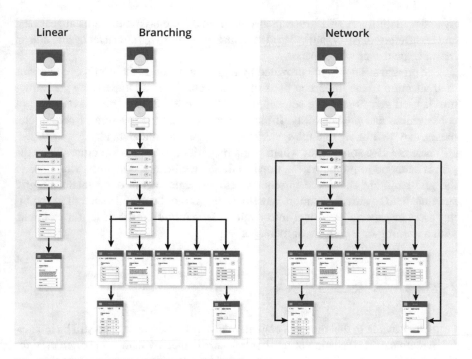

Fig. 12.7 Linear, branching, and networked architectures

12.6 Screen Layout

The process of arranging items on a screen influences the user's ability to locate items, understand the meaning of the items, understand how items relate to one another, and frankly provide an attractive, well organized, professional device. As indicated in Chap. 7, people naturally and spontaneously categorize things they encounter. Default categorization schemes tend to take on hierarchical qualities; and reflecting these logical categories in UIs is important. The most critical items should be placed in the most conspicuous position on the screen. In large screens such as tablets, laptop and desktops, this tends to be toward the upper left. On smaller screens, we tend to focus more on the middle. This process requires substantial skill and iteration, but the sections below provide a few tips.

Grid

A grid is a structure used to break-up an area into units. Grids of various shapes and sizes can be used to create a systematic method for organizing and arranging content. A grid helps provide consistent and predictable alignment of items. For text, alignment is particularly important because a structured layout can increase legibility and readability. A grid structure also helps make the layout of items appear neat and less cluttered. An organized grid structure also improves consistency and aids in the strategic use of white space.

The organization scheme provided by a grid helps guide the user's eye and thus their attention to the important parts of the screen. When an underlying grid structure is used across multiple screens within the same UI, this provides a sense of cohesiveness and predictability. If the user can predict the location of certain elements, they will likely find the system easier to use and understand.

There are also some cases where items may violate the grid structure. Since the grid is intended to promote neat, organized, and well-aligned layouts, violations to the grid structure should be infrequent and strategic. Items that violate the grid structure tend to attract attention, therefore, they might be useful as a tool for pointing users to especially critical information. This might be the case, for example, with a warning or pop-up error message.

Columns

Grids are flexible in that they can contain many, or few, columns and still be impactful. A simple grid consisting of one column surrounded by margins is often used for simple text, such as a legal document printed on letter sized paper. A complex grid layout consisting of multiple columns is often used for more complex content. For instance, think about a newspaper consisting of many different articles, images of varying size, and advertisements, many columns are used to provide structure.

The orientation of columns within the grid guide the structure of visual information. Tall and narrow columns of information are often easier for people to read than short wide rows. This may be due to the fact that many writing systems are written left-to-right and the distance required for the eye to travel is shorter when reading tall and narrow columns of information. For instance, in Fig. 12.8 (top right) the text consists of four groups of text. Group 1 consists of the same text as group 3, and group 2 the same as group 4. Groups 1 and 2 are tall and narrow in shape, while groups 2 and 4 are long and narrow.

A grid can also be leveraged to emphasize information hierarchy. Content can be contained by a single grid unit or it can span several columns. Space, or graphical real estate, can be used to represent importance. For example, a line of text spanning all columns at the top of the grid will often stand out as the title if the rest of the text conforms to a smaller grid.

Importantly, columns break up the grid into smaller units. However, all of the units do not need to be filled with content. In fact, it is equally as important to fill the grid with blank space.

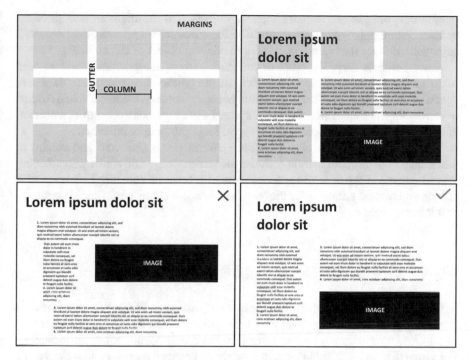

Fig. 12.8 A three column grid structure (top left). An example of how content fits into the grid structure (top right). When content does not confirm, it appears disorganized (bottom left), when content does conform, it has more structure (bottom right)

Blank Space

Adequate blank space makes screens easier to read, more organized, and more attractive than needlessly dense screens. Use white space to separate blocks of information. Information on the screen should be distributed so that neither the overall screen nor individual sections of the screen appear dense. This helps avoid any one part of the screen looking too dense or too sparse (AAMI, 2018).

Gutters, Margins, and Padding

Content that is too densely displayed can be difficult to read. Gutter and margins are separation elements that allow a page to "breathe," so to speak. The vertical and horizontal lines that create can also be used to lead the user through the information, effectively "guiding" them through the content.

With influence from print design, gutters and margins are intentional spaces used to organize elements on a page. Margins are used to create gutters between columns. Padding can be used to create even more blank space between objects and text. A margin falls outside two or more adjacent objects, it is the space around all of the content. While padding is used to describe the space within an individual object, it essentially defines the bounds of content inside that object as shown in Fig. 12.9.

Since columns form the basis of the grid, gutters, margins and padding help maintain the structure of the content within the grid system and create the blank space. Blank space provides separation between content elements and prevents overcrowding. All elements of the grid system work together to construct a predictable layout which in turn means users spend less time finding information.

You can improve readability by adding several pixels (in width and height) of white space as a gutter or margin between the screen's edge and content. Readability is also enhanced when white space separated blocks of content. This reinforces natural and logical grouping for the user. In essence it chunks information better so that it can be held in WM.

Grouping

In Chap. 5 we discussed the Gestalt principle of proximity. Placing content elements together in close proximity creates a relationship between those elements. Because of the close proximity it is assumed that those elements contain related information. For visual displays, related information might be grouped together and separated by blank space to put distance between different groups of information making it easier for users to navigate the page.

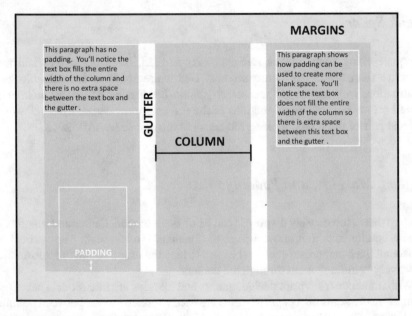

Fig. 12.9 Gutters, margins, and padding

12.7 Legibility

Misreading a display could be tragic in healthcare, leading to errors in diagnosis and treatment. As a result, legibility is paramount. This requires strategically selecting font style, font size, contrast, resolution, and other factors related to visibility and legibility. Sans serif fonts, those that do not have extra details like hats or feet on them, tend to be easier to read on computerized displays. These include Arial, Helvetica and a variety of others. To improve legibility, it's also important to use scalable fonts, with a moderate degree of antialiasing to provide a smoother appearance.

Text Size

Text size was described in Chap. 5, but as a reminder a few suggestions are provided here. Choosing the appropriate text size depends on the user's distance from the display. Information on a watch-based interface will be smaller than on a screen viewed from across the room. The watch is mere inches from your face. A good rule of thumb found in (AAMI HE75:2009/(R)2018) is for the height of critical information to be 1/150th of the viewing distance, and for important but noncritical information to be about half that height. AAMI's HE 75 points out that a key value on a patient monitor viewed from 3 feet away should then be at least 1/8-in. high (a 9-point font: 1 point equals 1/72.27 in.).

All Capitals

Avoid using all capital letters in your text, especially in long strings of text. Capital letters can draw attention to important information in short strings, but take up more space and take longer to read in long strings. Also, from a cultural perspective, it seems like yelling.

Contrast

Legibility relies to a large degree on sufficient contrast between foreground and background. Further, text with especially high contrast suggests importance because it attracts attention so easily. As a result, especially important or critical information should use high contrast, particularly if viewed from far distances. Black text on a white background produces the greatest contrast and usually the easiest readability. Less critical information could use slightly less contrast however. And, as mentioned in Chap. 5 avoid saturated blue and red together, since it produces chromostereopsis (Fig. 12.10).

Ensure that lines of text are separated by at least one or two pixels (AAMI: HE75) between the letters with ascenders and the letters with descenders. Often providing additional spaces between lines (leading) provides a more legible and less crowded appearance.

Text Justification

Generally, text should be left justified with a ragged right margin. People can read faster, and find information faster, if they know where to expect to start scanning. English text is read left to right, so a left justified consistent place to start reading makes reading more efficient. Though full justification (block justified) provides a similar left line to begin reading, it tends to space out words in inconsistent and unpredictable ways and can result in large spaces between words (Fig. 12.11).

12.8   Color

Color conveys meaning, helps separate blocks of information, and provides visual interest. It can have numerous other benefits, including:

- Making displays more interesting
- Eliciting emotional responses
- Draw attention to important items
- Organizing displays

The key is to use color to convey meaningful information, but to avoid using color as the sole means of conveying information. Instead, use redundant coding. For example, an icon indicating danger should not just be an amorphous red blob, but should provide a label that reads "Danger." It also might have an exclamation point in the middle of the icon, and perhaps the icon would be triangular in shape. It would look like Fig. 12.12. Each code, the red color, the "Danger" label, the

Fig. 12.10 Black text on a white background creates high contrast and makes the text easier to read

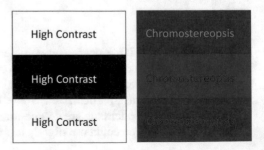

Left Justified Text	Block Justified Text
Generally, text should be left justified with a ragged right margin. People can read faster, and find information faster, if they know where to expect to start scanning. English text is read left to right, so a left justified consistent place to start reading makes reading more efficient. Though full justification (block justified) provides a similar left line to begin reading, it tends to space out words in inconsistent and unpredictable ways and can result in large spaces between words.	Generally, text should be left justified with a ragged right margin. People can read faster, and find information faster, if they know where to expect to start scanning. English text is read left to right, so a left justified consistent place to start reading makes reading more efficient. Though full justification (block justified) provides a similar left line to begin reading, it tends to space out words in inconsistent and unpredictable ways and can result in large spaces between words.

Fig. 12.11 Comparison of left justified and block justified texts

Fig. 12.12 Icon indicating danger

triangular shape, and the exclamation point reinforce the main message. Further, if your user is color deficient, they will not need to rely solely on color to get this important piece of information. In fact, this practice is required by Sect. 508 of the Americans with Disabilities Act (1990).

Color Guidelines

Many effective UI designers create simple, clean aesthetics by using color sparingly. They limit their designs to a few complimentary colors while focusing most on legibility. Here are a few tips for using color effectively:

- Design for monochrome first
- Use color conservatively, by limiting the number and amount of colors
- Use color coding to support tasks
- Consider the needs of color-deficient users
- Be consistent in color coding
- Adhere to common expectations about color codes
- Be alert to problems with color pairings
- Use color changes to indicate status changes

Table 12.2 describes some conventional color meanings in the United States. It's important to remember that colors can have different meanings in different cultures.

Table 12.2 Color codes for medical applications in the United States (Association for the Advancement of Medical Instrumentation, AAMI, 2009)

Color	Meaning
White	Conventional practice: Primary information on a black background
Red	Alarm condition: High priority (medical and nonmedical) Hazard: Danger (an associated hazard will be deadly or will cause property damage) Conventional practice: Arterial (oxygenated) blood pressure Conventional practice: OFF, power OFF Conventional practice: Stop, emergency stop Conventional practice: Fault condition Conventional practice: Energy being delivered (e.g., laser firing) Conventional practice: Stay clear Association: Warm, hot
Orange (amber)	Alarm condition: Medium priority (nonmedical) Hazard: Warning (an associated hazard could be deadly or injurious or cause property damage)
Yellow	Alarm condition: Low priority (nonmedical) Alarm condition: Medium or low priority (medical) Hazards: Caution (an associated hazard could be injurious or cause property damage gas: Air Conventional practice: Slow Conventional practice: Pulmonary artery blood pressure Association: Warm, sunny
Green	Conventional practice: ON, power ON Conventional practice: Go or continue Conventional practice: All OK (normal) Conventional practice: Ready (available for use) Conventional practice: Could be coded to other physiological variables Gas: Oxygen Association: Good Association: Environmentally friendly
Blue	Conventional practice: Secondary information on a white background association: Frozen, cold Conventional practice: Deoxygenated lungs or blood Conventional practice: Central venous (deoxygenated) blood pressure gas: Nitrous oxide
Cyan	Alarm condition: Low priority (medical)
Gray	Conventional practice: Unavailable or nonapplicable option or information Gas: Carbon dioxide
Brown	Gas: Helium
Black	Conventional practice: Primary information on a white background Gas: Nitrogen

Data Visualization and Graphics

Graphics can convey a great deal of information in a small space. They can demonstrate how something should be done, and provide examples. It is important, however, to test graphics for their understandability, and to make sure they convey information better, faster and potentially more accurately than a textual description.

Table Design

In tables, people read columns of data easier than long rows. Shading alternate rows (Fig. 12.13) helps the user keep track of where they are when scanning long rows of data.

Visualizing and making sense of trends in data are critical to diagnostic and status monitoring activities. As a result, many devices, including patient monitors, therapeutic devices, and diagnostic devices, show this type of data. Trend displays show how measures change over time. It is critical that users understand how to read these data and what they mean. Users need to interpret the data, discriminate between historical data and current data, recognize the units of measure, distinguish older values from newer ones, determine the units of measure, and read the appropriate level of detail.

Fig. 12.13 Striped rows to improve reading

To this end, take care to present trend data at the most appropriate time resolution. For example, EEG data might rely on very high-resolution temporal data, whereas patient temperature over time does not need to be high resolution at all. It may be helpful to enable the user to adjust the time frame to meet their needs. If this is done, the time frame should be clearly indicated.

Small Displays

Small devices (Fig. 12.14) have very focused functionalities and few selectable areas. Because of this, discoverability can be an issue.

Display technology continues to allow progressively smaller screen sizes without compromising resolution, contrast, or brightness. The popularity of fitness watches and other types of wearables has championed much of this change in the consumer world.

Design tips for small displays include:

- Simplify the design, "less is more"
- Reduce or eliminate data entry
- Consider use frequency and importance
- Make clear what is selectable and what is not
- Leave room for scroll and swipe gestures to avoid inadvertent actions
- Consider relegating less important functions to other platforms such as desktop computers

Fig. 12.14 Small LCD and OLED displays. (Image by Vectorizus/shutterstock.com)

Resources

- Hartson, R., & Pyla, P. S. (2018). *The UX book: Agile UX design for a quality user experience*. Burlington, MA: Morgan Kaufmann.
- Johnson, J. (2013). *Designing with the mind in mind: Simple guide to understanding user interface design guidelines*. Amsterdam: Elsevier.
- Shneiderman, B., Plaisant, C., Cohen, M., Jacobs, S., Elmqvist, N., & Diakopoulos, N. (2016). *Designing the user interface: Strategies for effective human-computer interaction*. New York: Pearson.
- Weinschenk, S. (2011). *100 things every designer needs to know about people*. London: Pearson Education.

References

Americans with Disabilities Act of 1990, Pub. L. No. 101-336, § 508, 104 Stat. 328. (1990).

Association for the Advancement of Medical Instrumentation. (2018). *ANSI/AAMI HE75:2009/(R)2018 human factors engineering—Design of medical devices*. Fairfax, VA: Association for the Advancement of Medical Instrumentation.

Hassenzahl, M. (2008, September). User experience (UX) towards an experiential perspective on product quality. In *Proceedings of the 20th conference on l'Interaction Homme-Machine* (pp. 11–15).

Hutchins, E. L., & Holland, J. D. (1986). Direct manipulation interfaces. In D. A. Norman & S. Draper (Eds.), *User centered systems design*. LEA.

International Standards Organization. (2008). ISO 9241-303:2008. Ergonomics of human-system interaction—Part 303: Requirements for electronic visual displays. Available at.

Lee, J. D., Wickens, C. D., Liu, Y., & Boyle, L. N. (2017). *Designing for people: An introduction to human factors engineering*. Scotts Valley, CA: CreateSpace.

Nielsen, J. (1994). *Usability engineering*. Burlington, MA: Morgan Kaufmann.

Norman, D. (2013). *The design of everyday things*. New York: Basic Books.

Rosenfeld, L., Morville, P., & Arango, J. (2015). *Information architecture: For the web and beyond*. Sebastopol, CA: O'Reilly.

Shneiderman, B. (1998). *Designing the user interface* (3rd ed.). Boston, MA: Addison-Wesley.

Springer, A., & Whittaker, S. (2019). Progressive disclosure: Empirically motivated approaches to designing effective transparency. In *Proceedings of the 24th international conference on intelligent user interfaces* (pp. 107–120).

Toffler, A. (1970). *Future shock*. Manhattan: Random House.

Chapter 13
Designing Instructions for Use(rs)

13.1 Definitions

Instructions for use (IFU) are essentially the user manual for a medical device. The term "IFU" is sometimes used interchangeably with "labeling"; however, labeling encompasses more than just the IFU, and can include other interfaces such as packaging, product inserts, or on-device labels. IFUs provide operational information such as:

- How to use the device
- Why and when the device should be used
- Technical specifications
- Maintenance requirements
- Troubleshooting information
- Cautionary or risk-related information

This chapter discusses the challenges of designing effective IFUs and the reasons for those challenges. It then discusses the considerable consequences of writing poor IFUs, including lost efficiency and increased likelihood of use-error. Finally, we provide guidance for designing IFUs that facilitate safe and effective use of your medical devices using the Find, Comprehend, and Apply framework.

13.2 Do We Need Instructions for Use?

You might wonder whether we even need IFU in the first place. Can't we make the device so easy to use that we render the IFU unnecessary? This would be ideal, but for most devices, impractical. Generally, users would need to know how to assemble, use, maintain, clean, store, dispose of, or reuse the device. This is a lot of information to convey through the design of the device itself. Further, the technological

© Springer Nature Switzerland AG 2021
R. J. Branaghan et al., *Humanizing Healthcare – Human Factors for Medical
Device Design*, https://doi.org/10.1007/978-3-030-64433-8_13

trend is away from simplistic medical devices and toward simultaneously increasing functionality and decreasing device size. So, it is challenging, if not impossible, to eliminate the IFU. Although we strive to make every aspect of the device intuitive, we rarely completely achieve that goal. Instead, we usually need to rely on IFUs to convey use-related information.

Alternatively, you might ask if we could simply provide training through sales reps, in-services, and online training. These approaches may also be necessary but are not sufficient. Two of the authors recall attending an in-service on a new device attended by half of a hospital department. One or two people at the front of the room could actually see what was going on and got an opportunity for hands-on practice with the device. The vast majority of those in attendance, including the half of the department not in attendance, could not have used that new device after the in-service. Further, this type of training is not realistic for medical devices developed for lay users. For example, manufacturers cannot provide in-person training to all potential users of an over-the-counter blood pressure cuff.

Undoubtedly online instruction has merit, but sometimes people just need a small piece of key information, and it is inefficient to sit through an online training session to find it. In fact, there is no guarantee that the information sought will even be covered in that video. Instead IFUs with a good index or search feature would do quite nicely in this situation.

Finally, even though people commonly fail to read instructions, manufacturers still have a responsibility to design effective documentation to ensure safe use of their device. People who fail to refer to instructions may overlook important warnings, or start using a device before they've read the entirety of the IFU. These factors can result in unsafe device use. Further, when something inevitably goes wrong— accidents and litigation happen—the medical device manufacturer's first line of defense is often the IFU, with adequate labeling and warnings (Hyman, 2014).

13.3   No Respect

Despite their necessity, we have noticed that, like the late comedian Rodney Dangerfield, IFUs get no respect. Little attention is paid to IFU design, because of an implicit—though misguided—belief that design begins and ends with the device itself.

IFU as User Interface

Often, people do not even consider IFU to be part of the user interface (UI). They think of the UI as the controls and displays on the device itself. But they are just plain wrong; recall from Sect. 1.3 that the UI is everything the user comes into contact with physically, perceptually, and conceptually. This includes packaging, website, customer support, and also, the IFU. The simple fact that the IFU is part of the UI—that people interact with them—means that they are important and deserve good design.

IFUs as Checking a Box

Sometimes IFUs are written solely to "check a box." For example, the FDA mandates that labeling must be provided with certain types of medical devices (21 CFR Part 801, n.d.). When manufacturers simply create IFUs to "check a box," they miss opportunities to ensure that content and design match end users' abilities and needs, and support their limitations.

IFUs as an Afterthought

Oftentimes, IFU are an afterthought and not developed in tandem with the medical device or written by personnel with instructional design training. As such, they do not receive the same timely attention as the device itself. Manufacturers may even write the IFU as the last part of the design, after they have already missed a few delivery deadlines. By the time writers can work on the IFU, their deliverable is already overdue, and they are faced with the task of eliciting information from stressed out, sleep deprived, engineers who are simply trying to meet their deadlines. As you can imagine, explaining, in great detail, how the design works to a technical writer is not at the top of the engineer's priorities.

Worse yet, in many cases, the IFU are not even written by those with experience in instructional design, but are cobbled together by engineering, quality assurance, regulatory, and legal teams who have little understanding of the end user. This results in disjointed and complex instructions that do not correspond well with use of the device. By the time the IFU is completed, they are little more than a parking lot for mitigating bad design with cautions, warnings, and disclaimers. Like we said—no respect.

But Nobody Uses the IFU Anyway

Many people say that nobody uses the IFU anyway, so why have them at all? Sadly, there is some truth to that statement. Some IFUs are so bad that people cannot use them. On the other hand, we have to wonder whether people refuse to use an IFU or do they refuse to use a poorly designed IFU? The literature would suggest the former, that is, people are not likely to read instructions in general. However, we argue that if people are going to use instructions, then people would make better use of an

IFU, and be more successful using the device, if the IFU is designed to be helpful. If anything, we know that people may not use an IFU until they really need them. After all, they are not the kind of document you read for fun; you read them as a last resort, because you are stuck, or something is going wrong. At that moment IFUs are essential and thus, so too is their design.

Rewriting IFU into Standard Operating Procedures (SOP)

Because IFUs often fail to meet user needs, medical facilities often translate them into a more "usable document" to guide their employees on how to use a device—a standard operating procedure (SOP; Hildebrand et al., 2010; Sinocruz, Hildebrand, Neuman, & Branaghan, 2011). We worked with a hospital that was thrilled to whittle down a 124-page operator manual for a device to a 33-page SOP. Of course, we applauded their success, but the IFU should have been strong enough so that they did not need to do this. And, the result was still an SOP that was 33 pages long.

Of concern is that, when facilities translate IFUs, the content can be compromised. Critical information may be omitted representing an obvious concern for patient safety. Further, tremendous time is wasted on this process. In a fast-paced, high risk environment such as a hospital, workers should not waste time creating workarounds for devices and IFUs that were designed poorly by manufacturers.

Ease of Use vs. Regulatory Standards

A further challenge is to design an IFU that can be easily used while still meeting regulatory and, as applicable, internal manufacturer requirements for document content and control. While there are some required regulatory standards and guidance documents governing the content that must be included in IFUs, there are no required standards for how information should be presented such as format, platform, style, and length. As a result, IFUs from one device to the next, even within the same device category and from the same manufacturer, can be wildly different.

13.4   Developing Instructions for Use(rs)

To ensure that IFUs get the respect they deserve, we first reiterate that the IFU is part of the medical device interface and thus deserves the same attention to design that the actual device would command. In fact, we recommend an amendment to the term "instructions for use," as it is a bit of a misnomer, and instead propose manufacturers think of the term "instructions for use(rs)" to help guide their IFU development journey.

Second, we reiterate utilizing human-centered design (HCD). Throughout this book, we have advocated a HCD approach with three tenets: (1) an early and constant focus on the users and their tasks, (2) data to help guide our design decisions, and (3) iteration. We espouse this approach because it works for all interfaces regardless of whether you are developing a physical device, or an IFU.

Start Designing Early

Two basic ingredients are needed to write an effective instructional device: time and information (Robinson, 2009). IFU design and development take time; that time needs to start early in the process of interface development. IFUs and the devices they correspond to should be designed in tandem, not one after the other. IFU design needs information so the HCD approach should be taken to ensure all necessary design inputs have been identified. Research has shown that "basic decisions at the beginning of the design lifecycle matter" and seemingly simple design choices (e.g., booklet vs. foldout format) can actually have an impact on the user experience (Williamson, 2019).

Develop User Profile

Instructions must be designed for a range of users, uses, and environments. Is your device used by healthcare providers? Patients with rare diseases? Caregivers? General lay users? Start by identifying all end users (primary users, secondary users, rare users such as maintenance technicians, etc.) and develop profiles by understanding the scope of their needs, capabilities, and limitations (see Sect. 2.16 for more information). In addition to identifying these user group profiles, consider how the IFU should accommodate different levels within those groups of users including:

- Beginners
- Experts
- Infrequent users
- Education level
- Literacy (important for lay user devices)

Identifying these inputs will begin to provide design requirements for your IFU. For example, special attention should always be given to devices developed for lay users. The FDA recommends that content created for lay users should accommodate at least a fourth to fifth grade reading level and no higher than an eighth grade reading level (FDA, 2001). Accordingly, instructions intended for lay users need to be written with simple, non-technical language that is easy to understand.

Develop Environmental Profile

The environment of use can greatly impact a user's interaction with the IFU. Developers need to consider how use is impacted by environmental elements such as lighting, temperature and humidity, space, noise or distractions, and access to technical support. Oftentimes, IFU formats are not conducive to their use-environment. For example, paper-based instructions for technicians who are reprocessing endoscopes are not useful because they work in wet environments and thus are not likely to reference the IFU at the point of use. Similarly, a small, paper-based booklet-style IFU may not be a good design for patients with rheumatoid arthritis, who have difficulty with manual dexterity, and are trying to learn how to administer a self-injection while at home alone. An IFU may be designed well with easy to find and understand information, but if the form factor doesn't support easy use in a given environment, then it doesn't really matter.

Further, while paper-based instructions are the standard format, this format is incredibly conducive to sitting on a shelf or getting lost in a stack of bills and papers at home, neither of which are ideal usability scenarios. IFU design needs to consider how the actual document form and its content fit into the use-environment.

Consider the User's Tasks

Consider using a task analysis to identify the individual steps and tasks that must be accomplished to use the device as intended. Chapter 2 outlines the methods for performing a task analysis. Then, consider the larger use-scenarios for the IFU. Will the user need to rely on the IFU during each use of the device? Will the IFU only be used as an adjunct to training? Where will the IFU likely be stored once it's been received? Will a user be referring to the IFU during a crisis? Answering these questions will provide insights and further define the requirements to the IFU design inputs.

Determine the Appropriate Format

The most common way to provide an IFU is through a paper-based copy. Given the rise in access to and availability of technology, there are trends of providing digital alternatives to paper-based copies. Paper-based IFU formats can range in type, length, and size. Different paper formats include but are not limited to:

- Booklets
- Pamphlets
- Foldable inserts
- Quick reference guides (QRGs)

- Leaflets

Digital IFU formats similarly can range in type, length, and also interactivity. Digital formats may include:

- PDF (can be an exact version of paper IFU or have interactivity (e.g., hyperlinks))
- Apps
- Electronic instructions for use (eIFU)

The size of a digital format will likely be dependent on the platform from which it is being accessed (e.g., more information can be presented on a computer screen as compared to a smart phone screen). Digital platforms may include desktop or laptop computers, smart phones, tablets, or the device itself. For example, some medical devices have embedded software which include an IFU.

Determining whether an IFU should be paper-based or digital will depend on some of the other design inputs identified, including user tasks and use-environment. Consideration should also be given to the benefits and drawbacks associated with each option. Table 13.1 below presents some of the pros and cons associated with paper-based versus digital IFUs.

Identify Appropriate Authors

Just as one wouldn't go about designing a building without an architect, one should not develop an IFU without the appropriate personnel. Those responsible for developing the IFU will vary from one manufacturer to the next; however, the key writer should be someone who is well vested in instructional design and human factors engineering. Another key author or contributor should be a subject matter expert (SME). One of our authors recently revised the IFU for an ocular anesthesia device. While we have the knowledge and experience to appropriately format information and ensure that content is easy to understand, we consulted with an ophthalmologist with retina speciality to ensure appropriate context, such as workflow and aseptic technique was integrated into the step by step instructions.

Consider the Regulatory Requirements

Before you begin developing the content for your IFU and medical device, do the necessary background research to understand regulatory requirements as well as non-regulated industry standards for your device. For example, in the United States, the FDA governs the labeling of medical devices. Regulations for developing IFUs or "labeling" for medical devices can be found in the Code of Federal Regulations (CFR), and are specifically covered in 21 CFR 801.1–801.437. Thus, 21 CFR

Table 13.1 Comparison of paper-based and digital IFUs

	Paper-based IFU	Digital IFU
Pros	*Consistency*: Most users are familiar with paper-based IFUs and are comfortable in searching for and identifying information needed. Additionally, each user will see the same materials as compared to digital IFUs which can be interpreted differently depending on their platform	*Interactivity*: Can be interactive and include video or audio files to further demonstrate and provide detail for concepts that are difficult to explain on paper. Can also facilitate prospective memory by including reminders or checklists
	Accessibility: Paper IFUs are typically included in the packaging with a device, making it easy for users to immediately access the IFU and find information. They don't have to look up information in a separate location (e.g., digital version)	*Accessibility*: Can be better than paper IFUs—which may get separated from the device. eIFUs can potentially be available anywhere
	Size and space: Paper IFUs can be bigger or smaller in order to control how much information is presented at one time. This can be difficult to do in digital IFUs which are limited in size to the screen they are presented on	*Size and space*: Digital IFUs can simplify processes by presenting pertinent information only as it's needed
		Easy to revise: Updates and revisions can be made quickly and easily
		Sustainable: Reduces costs associated with printing and environmental footprint
Cons	*Durability*: Paper IFUs may not be sustainable for their use in that they can be prone to tears and are not likely to hold up to wet environments	*Accessibility*: Depending on the environment of use, users may have delayed or no access to a digital IFU when it is actually needed. Further, given that power can be lost at any time, a paper-based copy may still be developed regardless as a backup
	Accessibility: Paper IFUs are easy to misplace and/or separate from the device	
	Interactivity: Users' interactions with paper IFUs are limited to reading text and comprehending images	*Requires Technical Ability*: Digital IFUs may cause difficulty for or deter users who are not technologically savvy

801.1–801.437 are mandatory requirements (Alper, Arndt, Borgardt, & Johnson, 2019).

In the US, there are several documents published by the FDA that provide guidance for developing and evaluating medical device labeling. The guidance documents include but are not limited to:

- Guidance on Medical Device Patient Labeling; Final Guidance for Industry and FDA Reviewers (CDRH, 2001)
- Guidance for Industry: Label Comprehension Studies for Nonprescription Drug Products (CDER, 2010)
- Labeling Regulatory Requirements for Medical Devices (CDRH, 1989)
- Guidance for Industry and FDA on Alternative to Certain Prescription Device Labeling Requirements (CDRH, 2001)

Perhaps the most important industry standard to adhere to when developing IFU content for medical devices is the ANSI/AAMI HE75-2009: human factors engineering—design of medical devices. This standard provides comprehensive guidance for designing IFUs for diverse populations and environments.

Alper et al. (2019) notes that "There are also non-recognized standards related to labeling, such as the ANSI Z535 series of standards, though the ANSI standards are typically applied to consumer products rather than medical devices. In this case, following the ANSI Z535 (2011) series of standards could result in labeling that is inconsistent with mandatory requirements, ultimately resulting in a mis-branded device." See Chap. 9 for additional information regarding regulatory requirements for medical device interfaces.

13.5   A Framework for Developing Good IFUs and a Model of IFU Use

In our experience, determining your IFU design inputs with regard to the users, uses, environments, formats, authors, and regulatory documentation is the first step. The next involves putting those requirements into action. Unsurprisingly, this step can be more daunting. Below, we introduce a model for how people interact with IFUs that in turn, provides a framework for approaching IFU design and evaluation. The model consists of three components: Find, Comprehend, and Apply. As we discuss this model as a framework for IFU design, we describe the pitfalls to avoid and the design principles to implement to ensure your IFU is not just usable but safe and effective as well. It's also important to note that many of the design solutions will support more than one component; however, for the purposes of simplicity, we'll list each principle just once.

Generally, people engage in three cognitive tasks, listed below, when using IFU (Fig. 13.1).

- *Find*—people look for the information they need by paging through the document, skimming high level phrases, conducting a keyword search, recalling where they have seen the information before, or guessing its location based on their current mental model. With any luck, and with good design, people recognize the content they are searching for. Of course, some information will be easier to recognize than others; conspicuous information (see Sect. 5.14), which

Fig. 13.1 A model of instructions for use (IFU)

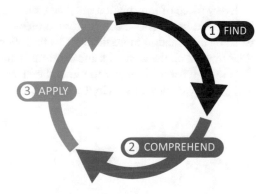

stands out from its surroundings is more likely to be recognized than information surrounded by a sea of sameness.

*   *Comprehend*—once users have found the information, they need to learn and understand and learn that information.
*   *Apply*—finally, people must be able to use the information to conduct some task or activity with the device. Often, simply understanding the information is not enough—people need support while they do the task. This might entail working with the device while simultaneously referring to the open IFU (either in text or electronic form).

Next, we discuss how to design the IFU to facilitate these three cognitive activities.

Finding Information

People need to be able to easily locate information teaching them how to use the device. The IFU should be designed to facilitate finding this information. Below we discuss a few ways to reduce cognitive effort while guiding the user's attention to the information they need.

Signal vs. Noise

Albert Einstein pointed out that everything should be made as simple as possible, but no simpler. So it is with IFUs. However, IFUs often fail to provide the right amount of detail. For one, they can be too long, wordy, and detailed. This adds visual noise to the document, making it difficult to locate any one specific piece of information. In Chap. 5, we discussed the Where's Waldo puzzles, in which your task is to find a character in a red and white shirt in a busy visual scene. The puzzle can be very challenging depending on how much visual noise is present. Finding information in an IFU is often similar to this—but without the fun.

Every piece of information must be examined critically to determine if it should be included in the IFU. Each additional piece of information may make some other piece of information more difficult to find. For example, we have observed many IFUs that list all the warnings and cautions before the user even gets to the first step. Given that a user is unlikely to remember 1 warning out of a list of 15 by the time they get to the relevant step in the IFU—this approach is not helpful at all.

Organization

The more organized materials are, the less effort is required to find what you are looking for (Parrish, 2007; Robinson, 2009). Organization provides the reader with clues that enable them to make educated guesses when looking for information. Recall from Chap. 7 that humans spontaneously categorize information, and that categorization is almost always hierarchical. Similarly, when looking for information, categorization is king. People tend to look for the category first, followed by the specific item. Since items within a category are more related to each than items in different categories, careful organization will guide the user, like breadcrumbs, to the information they need. Dickinson et al. (2010) created two different presentations of identical instructions for a fictional drug product Atenofen (see Fig. 13.2), one with categories and one without. Results from usability testing with both formats found that people found information in IFU with categories (Fig. 13.2B) than the one without (Fig. 13.2A). Clearly, good design makes a difference in assisting readers as they search for information.

There are three simple rules to smart categorization: (1) put related information together in close proximity (Miller, 1956; Perlman, Pothos, Edwards, & Tzelgov, 2010; Robinson, 2009), (2) separate that information from unrelated information using blank space (aka "whitespace"), or borders, and (3) reflect the categorical hierarchies using consistent headers, font size, font weight, font capitalization, or indentation. Presenting content in a consistent manner enables users to predict the location of, and immediately access, information (Sinocruz et al., 2011). Consistency also enables users to learn rules once and apply them repeatedly.

To adhere to the first rule, make sure to place images adjacent to its related text, and do the same for warnings and cautions. Do not place warnings and other cautionary information ad nauseum at the beginning of the IFU, where readers will either ignore it or forget it by the time it is needed.

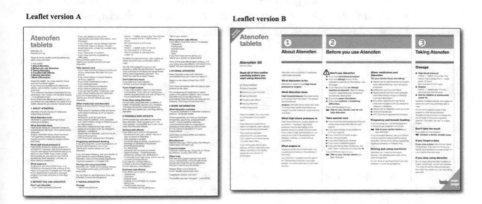

Fig. 13.2 Two IFU layout options. (Image courtesy of David Dickinson at Consumation, Ltd. with permission)

Comprehending Information

Once users find the information, they need to be able to make sense of what they have found. Thankfully, several factors that make IFU information easy to find (e.g., categorization, consistency and hierarchical organization), also make it easy to comprehend.

Cognitive Load Theory

One way to improve comprehension was described by Sweller (1988) in a Cognitive Science article on cognitive load theory. You can think of cognitive load as the amount of working memory being applied at any particular time. While learning, cognitive load depends on three variables: (1) the difficulty of the topic, (2) the difficulty of using the instructions on the topic, and (3) the work being done to actually learn—to build knowledge. These cognitive loads are called intrinsic, extraneous, and germane, respectively.

Intrinsic cognitive load depends on the topic: the more difficult the topic, the greater the intrinsic load. Unfortunately, there is little we can do about intrinsic load, it will depend on the complexity of the device. Extraneous load, on the other hand, is influenced by the way information is presented to the user. Good instructional design reduces extraneous load, by making the topic easy for the user to learn, leaving more working memory available to actually learn the material. Poor design makes extraneous load too much to bear. Finally, germane cognitive load is the cognitive work conducted when we learn new material, committing it to long term memory. This is the reason you read the IFU to begin with. You hope to know more after reading then you did when you started.

Promoting comprehension requires reducing extraneous load so more working memory can be dedicated to germane load (Sweller, van Merrienboer, & Paas, 1998). Let's look at a few ways of doing this. There are several design strategies that can be implemented into IFUs to combat extraneous load.

Chunking

Recall from Chap. 7 on working memory that chunking—combining related things into one—frees up working memory by making use of meaningful items already stored in long term memory. Designers who chunk items appropriately for learners help accomplish the same thing. In essence they reduce extraneous load so that readers can learn more. One way to chunk is to describe tasks or steps with short, action-oriented sentences.

IFUs are often chunked poorly. For example, IFUs will commonly present all of the disclaimers, warnings, and cautionary language at the front of the document. This information may not be remembered by the time a user needs to apply it at a later step in the process. Further, IFUs can make the mistake of requesting users to jump around in the document to locate pertinent information.

Meaning

Meaning is related to chunking. Long term memory (part of which can be activated to serve working memory) is organized according to meaning. As a result, meaningful information is easy to remember because it already resides in long term memory. Arbitrary information, on the other hand, is difficult to remember. Accordingly, adding context, rationale, and/or purpose to instructions will promote understanding by creating meaning. If users understand why they are doing something, they are much more likely to remember it. For example, instead of simply telling a reprocessing technician to brush a port on an endoscope, you can tell them to brush a port to remove debris.

Familiarity

Using familiar words and icons can reduce extraneous load. Using jargon and unfamiliar acronyms, on the other hand, can increase extraneous load. For example, in Fig. 13.3, a "channel cleaning brush" and a "channel opening cleaning brush" can instead be called the "long brush" and the "short brush" based on their recognizable features. These familiar terms reduce cognitive load and are easier to remember instead of hard-to-remember technical names or product codes in the IFU.

Fig. 13.3 Familiar and descriptive terms for device accessories

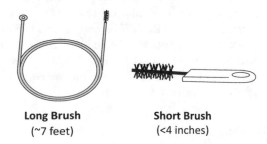

Long Brush
(~7 feet)

Short Brush
(<4 inches)

Conciseness

Being concise helps simplify instructions. We have experienced too many IFUs that are lengthy due to repetitive information. To be concise in procedural tasks, begin sentences with an action verb (Robinson, 2009). For example, instead of writing, "the connector cap should be attached," write, "Attach the connector cap." On the other hand, we have experienced the opposite as well; manufacturers can provide instructions that are too short. IFUs that try to oversimplify the process can be bad because they may not contain all necessary information to complete tasks.

Facilitating Learning

Learning tends to be improved with (1) organization, (2) metaphor and analogy, and (3) illustration. We discuss each of those below. Learning is improved through organization simply because it enables us to make use of knowledge structures (Chap. 7) already available in our minds.

Metaphor and analogy do something similar. That is, they describe a new topic by pointing out its similarity to a different topic with which we are already familiar. Then, we just apply that existing knowledge to this new topic.

Illustration is effective as well. In our experience, IFUs lack effective images. In truth, most IFUs lack images completely. Using graphics and illustrations reduces memory demand, provides examples, and helps people remember information especially when used in combination with written instructions (Houts, Doak, Doak, & Loscalzo, 2006; Katz, Kripalani, & Weiss, 2006; Kools, van de Wiel, Ruiter, & Kok, 2006). Steps that lack images or illustrations may be overlooked or interpreted as optional (Sethumadhavan, Cherne, & Shames, 2017).

13.6   Applying Information

Once people have reviewed the IFU, found the information they need and understand it, they need to successfully apply that information to use the device. Sometimes people cannot remember the exact steps required to complete a task without having instructions in front to them. In this situation, people need to use and actively reference the IFU for support while they perform the task. This might entail working with the device while simultaneously referring to the open IFU (either in text or electronic form). Other times, users may not be able to review the IFU while using the device. In either situation, there are several ways to support this activity of applying information.

Sequencing

Instructions should be arranged into tasks and subtasks and should be outlined with numbers or bullets appropriately. Steps that need to be carried out in serial order should be outlined using sequential numbers, and bullets should be used when stating or introducing information. Using numbers for sequential steps will help to ensure instructions are applied in the correct order.

Help Readers Save Their Place

While applying instructions, users may need to read a step, then conduct that step, only to return to what they were reading. We do not want the reader to lose their place in the instructions. Bullets, numbered tasks, or checklists enable users to do that more easily than large paragraphs of unstructured text.

Provide Feedback

Provide information about what users should expect to see, hear, and feel when they do something correctly. Use descriptive language of visual, tactile, and auditory feedback cues. A visual example would be, "Ensure there is no gap between the cap and the tip of the device." An auditory example would be, "Listen for the beep to signal that the task is complete." A tactile example might be, "Insert the device until it clicks into place." Including feedback not only allows users to check their work, but ensures that they are aware of their progress in using the device.

13.7   IFU Iteration and Evaluation

Using the Find, Comprehend, and Apply framework, you have now developed your first IFU draft. Now what? Well, following the HCD process, you will want to evaluate and iterate. It's important to conduct usability testing of your IFU early and often to assess its efficacy and ensure you are moving in the right direction and you can continue to iterate on the design until it has reached a sufficient level of usability. There are multiple methods that can be used to assess the usability of an IFU including traditional usability testing, cognitive walkthroughs, and heuristic evaluations (see Chap. 4 for more information on these methods).

When performing usability testing for IFUs, there are some nuances to consider. There are essentially three concepts that need to be evaluated and they are, not surprisingly, tied to the framework we introduced above:

- Search: Can people find and see the necessary information?
- Comprehension: Do people understand the information?
- Application—Do people do the step appropriately? (Fig. 13.4).

Given that we are talking about IFUs for medical devices, it is always important to understand any additional regulatory requirements for evaluating your IFU. For example, label comprehension studies are commonly requested by regulatory agencies for over-the-counter products (see Chap. 9 for more information on regulatory processes).

Resources

- 21 CFR 801 Medical device labeling.
- 21 CFR 809 In vitro diagnostic labeling.
- ANSI/AAMI HE75:2009/(R)2013 Human factors engineering—Design of medical devices.
- AAMI TIR49: 2013 Technical Information Report Design of training and instructional materials for medical devices used in non-clinical environments.
- Guidance on Medical Device Patient Labeling; Final Guidance for Industry and FDA Reviewers (CDRH, 2001).
- Guidance for Industry: Label Comprehension Studies for Nonprescription Drug Products (CDER, 2010).
- Labeling Regulatory Requirements for Medical Devices (CDRH, 1989).

Fig. 13.4 Example of formative usability testing to evaluate endoscope reprocessing IFU

- Guidance for Industry and FDA on Alternative to Certain Prescription Device Labeling Requirements (CDRH, 2001).
- Black, A., Luna, P., Lund, O., & Walker, S. (Eds.). (2017). *Information design: Research and practice*. Milton Park: Taylor & Francis.
- Coe, M. (1996). *Human factors for technical communicators*. Hoboken, NJ: Wiley.

References

21 CFR Part 801. (n.d.). 21 CFR 801.1–801.432 Medical device labeling.

21 CFR Part 809. (n.d.). 21 CFR 809.10–809.11 In vitro diagnostics products for human use labeling.

Alper, S. J., Arndt, S., Borgardt, J., & Johnson, K. (2019, September). Human factors and medical device instructions for use: It's not just good business and good science, it's the law. In *Proceedings of the international symposium on human factors and ergonomics in health care* (Vol. 8, No. 1, pp. 213–216). Sage, CA/Los Angeles, CA: SAGE Publications.

American National Standards Institute (ANSI). (2011). Product safety signs and labels. ANSI Z535.4-1991, 1998, 2002, 2011.

Hildebrand, E. A., Branaghan, R. J., Wu, Q., Jolly, J., Garland, T. B., Taggart, M., et al. (2010, September). Exploring human factors in endoscope reprocessing. In *Proceedings of the human factors and ergonomics society annual meeting* (Vol. 54, No. 12, pp. 894–898). Sage, CA/Los Angeles, CA: SAGE Publications.

Houts, P. S., Doak, C. C., Doak, L. G., & Loscalzo, M. J. (2006). The role of pictures in improving health communication: A review of research on attention, comprehension, recall, and adherence. *Patient Education and Counseling, 61*(2), 173–190.

Hyman, W. (2014, August 11). Medical devices: Who needs to read device instructions. Patient Safety & Quality Healthcare. Retrieved August 27, 2020, from https://www.psqh.com/analysis/medical-devices-who-needs-to-read-device-instructions/.

Katz, M. G., Kripalani, S., & Weiss, B. D. (2006). Use of pictorial aids in medication instructions: A review of the literature. *American Journal of Health-System Pharmacy, 63*(23), 2391–2397.

Kools, M., van de Wiel, M. W., Ruiter, R. A., & Kok, G. (2006). *Pictures and text in instructions for medical devices: Effects on recall and actual performance*. Patient.

Miller, G. A. (1956). The magical number seven, plus or minus two: Some limits on our capacity for processing information. *Psychological Review, 63*(2), 81.

Parrish. (2007). Aesthetic principles for instructional design. *Educational Technology Research and Development, 57*, 511–528.

Perlman, A., Pothos, E. M., Edwards, D. J., & Tzelgov, J. (2010). Task-relevant chunking in sequence learning. *Journal of Experimental Psychology: Human Perception and Performance, 36*(3), 649.

Robinson, P. A. (2009). *Writing and designing manuals and warnings 4e*. Boca Raton, FL: CRC Press.

Sethumadhavan, A., Cherne, N., & Shames, A. (2017). A guide to developing usable instructions for use for medical devices. *MED DEVICE ONLINE*.

Sinocruz, J. Q., Hildebrand, E. A., Neuman, B. L., & Branaghan, R. J. (2011, September). Human factors implications for standard operating procedure development and usability in reprocessing safety. In *Proceedings of the human factors and ergonomics society annual meeting* (Vol. 55, No. 1, pp. 803–807). Sage, CA/Los Angeles, CA: SAGE Publications.

Sweller, J. (1988, April). Cognitive load during problem solving: Effects on learning. *Cognitive Science, 10*(2), 257–285.

Sweller, J., van Merrienboer, J. J. G., & Paas, F. G. W. C. (1998). Cognitive architecture and instructional design. *Educational Psychology Review, 10*(3), 251–296.

U.S. Food and Drug Administration (FDA). (2001). *Guidance on medical device patient labeling: Final guidance for industry and FDA reviewers*. Rockville, MD: FDA.

Williamson, R. (2019, September). How does it all unfold? The role of format layout in the user experience of medical instructions for use. In *Proceedings of the international symposium on human factors and ergonomics in health care* (Vol. 8, No. 1, pp. 248–251). Sage, CA/Los Angeles, CA: SAGE Publications.

Chapter 14
Reusable Medical Devices, Reprocessing, and Design for Maintenance

14.1 Introduction

When we think about any medical procedure, two people immediately come to mind: the person giving the therapy and the person receiving the therapy. These are the primary users, and they are involved in the primary procedure—the giving and receiving of therapy. As a result, when developing an improved cardiac catheter for better control when placing implants, we think of the primary user as the surgeon and placing the implant as the primary procedure. Similarly, when designing a wearable robotic device to assist in stroke rehabilitation, we think of the primary users as the therapist who fits the device to the patient and the patient who uses it during rehab activities.

This process of considering the users and their main tasks is exactly what we have advocated throughout this book, calling it human-centered design. The problem is that there are typically many other forgotten users and forgotten tasks associated with the device. For example, someone had to identify, pick, and retrieve the implant from central supply. Someone unpackaged it and entered the serial number into the EHR. Someone disposed of it after the procedure, or in the case of reusable medical equipment (RME), someone maintained, cleaned, disinfected, and sterilized the equipment.

The safety and effectiveness of that device depend as much on those users as on the primary users. This is especially true for medical devices that are reusable. For instance, a reprocessed device with bioburden left over from the previous procedure is of no use to anyone. Indeed, it is worse than useless; it is dangerous.

Reusable devices range from simple, such as a portable Sp02 monitor, to incredibly complex, such as a flexible video duodenoscope (Fig. 14.1). Despite the extensive variety in reusable devices, each requires some form of maintainability, which can range widely in itself and may involve multiple end users to ensure they remain functional and clean between patient uses. Failing to consider all end users and all of the various uses a device will encounter can have drastic consequences.

© Springer Nature Switzerland AG 2021
R. J. Branaghan et al., *Humanizing Healthcare – Human Factors for Medical Device Design*, https://doi.org/10.1007/978-3-030-64433-8_14

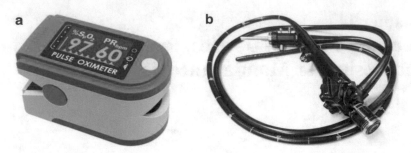

Fig. 14.1 (**a**) Reusable pulse oximeter. (Image by natatravel/shutterstock.com.) (**b**) Reusable endoscope. (Image by tadaki crew/shutterstock.com)

In early 2009 the Veterans Health Administration (VHA), the largest integrated healthcare system in the United States, sent out over 10,000 letters to patients to notify them that they may have been exposed to or infected with bloodborne pathogens, including Hepatitis B, Hepatitis C, or HIV. The VHA explained that the culprits for this potential exposure were improperly reprocessed endoscopes from procedures performed between 2003 and 2008 (ECRI, 2009, 2010). Most incident reports blamed reprocessing technicians, the medical professionals responsible for cleaning the devices between patient uses, and cited an inability to adhere to the manufacturer's cleaning instructions for use (IFU) as one of the causes for these systemic failures.

In the months and years that followed, lawsuits were issued against healthcare facilities and manufacturers. The media published headline after headline that appropriately shared news of these problems with the public, but likely also diminished the public's trust in their healthcare providers and these medical devices. Worst of all, patients that were infected not only experienced severe complications, but some ultimately died from their complications.

Since 2009, there have been concerted efforts to reduce the likelihood for these types of adverse events by encouraging the implementation of human factors engineering (HFE) into the design of reusable medical devices. For example, AAMI and FDA have both published guidance documents aimed at providing recommendations for the development and evaluation of reprocessing IFUs for reusable medical devices (AAMI, 2010; FDA, 2015). However, issues with reprocessing continue to occur. Using endoscopes again as an example, in 2014, Seattle Children's Hospital had to notify over 100 patients, ranging from toddlers to teenagers, that they may have been exposed to Hepatitis or HIV after discovering that there were lapses in their colonoscopes cleaning procedures (Bartley, 2014). Again, in 2017, almost a decade after the initial issues with endoscope reprocessing, the VHA had to send out another 500+ letters to veteran patients, to notify them of a potential infection risk associated with improperly sterilized endoscopes (Davis, 2017). Perhaps more alarming, a study performed in 2018 evaluated endoscopes deemed "ready for use" across three separate hospitals and found that 71% tested positive for bacteria, indicating that the reprocessing procedure was not effective (Ofstead, Heymann, Quick,

Eiland, & Wetzler, 2018). Thus, it appears that simply requiring IFUs designed with HFE considerations may not be enough.

These problems with endoscope reprocessing are a prime example of failing to consider all end users during development of the device. There is no doubt that the endoscopes involved in those adverse events were well designed to be used by physicians performing specialized medical procedures. It is unlikely that a proportionate amount of research and design efforts focused on the needs of the reprocessing technicians who perform cleaning between each patient use. These ongoing adverse events demonstrate that there is a real lack of progress and innovation by manufacturers in designing reusable medical devices that are easy to clean, maintain, and thus safer for use.

14.2   Reusable Medical Devices and Designing for Maintenance

Reusable medical devices are any devices that are reprocessed and reused on multiple patients. They are used across all spectrums of healthcare. In fact, it is impossible to provide most aspects of healthcare without the use of a reusable medical device including preventative and diagnostic care, patient monitoring, therapy, and treatment (see Table 14.1). Reusable medical devices are unique in that, unlike other types of medical technology (e.g., biological implants) or medications, medical devices require ongoing maintenance, both scheduled and unscheduled, throughout their lifecycle (Jamshidi, Rahimi, Ait-kadi, & Bartolome, 2014).

Maintenance of medical devices is only one component of the medical device lifecycle (see Fig. 14.2), but it is an important component for various reasons. First and foremost, maintenance activities ensure that reusable medical devices are safe to be reused on patients and remove or reduce the likelihood for device failures, cross-contamination, and spread of infections. Second, from a cost perspective, well-designed and efficient maintenance activities are critical for a customer's bottom line. More money may be spent on maintaining the device than on its initial procurement (Dhillon, 2011; Jamshidi et al., 2014). In fact, some medical devices may spend the majority of their lifecycle time in the maintenance phase. Finally, reusing devices provides a "green" solution to the medical device community. It

Table 14.1 Examples of reusable medical devices by healthcare area

Area in healthcare	Examples of reusable devices
Preventative and diagnostic care devices	Endoscopes, blood pressure cuffs, dental teeth cleaning tools, ultrasound, and MRI machines, surgical instruments, etc.
Patient monitoring devices	Infusion pumps, surgical tables, pulse, oximeters, etc.
Therapy devices	Assistive robotic devices, vibration devices, medical treadmills, etc.

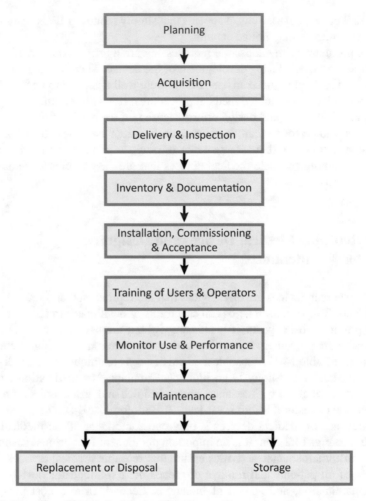

Fig. 14.2 Overview of the medical device lifecycle

helps to reduce the amount of medical waste. Thus, integrating maintenance into the design for medical devices, or simply designing for maintainability, is a critical consideration for developing optimal reusable medical devices.

What does it mean to integrate maintenance into the design of a device? It means that the users who perform maintenance, the maintenance activities themselves, and the maintenance environments need to be considered as inputs to the design process. Formally, designing for maintenance refers to a device's ability to be maintained or restored to an effective usable condition (Weininger, Kapur, & Pecht, 2010). There are multiple tasks that can be classified as maintenance for reusable medical devices; but they typically fall into two categories:

1. Scheduled, which includes routine and preventative maintenance.
2. Unscheduled, which includes repairs and corrective maintenance.

Maintenance that is scheduled or preventative may include tasks such as replacing batteries in portable electronic devices, lubricating gears and motors in mechanically powered devices, or inspecting for sharpness and alignment in surgical instruments. The list goes on. Preventative maintenance is performed routinely and is aimed at reducing risks related to device failure or cross-contamination.

Meanwhile, unscheduled maintenance includes tasks such as repairing broken components, performing software updates, and troubleshooting device failures. The users who perform these tasks range widely in education and training, and may include biomedical engineers, clinical engineers, technology managers, and reprocessing staff and technicians. The environments in which maintenance is performed varies widely as well; clinical lab spaces, procedure rooms, and sterile processing departments are common examples.

Although all maintenance activities are important for improving the usability of a reusable medical device, moving forward, this chapter will focus on one type of maintenance activity: reprocessing. Why reprocessing? As is evident from case studies discussed in the beginning of the chapter, reprocessing has become somewhat of a poster-child for issues with regard to maintaining reusable medical devices.

14.3 Reprocessing and Designing for Maintenance

What Is Reprocessing?

Reprocessing is a subcomponent of maintainability and has been formally defined as the process that renders a clinically used device safe and ready for its intended reuse (AAMI ST91). In the 1960s, the Spaulding classification system was used to assign reusable medical devices into three categories, based on risk of infection associated with reuse of the device (AAMI, 2010, 2017; FDA; Ramakrishna, 2002) and the type of reprocessing procedure that they require. FDA and AAMI have adopted these categories as well, which include:

1. Critical devices are those that penetrate skin or mucosa and carry a high risk of infection. Critical devices should be sterilized between uses. Examples include biopsy forceps.
2. Semicritical devices do not penetrate intact mucous membranes and should be sterilized or high-level disinfected. Endoscopes are an example of semicritical devices. They can be sensitive to heat, thus high-level disinfection is a commonly practiced method of reprocessing.
3. Noncritical devices, such as blood pressure cuffs or patient carts, only contact intact skin and have very little risk of transmitting infection. Non-critical devices do not need to be sterilized and may be cleaned using low-level disinfection.

Depending on the classification of a device, different resources will be required to ensure the device is ready for reuse. Figure 14.3 illustrates the various pathways that critical, semicritical, and noncritical devices follow during their reprocessing maintenance lifecycle.

Reprocessing of reusable medical equipment is an activity that is typically performed by the role of a reprocessing technician or central service/sterile processing professionals. However, in some healthcare facilities, nurses or other technicians may perform some of the tasks. Medical device reprocessing can occur in a facility's own Sterile Processing Department or be outsourced and completed by third-party vendors. Regardless of where reprocessing is performed, the steps involved are largely the same. Reprocessing typically includes six major tasks (see Fig. 14.4):

1. Precleaning or point of use cleaning: This cleaning typically happens immediately after use in a patient procedure, while the device is still in the procedure room. Precleaning typically involves soaking or spraying devices with a water or detergent solution. The purpose is to remove gross contaminants from the devices before it dries, which make it more difficult to remove during manual cleaning. Precleaning will typically be performed by a nurse or a technician who predominantly work in the procedure rooms.
2. Maintenance testing: This may occur prior to or after manual cleaning. This includes evaluating devices to determine if there is damage and whether the

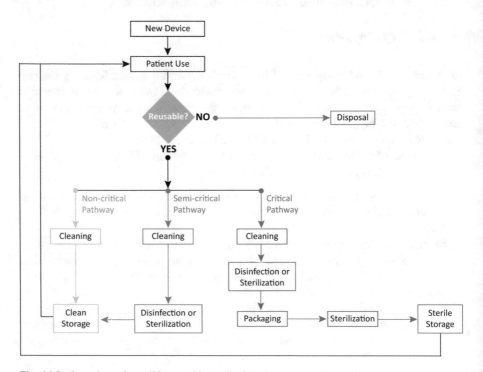

Fig. 14.3 Overview of possible reusable medical device reprocessing pathways

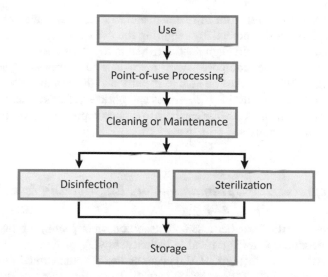

Fig. 14.4 Overview of major reprocessing tasks for reusable medical devices

device is mechanically sound to ensure effectiveness in either the subsequent cleaning process or patient procedure. This step, and all remaining steps, is likely to be carried about by reprocessing technicians in a sterile/central processing department.

3. Manual cleaning: Devices are typically soaked in or wiped down with detergent solution and any remaining gross contaminants removed by manually and thoroughly wiping or brushing or flushing. Devices are often rinsed during this phase as well to remove residual cleaning solution.
4. Disinfection or sterilization: The choice to perform high-level disinfection or sterilization is made based on manufacturer instructions and recommendations. Whether or not a device can be sterilized will often depend on its materials and ability to withstand different exposure to sterilization methods. High-level disinfection can be conducted manually and often involves soaking or flushing devices with a germicidal agent. Alternatively, reprocessing machines can perform those same steps automatically. If high-level disinfection is performed, another rinsing will typically take place to remove residual solution.
5. Drying: After high-level disinfection or sterilization has been completed, reusable medical devices are often required to complete a dry time to ensure that all remaining liquid has time to dissolve or is physically removed through the use of forced air to promote drying.
6. Storage: Proper handling of reusable medical devices during storage is important to ensure that contaminants are not reintroduced to the sterilized or disinfected equipment. Typically, healthcare facilities will have different procedures and physical locations for sterile vs. clean storage.

Depending on the type of reusable medical device and the manufacturer, the steps above can vary widely, even for the same classes and types of devices. For example, two endoscopes from different manufacturers may have very different reprocessing procedures, potentially requiring the use of different equipment and cleaning agents. Further, although not addressed in the steps above, some reusable medical devices come equipped with disposable and reusable accessories. Reprocessing technicians must be able to identify and discard disposable accessories, then reprocess the reusable accessories appropriately. Incorporating HFE into the design of these medical devices can help to support reprocessing users in accomplishing these tasks safely and effectively.

Why Is Reprocessing a Human Factors Engineering Issue?

Reprocessing is a HFE issue because poor usability can impact the ability to control infection. Infection control is critical to patient safety. Approximately two million healthcare-associated infections (HAIs) occur annually in the United States, resulting in approximately 90,000 deaths (Stone, 2009). Reprocessing medical devices has consistently been identified as a top health technology hazard by the ECRI Institute over the last 10 years (ECRI, 2018). More specifically, many HAIs have been traced to errors in reprocessing procedures resulting in contaminated reusable medical devices. Failure to properly reprocess a medical device carries the risk of person-to-person transmission of viruses such as HIV, Hepatitis C, Hepatitis B, or other bacteria, as well as the transmission of environmental pathogens such as pseudomonas (Mehta et al., 2006; Nelson, 2003; Weber & Rutala, 2001).

If reprocessing is so risky, someone might ask why continue to pursue it? Well, there has actually been a shift, wherever possible, toward single-use devices (SUDs). SUDs are devices that are used on patients and then discarded after one use. However, SUDs are not necessarily the best answer in all scenarios because they have their own drawbacks. SUDs increase costs; the cost of materials and the cost to manufacture so many of them increase downstream costs to customers. SUDs also result in more waste; waste of used materials and also waste of storage space, which is a precious commodity in healthcare facilities. Further, SUDs cannot replace all medical devices. So, while SUDs can help to reduce the number of HAIs, unfortunately, not all medical devices can be made disposable. For example, endoscopes are complex devices that cost thousands of dollars to manufacture and train physicians on how to use, so making them disposable would be prohibitively expensive. As a result, devices like endoscopes must continue to be reprocessed, or cleaned and sterilized after each use.

So, if reprocessing is here to stay, how can medical devices be improved to better meet the needs of all end users and reduce the likelihood for infection? Well, first, it helps to understand where usability often falls short in the design of reusable medical devices. The most obvious problem is that reusable medical devices are designed with patient procedures in mind first, and reprocessing is an afterthought. Reviews of existing literature on the causes of reprocessing errors generally indicate that the blame is almost always placed squarely on the human operator and "user" error, citing noncompliance with instructions as the primary culprit of errors (Nelson, 2003; Ofstead, Hopkins, Buro, Eiland, & Wetzler, 2020; Ramakrishna, 2002). However, it is difficult to understand how reprocessing users can be blamed when devices weren't designed well for *their* tasks. These analyses rarely take into account the poor interaction between the reprocessing user and the actual device design, nor do they address user interactions with other device interfaces including the instructions-for-use (IFU) and training (Hildebrand et al., 2010).

HFE practitioners recognize that assigning blame to the user does nothing to mitigate problems. Many problems classified as human error are actually the result of a bad match between users and device design. Put simply, reusable medical devices that were not designed with cleaning in mind often result in devices that are difficult to clean. Consequently, their cleaning processes are long and complex. For example, reprocessing a flexible endoscope can require over 200 individual steps and the use of over 20 different cleaning accessories (Hildebrand et al., 2010; Jolly et al., 2012). Moreover, facilities often reprocess as many as 250 endoscopes a week (Hildebrand et al., 2010, 2011; Ofstead et al., 2020). If our math is correct, cleaning 50 scopes per day amounts to conducting 10,000 steps per day. As human attention and working memory have limited capacities, each additional step in a cleaning process increases the likelihood of a lapse or omission in the process. That's 10,000 opportunities for an error, with the end result being an endoscope that is not cleaned improperly (per day!). The logic further follows that if a device and cleaning process are complex, the materials that support this process, such as IFUs and training, are likely to be complex as well. There is converging evidence in the literature that confirms these assumptions by identifying three HFE issues within reprocessing devices interfaces that lead participants to commit errors: high memory demands, lack of visibility, and insufficient feedback (Hildebrand et al., 2010, 2011; Jolly et al., 2012).

The final section in this chapter will discuss how to improve medical device design for the most common reprocessing device interfaces: the reusable medical device, reprocessing instructions for use (IFU), and training. It also provides design principles and guidelines for addressing the HFE issues that have been identified in the literature: memory, visibility, and feedback. Further, we provide resources to ensure regulatory compliance with the design of any reusable medical devices that involve reprocessing.

14.4   Designing Reusable Medical Devices to Optimize Reprocessing

Manufacturers must design reusable medical devices that account for all reprocessing users' needs, capabilities, and limitations. Reprocessing users include staff that are performing precleaning, which could be a nurse in the OR or reprocessing technicians that performs manual cleaning, disinfection, and sterilization. Utilizing the human-centered design process discussed throughout this book will result in better usability, fewer errors, and in the case of reusable medical devices in general, fewer infections.

Interface 1: Reusable Medical Device

First and foremost, reusable medical devices need to be easy to clean. The challenge is to also design reusable medical devices to withstand multiple uses as well as cleaning, disinfection, and potential sterilization processes between uses. Accordingly, these devices must be made of materials that can endure repeated reprocessing, including manual brushing and the use of chemicals. Once those materials have been determined, designers should look for opportunities to incorporate cleaning affordances and eliminate features that are difficult to clean. The FDA and AAMI (AAMI TIR12, 2010; FDA, 2015) have identified features that are prone to retaining debris, including but not limited to:

- Lumens; long, narrow interior channels
- Sharp internal corners and angles
- Dead-end channels and zones
- Springs, coils, and twisted or braided wires
- Hinges
- Sleeves surrounded rods, blades, activators, and insertors
- Adjacent device surfaces between that debris can be forced or caught during sue
- O-rings
- Valves that regulate the flow of fluid through a device
- Devices that have features that cannot be disassembled or require extensive disassembly and reassembly

From an HFE perspective, there are a number of strategies that can be employed during design to support reprocessing including, but not limited to:

- Make things visible

    Ensure the device design does not have any hard to see or hidden components, such as interior channels. When components have to be internal, use clear materials so that users can better identify components that need to be clean and ensure they are cleaning the correct component. Making components visible also supports user memory (i.e., the user can see and remember what they need to clean) and user feedback (i.e., the user can see that there is no remaining bioburden).
- Make things memorable

    Many devices use labeling to assist primary users (Fig. 14.5). Consider using text and icons to support reprocessing users' identification and recognition of components that need to be cleaned.

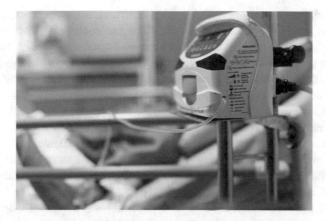

Fig. 14.5 An infusion pump is an example of a reusable device that provides on-device labeling to support primary users in patient-related tasks. (Image by The Five Aggregates/shutterstock.com)

Utilize color coding to support user memory by identifying components on devices that need to be cleaned, disassembled, or discarded. For example, one endoscope manufacturer uses the color green to identify single-use components.

– Provide feedback

In addition to developing the device interface, consider ways in which the design can support users' understanding of when cleaning has been completed and is sufficient. Further, devising methods that can provide immediate feedback and/or alert reprocessing technicians of failures could be critical for preventing potential errors. For example, rapid cleaning monitors could facilitate the recognition of improperly cleaned medical devices (Alfa, 2016).

As further confirmation that re-envisioning the design process for reusable medical devices route will limit potential disease transmission, FDA has recently recommended a transition to innovative designs for reprocessing procedures (FDA, 2020).

Interface 2: Reprocessing Instructions for Use

If a device is easy to physically clean, it is likely to have an easy cleaning process as well. The more manufacturers can do to simplify the cleaning process (i.e., by designing easy to clean devices), the better they will be able to support users' working memory. Fewer steps mean easier to remember.

Typically, when new equipment is purchased by a healthcare facility, reprocessing IFUs may be provided by the manufacturer. These IFUs detail the required process to follow in order to effectively clean, disinfect, or sterilize. In general, reprocessing IFUs tend to be difficult to use because they are not designed with end users in mind, they are accommodating for a difficult cleaning process, and they are dealing with the fallacy that reprocessing IFUs are documents that just "sit on a shelf." This presents little motivation to expend resources needed for adequate design to optimize usability. However, in 2015 the FDA published guidance that requires manufacturers to provide reprocessing instructions that instruct users how to safely reprocess a device. The guidance recommends that those instructions must be validated through usability testing to ensure that they can be understood by representative end users. (Alfa, 2016; FDA, 2015). As a result, manufacturers have recently increased their efforts to create IFUs that are easy to use.

Alas, even the best reprocessing instructions are not always helpful. For example, it may be quite difficult for users to reference a paper booklet in the wet environment of a reprocessing department. To continue evolving and improving reprocessing practices, innovation beyond paper IFUs is needed. Design considerations for alternative methods of supporting and guiding users through reprocessing procedures will be important. Chapter 13 goes into detail about how to design IFUs to optimize usability. All of the strategies and recommendations discussed there apply for reprocessing IFUs as well.

Interface 3: Training

For reprocessing procedures that are more complex, manufacturers may decide to provide additional training beyond the IFU. Most manufacturers accomplish this with in-service training for the reprocessing staff. Regrettably, it is difficult (and frankly unlikely) to assemble all of the reprocessing staff at once for training. As a result, those who were present are left to train those who could not attend. If training occurs at all, it can be wildly inconsistent under this model. At that point, the healthcare facility bears the burden for ensuring that staff is trained to complete the reprocessing procedure as intended. From a design standpoint, unless training can be guaranteed across all facilities, it should not be relied upon as a method of mitigating risk when considering reprocessing in the device design process.

If manufacturers intend to provide training, they should approach its development with the same user-centered design approach applied to the device development. ANSI/AAMI ST91 provides guidance on implementing training and educational programs. When developing an in-service, educational video, or visual aids, manufacturers should apply the same HFE approach to the design of that method.

Literature suggests that medical professionals may benefit most from keeping training "short, active, and relevant" (Rideout, Held, & Holmes, 2016). Keeping training concise will help users to retain information. Keeping training active by including didactic and hands-on practice accommodates different learning styles. And finally, keeping training relevant by including the clinical relevance of reprocessing activities can help users to stay engaged (Rideout et al., 2016).

Regardless of the approach taken, designers should ensure that information across all training mediums and the IFU are consistent. They should not promote contradictory ideas. Finally, as with IFUs, manufacturers should seek to validate their training materials through simulated-use testing with representative users to verify that the training yields the intended effect.

14.5 Conclusion

As the medical device industry continues to grow and the need for complex reusable medical equipment increases, the HFE of reprocessing needs to be at the forefront of the design process. Reusable medical devices are only as effective clinically as their maintenance and cleaning procedures afford; that is, if errors are made during reprocessing, the devices may do more harm than good. Implementing HFE in the design of reusable medical equipment and reprocessing procedures will result in safer, easier, and more effective devices.

Defining reprocessing as a design requirement during early product development is the best way to incorporate human factors engineering. Design inputs should include reprocessing users capabilities, limitations, environments, and expectations will help to create an easier reprocessing procedure overall. The logic here is that if the design of the device is easy to clean, the process for performing cleaning should be easy as well. If the process is easy, then the interfaces that support reprocessing, such as the instructions and training, should be simplified as well. Reusable medical device design teams should contain at least one member who is responsible for ensuring the HFE of reprocessing is appropriately considered. To ensure compliance with regulatory standards, reprocessing device interfaces (including training and IFUs) should be evaluated through simulated-use usability studies with representative users as early as possible (AAMI TIR12; FDA; IEC).

Resources

- The following standards and guidance documents provide specific methods applying HFE approach to reprocessing of reusable medical devices. Additional information is provided to help guide manufacturers in choosing materials and chemicals to optimize cleaning and designing cleaning processes that afford usability.
- AAMI TIR12: Designing, testing, and labeling reusable medical devices for reprocessing.
- AAMI TIR55: Human factors engineering for processing medical devices.
- AAMI HE75: Human factors engineering - Design of medical devices.
- ANSI/AMMI ST91: Flexible and semi-rigid endoscope processing in health care facilities.

- ASTM F3357-19 Standard Guide for Designing Reusable Medical Devices for Cleanability.
- FDA guidance: Applying Human Factors and Usability Engineering to Medical Devices.
- FDA guidance: Reprocessing Medical Devices in Health Care Settings: Validation Methods and Labeling.
- IEC/ISO TR 62366: Application of usability engineering to medical devices.

References

Association for the Advancement of Medical Instrumentation (AAMI TIR12). (2010). *Designing, testing, and labeling reusable medical devices for reprocessing in health care facilities: A guide for medical device manufacturers, TIR12.*

Association for the Advancement of Medical Instrumentation (AAMI TIR55). (2017). *Human factors engineering for processing medical devices, TIR55.*

Bartley. (2014). Children's patients may be at risk of infection after colonoscopies. *The Seattle Times.* January 22, 2014.

Davis, H. L. (2017, August 16). VA medical center warning 526 patients of infection risk from scopes. *The Buffalo News.*

Dhillon, B. S. (2011). Medical equipment reliability: A review, analysis, and improvement strategies. *International Journal of Reliability Quality, and Safety Engineering, 18*(4), 391–403.

ECRI Institute (ECRI). (2009). *U.S. Veterans Health Administration announcements highlight need for comprehensive endoscopy-reprocessing protocols [special report].* Health Devices Alerts 2009 Apr 17. Accession no. S0193.

ECRI Institute (ECRI). (2010). Clear channels: Ensuring effective endoscope reprocessing. *Health Devices, 39*(10), 350–359.

Food and Drug Administration. (2020). *The FDA is recommending transition to duodenoscopes with innovative designs to enhance safety: FDA safety communication.*

Food and Drug Administration (FDA). (2015). *Reprocessing medical devices in health care settings: Validation methods and labeling.*

Food and Drug Administration (FDA). (2018). *FDA warns duodenoscope manufacturers about failure to comply with required postmarket surveillance studies to assess contamination risk.* https://www.fda.gov/news-events/press-announcements/fda-warns-duodenoscope-manufacturers-about-failure-comply-required-postmarket-surveillance-studies.

Hildebrand, E. A., Branaghan, R. J., Neuman, B. L., Jolly, J., Garland, T. B., Taggart, M., et al. (2011). An expert perspective of errors in endoscope reprocessing. In *Proceedings of the human factors and ergonomics society annual meeting* (Vol. 55, No. 1, pp. 748–752). Sage, CA/Los Angeles, CA: SAGE Publications.

Hildebrand, E. A., Branaghan, R. J., Wu, Q., Jolly, J., Garland, T. B., Taggart, M., et al. (2010). Exploring human factors in endoscope reprocessing. In *Proceedings of the human factors and ergonomics society annual meeting* (Vol. 54, No. 12, pp. 894–898). Sage, CA/Los Angeles, CA: SAGE Publications.

Jamshidi, A., Rahimi, S. A., Ait-kadi, D., & Bartolome, A. R. (2014). Medical devices inspection and maintenance; a literature review. In *IIE annual conference. Proceedings* (p. 3895). Institute of Industrial and Systems Engineers (IISE).

Jolly, J. D., Hildebrand, E. A., & Branaghan, R. J. (2013). Better instructions for use to improve reusable medical equipment (RME) sterility. *Human Factors, 55*(2), 397–410.

Jolly, J. D., Hildebrand, E. A., Branaghan, R. J., Garland, T. B., Epstein, D., Babcock-Parziale, J., et al. (2012). Patient safety and reprocessing: A usability test of the endoscope reprocessing procedure. *Human Factors and Ergonomics in Manufacturing & Service Industries, 22*(1), 39–51.

Nelson, D. B. (2003). Infectious disease complications of GI endoscopy: Part II. Exogenous infections. *Gastrointestinal Endoscopy, 57*, 695–711.

Ofstead, C. L., Heymann, O. L., Quick, M. R., Eiland, J. E., & Wetzler, H. P. (2018). Residual moisture and waterborne pathogens inside flexible endoscopes: Evidence from a multisite study of endoscope drying effectiveness. *American Journal of Infection Control, 46*(6), 689–696.

Ofstead, C. L., Hopkins, K. M., Buro, B. L., Eiland, J. E., & Wetzler, H. P. (2020). Challenges in achieving effective high-level disinfection in endoscope reprocessing. *American Journal of Infection Control, 48*(3), 309–315.

Ramakrishna, B. S. (2002). Safety of technology: Infection control standards in endoscopy. *Journal of Gastroenterology and Hepatology, 17*(4), 361–368.

Rauwers, A. W., Kwakman, J. A., Vos, M. C., & Bruno, M. J. (2019). Endoscope-associated infections: A brief summary of the current state and views toward the future. *Techniques in Gastrointestinal Endoscopy, 21*(4), 150608.

Rideout, M., Held, M., & Holmes, A. V. (2016). The didactic makeover: Keep it short, active, relevant. *Pediatrics, 138*(1), e20160751.

Stone, P. W. (2009). Economic burden of healthcare-associated infections: An American perspective. *Expert Review of Pharmacoeconomics & Outcomes Research, 9*(5), 417–422.

Weininger, S., Kapur, K. C., & Pecht, M. (2010). Exploring medical device reliability and its relationship to safety and effectiveness. *IEEE Transactions on Components and Packaging Technologies, 33*(1), 240–242.

Chapter 15
Home Healthcare

15.1 Introduction

The bulk of healthcare is migrating from professional healthcare facilities to patient homes. In fact, the home has become the most common site for healthcare administration (Zayas-Cabán & Valdez, 2012). This reflects a broader trend in which people become more responsible for their own health, and individuals take more active roles in managing their healthcare (National Academy of Sciences, 2010). And, all indications suggest that this trend will continue for some time. A big part of this transition is the successful adoption of home-use medical devices.

According to the FDA (2014), a home-use medical device is labeled for use in any environment outside a professional healthcare facility, including outdoors, offices, schools, vehicles, emergency shelters, and independent living homes. Numerous devices meet these criteria, so Story (2010) provided this helpful taxonomy of home health devices:

1. Medication administration equipment, such as syringes, cups, eye-droppers, sprays, patches, and syringes.
2. Test kits, from pregnancy and allergy kits to cholesterol and hormone tests.
3. First aid equipment, such as bandages, traction equipment, ostomy care, and defibrillators.
4. Assistive technologies, such as glasses, hearing aids, prostheses, orthotics, crutches, wheelchairs, and mobility aids.
5. Durable medical equipment, including beds, specialized mattresses, specialized chairs, lift equipment that may be either ceiling-mounted or portable, commodes, urinals, and bedpans.
6. Meters and monitors, such as thermometers, blood glucose meters, electrocardiogram monitors, and fetal monitors.
7. Treatment and therapy equipment, such as infusion pumps, dialysis equipment, transcutaneous electrical nerve stimulation equipment, and intravenous equipment.

© Springer Nature Switzerland AG 2021
R. J. Branaghan et al., *Humanizing Healthcare – Human Factors for Medical Device Design*, https://doi.org/10.1007/978-3-030-64433-8_15

8. Respiratory equipment, such as ventilators, forced airway devices, oxygen, masks, and suction.
9. Feeding equipment, such as feeding tubes and food pumps.
10. Voiding equipment, catheters, and colostomy gear.
11. Infant care equipment, such as incubators, warmers, bilirubin lights, and apnea monitors.
12. Telehealth equipment, such as cameras, sensors, and computers.

The taxonomy illustrates the wide range of home devices, but there is also a wide range of reasons for the surge in home-use devices. The first, and perhaps most obvious, is the aging world population. Due to improved medical treatment, pharmaceuticals, surgical procedures, nutrition, and hygiene, people are living longer. For example, in 1935, the average life expectancy in the US was 61.7 years. Now, it is closer to 80 years. By 2030, 72 million Americans will be over 65 years old, and many are living longer with chronic medical conditions that can be managed from home.

At the same time, people are having fewer children as a result of improved birth control technology, resulting in the older population being a greater percentage of the total. Figure 15.1 illustrates population pyramids from 1950 to 2050. Notice that the 1950 charts for the world and for the US show the pyramidal shape that gives the population pyramid its name. Now look at how the shape becomes more rectangular and evenly dispersed in 2019 and 2050. Because life expectancies are increasing, it's evident that the aging population will benefit from home use devices.

Second, as people age, they develop more chronic diseases including heart disease, stroke, cancer, diabetes, overweight, and autoimmune disorders (Schulz & Tomkins, 2010). Third, it is dramatically less expensive to care for yourself (or be cared for by family or other caregivers) at home. In fact, Charness (2010) argues that the best way to reduce healthcare costs is to eliminate, as much as possible, unnecessary visits to hospitals and unnecessary visits with physicians.

Actually, these visits are becoming more difficult to get anyway. The shortage of healthcare facilities and skilled personnel such as nurses and physical therapists means that therapy must be provided on an outpatient basis. As a result, patients are being released from hospitals and other facilities while they still need care (Talley & Crews, 2007). Additionally, medical insurance companies often insist on less time spent in professional facilities and more time spent at home.

A fourth reason for home device use is that people would rather be at home than in a clinical facility. Home is more comfortable, more familiar, more predictable, and that's where family and friends are. Because hospitals are not as safe as we would like, a home environment can be safer, with a reduced risk of developing life-threatening illnesses like Methicillin-resistant *Staphylococcus aureus* (MRSA) and *Clostridium difficile* (C-diff).

The final reason for this migration to home healthcare is that it is now technologically and economically feasible. The cost of devices and information technology has dropped to a point where home care makes sense. For example, today's $10 sensor would have cost thousands of dollars in the 1990s, and will cost much less in

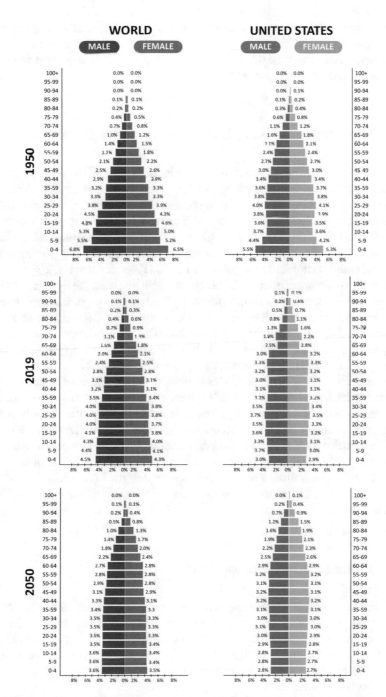

Fig. 15.1 Population pyramid-1950 (*top*); Population Pyramid-2019 (*middle*); Population Pyramid-2050 (*bottom*)

the near future. Similarly, today's mobile phones provide the computing power of a 1990s supercomputer. These changes allow for remote monitoring, data collection, and communication with healthcare providers, enabling patients to live and heal at home rather than at a professional facility.

15.2   Challenges of Home Use Medical Device Design

That's not to say there aren't any challenges with home use. Many complex devices, including ventilators, infusion pumps, dialysis machines, and artificial hearts, are now being deployed in patients' home environments. These are frequently operated by lay users, even though they were not designed for, and were not specifically labeled for them (Story, 2010). Regardless, individuals using medical devices at home should be able to use the devices safely and effectively and without making errors that could compromise the patient's health (Kaye & Crowley, 2000).

There are two main challenges to designing for home-use; the users are different and their environments of use are different. Designing for professional use in a healthcare facility enables designers to make simplifying assumptions about both the users and the use-environment (Branaghan, 2017). For example, device users in professional healthcare facilities are likely to be intellectually bright, able-bodied, relatively young (at least young enough to be in their working years), and adequately trained medical personnel. This is usually not true of users at home.

Also, professional healthcare settings themselves enable us to make simplifying assumptions about the environment of use. For example, the professional healthcare environment is likely to be consistent, well-lit, sanitary, configurable, with easy access to electrical power, and a reasonable amount of space. These assumptions are not valid in-home environments. Homes are designed for comfort, intimacy, entertainment, and socialization. They are not usually designed for medical care. Next, we provide more detail about home users and home-use environments, and highlight some design challenges.

15.3   Users of Home-Use Devices

Charness (2014) points out that the people who use home medical devices may be professional or lay caregivers, or the care recipients themselves. Indeed, the entire world population, from small children to older adults, are potential home healthcare users. This of course makes design difficult. Below, we discuss three broad groups of users: the patients themselves, formal caregivers, and informal caregivers, each with different characteristics.

It is no surprise that patients of all ages can have substantial physical, sensory, cognitive, and emotional impairments since they are being treated for medical conditions. These problems can be caused by a variety of factors such as age, illness, injury, surgical recovery. We may think first of older individuals who need care, but 14% of children under the age of 18 have special healthcare needs (Strickland et al., 2009).

The second group of users includes formal caregivers who are trained healthcare practitioners providing in-home services (Humphrey & Milone-Nuzzo, 1996). They can be employed by a home care agency, social service agency, or for-profit provider. Typically, they are divided into two categories: professionals and direct care workers.

Professional caregivers include nurses, physicians, therapists (physical therapists, speech and language therapists, occupational therapists), dieticians, and social workers. Examples of direct care workers include home health aides, home-makers, companions, and patient care attendants. Their responsibilities often include personal care, housekeeping, companionship, and assistance with activities of daily living. Direct-care workers in general, even those in formal home care agencies, receive little formal training before beginning employment and have high school diplomas or less education.

A third group of home care providers includes people who are hired by individuals in the home to help with healthcare. They often work without training or supervision, are hired without background checks, and arrive with unknown skill levels.

Probably the largest group of home caregivers are considered informal. According to the National Alliance for Caregiving (2020), 44 million US households, more than one-third, have at least one unpaid family medical caregiver. Usually these are family members, friends, neighbors, fellow church members, and so on. According to Schulz and Tomkins (2010), most people will serve this role at one time or another in their life, and some serve this role for months or years.

Story (2010) points out that the characteristics of individuals who use medical devices in the home are not well known, but a user's ability to operate a medical device depends on characteristics including:

- Physical size, strength, and stamina
- Physical dexterity, flexibility, and coordination
- Sensory capabilities (i.e., vision, hearing, tactile sensitivity)
- Cognitive abilities, including memory
- Comorbidities (i.e., multiple conditions or diseases)
- Literacy and language skills
- General health
- Mental and emotional state
- Level of education and training relative to the medical condition involved
- General knowledge of similar types of devices
- Knowledge of and experience with the particular device
- Ability to learn and adapt to a new device
- Willingness and motivation to use a new device.

Using this handy list as a rough framework, many of the considerations of design for home use are elaborated below. For the sake of brevity, we won't address each one in full, but we'll cover the highlights.

15.4   Physical Size, Strength, and Stamina

Due to underlying disease, atrophy from injury, aging, or lack of use, patients often have reduced muscle mass and strength (Visser et al., 2005). This tends to be worse in women than men, but both suffer from this problem. In some cases, grip strength also diminishes so that even opening a childproof prescription drug container is impossible. This loss of muscle can make it difficult to lift or move even modest amounts of weight.

Additionally, certain movements may become difficult. For example, patients may walk slower and be less sure-footed. Contributing to this difficulty, about two-thirds of US adults are overweight or obese, making it even more difficult to ambulate when other medical problems are present.

Caregivers may struggle with these problems as well. As the overall population ages, so does the age of both professional and informal caregivers. Approximately 50% of the registered nurse workforce will reach retirement age in the next 15 years. For older and/or smaller caregivers there can be serious musculoskeletal strains caused by lifting, positioning, and moving patients.

Importantly though, it is not just older caregivers who are at risk. As many as 1.4 million children in the United States between the ages of 8 and 18 provide care for an older adult (Levine et al., 2005). Many of these children do not have the strength to move equipment or patients either.

15.5   Dexterity, Flexibility, and Coordination

Patients often have diminished dexterity and coordination. For example, essential tremor, a neurological disorder associated with involuntary shaking, often leads to trembling hands, even when doing simple tasks. Although this can occur at any age, it is most common in people over 40 and increases with age (Benito-Leon & Louis, 2006). This condition makes it difficult to execute fine motor tasks, such as giving injections or handling pills. It can also reduce the precision of physical movements, such as moving a cursor with a mouse.

In general, as we age, our ability to perform motor tasks decreases due to reduced gray matter in our brains (Seidler et al., 2010). Additional problems related to motor movements are presented by conditions such as stroke or Parkinson's disease, making it difficult to select items on a keyboard or press buttons.

15.6 Sensory Capabilities (Vision, Hearing, Tactile Sensitivity)

Patients have different sensory abilities (vision, hearing, touch, smell, proprioception) than professional healthcare providers. Since healthcare workers are typically young and able-bodied, their senses are acute. This is often not true for patients or caregivers at home. Ill, injured, or aging patients may have impairments in some or even all of their sensory and perceptual systems.

For example, patients may have temporary or permanent visual impairment. Even during healthy aging, the lenses in our eyes become rigid and less responsive to our ciliary muscles. In fact, the ciliary muscles themselves weaken, making it difficult to achieve sharp focus. Even the general population over the age of 45–50 is not likely to be able to focus the lens of the eye on fine visual details without corrective lenses.

Additionally, disease or age can cause our lenses to yellow and become more opaque, allowing less light to enter the eye. This yellowing makes it difficult to discriminate fine blue text, and causes some loss of color sensitivity at short wavelengths, so that discriminating violet, blue, or green is more difficult.

Moreover, the pupil itself shrinks, allowing less light to enter the eye, and the pupil's response to dim light diminishes, making it difficult to see in low light conditions. Older people need more contrast to distinguish text and images. As a result, patients at home can have great difficulty reading displays, medication labels, or instructional materials. Finally, tiny imperfections on the lens of the eye can cause increased light scatter and glare.

Cataracts can be problematic for home device users. Cataracts, a clouding of the lens in the eye, are common among aging individuals. In fact, by 80 years old, more than 50% of all Americans have a cataract or have had surgery to correct a cataract (National Institute of Health, 2015). Vision affected by cataracts is often blurry, and slightly yellow or even brown, so objects are not as sharp and crisp as they once were. In terms of color perception, cataracts make it difficult to distinguish between dark shades (i.e., short wavelength colors) such as blues, purples, and black (National Institute of Health, 2015). Although cataracts may begin to develop in one's 40s or 50s, significant issues with one's vision may not be noted until the person is in their 60s.

Another concern is Age-Related Macular Degeneration (AMD), a slow-progressing disorder, often emerging around the age of 50 years (National Institute of Health, 2015). AMD causes a blurred spot in the center of your field of view (FOV), making it difficult to see what is directly in front of you. Over time, this spot grows, causing "blank spots" in vision (National Institute of Health, 2015).

Home use device users are likely to have challenges in hearing as well as vision. Hearing loss is present in about 44% of adults ages 60–69, with prevalence increasing with age (Cruickshanks et al., 1998; Pratt et al., 2009). Hearing is particularly impaired in the presence of background noise or when communication is rapid. With hearing loss, trying to understand what people say requires substantial perceptual and cognitive effort, detracting from other aspects of cognitive performance (Wingfield, Tun, & McCoy, 2005).

Approximately 10% of all middle-aged adults in the United States experience some form of hearing loss or impairment that affects how they communicate with others (Fisk, Rogers, Charness, Czaja, & Sharit, 2004). Not surprisingly, this percentage jumps to nearly 50% of men and 30% of women as they age beyond 65 years. This gradual decline poses an additional hurdle when designing medical devices for older adults. It limits one of the main sensory pathways—arguably, second only to vision—designers appeal to when communicating how a product or system is intended to be used.

Presbycusis is one of the most common degenerative hearing issues in older people. Those affected by presbycusis perceive low-pitch sounds successfully, but struggle to hear high-pitched sounds. This is of particular importance when communicating with older individuals, since language makes use of a wide range of pitches. Words are based on phonemes—perceptually distinct units of sound. Some phonemes are low pitched, occupying a range of about 125–50-0 Hz (e.g., "Z," "V," "J," "M," "D," "B," "G," "N," and most vowel sounds). Other phonemes are much higher pitch (e.g., "SH," "CH," "K," "F," "S," and "TH"), occurring between 700 and 5000 Hz depending on the person (Cruttenden, 2014). A person affected by presbycusis can hear only bits and pieces of words in a conversation. That is, they understand the low pitches—for example, vowel sounds—but miss out on the detailed, high pitch sounds that are essential to distinguish words from one another (e.g., "sat" might be confused with "fat" or "chat").

Beyond communication issues, the deterioration of higher sound registers can also affect an individual's ability to successfully interact with technology. For instance, elderly individuals with presbycusis might struggle hearing certain high-pitched alarm tones (Gates & Mills, 2005) (Fig. 15.2).

Another problem, noise-induced hearing loss (NIHL) is caused by loud, frequent, and sustained noises in one's environment. Indeed, NIHL is most frequently seen among those who worked in or near noisy environments, such as concert halls, construction sites, aviation crews, or machine shops. However, NIHL may also occur due to singular events, such as being too close to a loud, percussive noise (e.g., fireworks, gunshots, train horns). Interestingly, however, most elderly individuals who experience hearing loss suffer from some combination of presbycusis and NIHL, rather than just one or the other.

Comorbidity plays a significant role in hearing loss. For example, Agrawal, Niparko, and Dobic (2010) observed a higher prevalence of hearing dysfunction in patients with diabetes. Hearing loss was worse when the individual (a) had diabetes for longer periods of time, (b) reported higher A1C levels, and (c) experienced other diabetes-related complications (Agrawal et al., 2010). Unlike presbycusis, however, hearing loss associated with diabetes appears to affect all registers of pitch perception. That is, it affects the elderly's ability to hear low, medium, and high-pitched sounds equally (Agrawal et al., 2010).

Older adults are less accurate at localizing sounds in an environment, particularly when sounds come from in front or behind them (Abel, Giguère, Consoli, & Papsin,

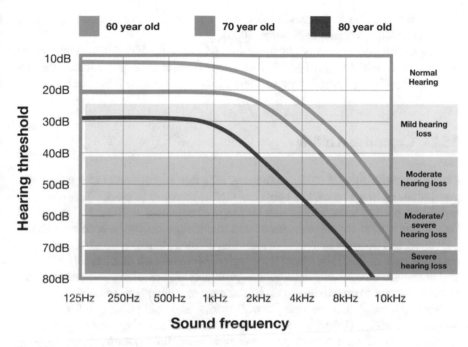

Fig. 15.2 Presbycusis and hearing loss

2000), or from above or below (Noble, Byrne, & Lepage, 1994). These issues are exacerbated in noisy environments (Boot, Nichols, Rogers, & Fisk, 2012). Even simple technologies such as ceiling fans, humidifiers, and computers can emit enough noise that older adults will experience at least some degradation in comprehension during conversations. What's more, when multiple people converse, older adults may struggle to identify who is speaking as speakers change quickly. This problem is particularly pronounced when there are several women participating in a conversation, due to their generally higher pitch speaking voices compared to males (Boot et al., 2012).

Now that we have discussed challenges with vision and hearing, let's turn our attention to haptic perception. Haptic perception is loosely defined as the sense of touch (Boot et al., 2012); however, the broader scope of haptics covers discovering objects through touch, detecting changes in temperature, and even grasping and applying constant force to an object (e.g., holding a small medical device). Haptic senses have been studied far less than vision and hearing in the aging population.

Home patients often have diminished haptic capabilities due to illness or medication use. For example, patients with diabetes or who are on HIV medication may experience peripheral neuropathy, causing weakness, numbness, and pain in their hands and fingers, making them less sensitive to tactile information. Further the density of nerve endings responsible for light touch and vibration detection decreases as humans age. Nerve conduction itself slows with age (Rivner, Swift, & Malik,

2001), probably due to decreased thickness of the myelin coating on nerve fibers (Verdú, Ceballos, Vilches, & Navarro, 2000).

Further, roughly half of adults older than 65 years old experience symptoms of arthritis, often with localized pain in their hands (Fisk et al., 2004). Medical devices with small parts requiring fine movements can be troublesome for home users.

15.7 Cognitive Abilities

Continuing with Story's (2010) taxonomy, let's turn our attention to the cognitive capabilities of home users. For home use medical devices to work effectively, they will likely need to accommodate a wide range of cognitive abilities. Home patients may struggle with cognitive issues related to attention, memory, problem solving, judgment, and decision making. For example, divided attention, which requires processing of two or more sources of information, is diminished dramatically with age, and often with illness or injury. Attention problems are especially pronounced when patients attempt to do two things at once or even to switch between tasks.

Similarly, many patients struggle with storing and manipulating information in working memory (Jastrzembski & Charness, 2007). Using working memory includes organizing, integrating, and manipulating information. When working memory is impaired, it causes difficulties for all kinds of tasks including decision making, learning, and goal directed behavior. It also means that users may have difficulty following multiple steps in succession.

Some parts of long-term memory can be affected by illness, aging, or injury as well. Recall that long-term memory is divided into semantic memory and episodic memory. Semantic memory is memory for meaning (including facts, knowledge and skills) and episodic memory is memory for events. Typically, home patients have little difficulty retrieving previously stored semantic memory (facts and skills), but they often do have difficulty creating new semantic memories (that is, learning).

Unfortunately, home patients, especially those with brain injuries or dementia, can have great difficulty remembering new events. For example, they might not recall if they did or did not program their device or take their medication. Finally, abstract problem solving, sometimes referred to as fluid abilities, tend to decrease sharply with age (Salthouse, 2010). Such declines are likely to impact the ability to troubleshoot new malfunctioning healthcare devices.

A particularly striking change in cognition is slowing in the learning rate. For instance, an older adult may require twice as much time as younger adults to learn to use software (e.g., Charness, Kelley, Bosman, & Mottram, 2001). As a result, device designers, for example, need to take learning time into account so that people don't become frustrated and give up on a device.

Taken together, these changes suggest that it may be difficult for patients (and some caregivers) to learn and remember relatively simple sets of activities, not to mention more complex step-by-step procedures like blood glucose monitoring (which might be 50–60 steps long; Rogers et al., 2001).

This may provide clues into problems of noncompliance and medical device abandonment. Some of the devices may simply be too complex. For example, Charness (2010) points out that some devices, such as hearing aids, have an abandonment rate as high as 30–50%. A big part of this problem may be that the devices do not match the capabilities and limitations of the home user.

15.8 Literacy and Language Skills

Although communication, and cognitive processing in general, tends to slow a bit, older adults usually continue to have strong speech and language abilities. Unfortunately, strokes, brain injuries, and surgery can reduce these abilities. Also, people speak different languages, come from different countries, have differing educations, and have varying levels of familiarity with health-related concepts (health literacy). Some will already be familiar with devices similar to yours, whereas others may have never heard of such a thing. Some may simply be upgrading to a new model, whereas others may be using this type of device for the very first time.

Reading abilities are a particular concern. For example, an estimated 11 million US residents have below basic literacy skills, which limit their ability to perform even relatively simple tasks (National Research Council, 2020).

15.9 Emotions and Motivation

Needing and providing healthcare are both stressful situations with major effects on health and well-being. According to Schulz and Tomkins (2010), for example, caregivers experience more health problems than noncaregivers. They have worse psychological well-being, health habits, physiological responses, psychiatric illness, physical illness, and mortality. Depression as a result of caregiving is especially worrisome because depression is the second leading cause of disability worldwide (Talley & Crews, 2007), and often adversely affects work performance as well as close relationships (Druss et al., 2009), reducing access to social support needed for effective care (Schulz & Tomkins, 2010).

Stress is a problem too, since it affects cognitive abilities. Medical conditions, changes in medical status, and learning new devices can all cause anxiety for the patient and caregivers, increasing the likelihood of use-errors. Laboratory research demonstrates the detrimental effects of stress on attention, memory, perceptual motor performance, and judgment and decision making (Staal, 2004). Chronic stress can decrease executive functioning, especially attentional control, and prospective memory. It also slows working memory processing, especially when attention is divided (Brand, Hanson, & Godaert, 2000).

15.10   Environment

Homes are different from professional healthcare environments. Healthcare facilities are designed specifically to apply technology to patient care. They are well-lit, temperature and humidity controlled, with good airflow. They have adequate access to electrical outlets and possess backup generators for power outages. They have smooth floors for transporting equipment and hang much of their equipment from the ceiling. Generally, they do not have children underfoot or pets roaming around. Further, much of the space is dedicated to reducing infection, providing space, procedures, and materials for cleaning and reprocessing medical equipment, and cleaning the environment.

Homes, on the other hand, are not designed for technology or patient care. They are less controlled and more varied. Lighting may be inadequate and inconsistent. Temperature, humidity, and airflow can be unpredictable. Space can be limited. Carpeting can make it difficult to move equipment, and cords and cables can get tangled and create tripping hazards. The environment can be messy, cluttered, and potentially unsanitary.

Although we call these home-use devices, these devices could be used in any number of places besides healthcare facilities. For example, they may need to operate in work environments, schools, automobiles, public transportation, hotels, and other locations. This can be problematic if any of these environments do not have appropriate power or communication. Further, depending on the device and the procedure, people may use medical devices in a private space, such as a bedroom, office, or restroom, or in a public space, such as an airplane, theater, or park (Story, 2010). In fact, it is important to provide labeling indicating that environmental conditions could raise safety or efficacy concerns.

These environments might not possess the utilities required for proper use. For example, the electrical power may go out for a variety of reasons and an emergency backup system may be needed, especially if a device is keeping a person alive (Story, 2010). Further, some devices, like dialysis machines, require water to operate. You should specify the type of water needed to operate the device. For example, specify if well water or treated tap water will do (FDA, 2014).

Finally, home use devices enable the patient to explore domestic or international travel or simply to visit friends and family. This means that medical devices may need to accommodate travel of all modes including automobiles, trains, and airplanes. For airplanes, this means the device will need to be inspected by security without undue hassle or embarrassment. This is especially tricky with devices worn on the body. Consider whether the device could be affected by X-ray or millimeter wave technologies (FDA, 2014). Some devices may require the patient to avoid automated screening technologies and opt for pat-down screening instead. Include this information in training, instructions for use, and, if at all possible, on the device itself.

Further, other countries may require different electrical supplies for different voltage rates. Information for this, as well as how to acquire adapters if needed, should be readily available on the device itself. Battery backup may be necessary in other countries. Finally, patients may need labeling indicating how to get help if the device malfunctions while they are away from home.

Storage is important as well. It is helpful to make sure patients and caregivers know how to store the device and its accessories safely. For example, the device may need to be stored in a cool dry environment. If so, that information should be clearly visible.

15.11   Design Considerations

Home use devices must fit into a wide variety of environments, at home and on the go. Further, they must be safe and easy to use by patients as well as caregivers who vary dramatically in physical size, strength, dexterity, mobility, stamina, and coordination.

Designers should consider simplifying devices by eliminating all nonessential functionality and reducing the number of features. This can make the device easier to learn (and relearn) by reducing clutter, minimizing choices, and simplifying information architecture. Also, consider which parameters home users have access to and which should be locked out. For example, recently we worked on a cardiac device. In that situation, it was important for users to be able to verify their heart rate, but not to be able to change it.

Where possible provide knowledge in the world in the form of operational cues, labels, and icons rather than forcing users to remember how the device works (i.e., retaining knowledge in the head). Providing knowledge in the world, as discussed in Chap. 7, provides information about how to use the device on the device itself, where it is readily available. This reduces demand on the user's memory. Reminders are related to knowledge in the world. Remind users when to conduct tasks. This is especially helpful for patients with dementia. Also, provide plenty of visible and audible feedback on the status of the device and its operations.

Make all stimuli on the device easy to see or easy to hear. As people get older, their hearing and vision often decline, to the point that labeling can be difficult to read. Remember that people will use the device under a variety of ambient light conditions. This can make it difficult for users to see and hear the information you would like to convey. Make sure text size is adequate, provide high contrast between foreground and background, and avoid colors that are close together on the visible spectrum since this reduces contrast.

As discussed in Chap. 6, there are plenty of hearing considerations to attend to as well. For example, many devices may need both visual and auditory warnings to alert older users to potential hazards. The design of alarms for the home has special challenges. Noise and distractions in the home and around town may make it difficult for the user to detect. Device alarm systems with high- or medium-priority alarm signals should be designed to be perceived in environments typically found in the home. Some medical conditions cause hearing difficulties. To minimize these problems, design alarms and alerts to utilize lower frequencies and higher volumes. Alarm and alert sounds should be markedly different from ambient sounds. Also, consider using multimodal alarms that combine visual and auditory stimuli to attract attention.

When it comes to operating the device, older adults need larger targets (e.g., large keys with substantial space between them). Take care to make buttons and controls a reasonable size, and avoid UIs that require fine motor skills. Making controls larger also helps people with diminished touch sensitivity, as does auditory and tactile feedback. Touch screens that use natural gestures for input can often be difficult for home users to take advantage of. Simply, fine movements like pinching can prove difficult to people with movement disorders.

Even though the best designers will attempt to design risk out of the system and make systems obvious to use, sometimes instructions for use or other training will still be required. Chap. 13 provides guidance for designing effective IFU; however, we provide a few reminders here. Make the IFU as simple, concise, and understandable as possible. Clearly describe contraindications, discussing why certain people should not use the device, as well as warnings and hazards. Where possible, write the IFU and training for lay users, rather than medical professionals, at an eighth-grade level or lower. Present information a little slower than you normally would, and consider providing it in multiple modes (e.g., on paper and on a recording). If possible, enable people to practice using the device. Limit jargon, and provide well-designed diagrams wherever possible.

Also, people using your device at home do not have easy access to the experts available at a healthcare facility. Make sure to provide some type of user support all day every day. This could be through a toll-free telephone number, email address, or website for questions and concerns about device use and maintenance. Also, if the device is life-supporting or life-sustaining, provide a person available 24 h a day to talk with users (Story, 2010).

Home devices are likely to be moved around quite a bit, at home, around town, and even when traveling, so they need to be rugged. For example, we conducted contextual inquiry research with previously critically ill patients using a particular device, and found that it was common for them to drop the device from the kitchen counter when they were making dinner. Now the good news is that these patients felt good enough to make dinner, which was testament to the effectiveness of this treatment. The bad news is, if the device broke upon being dropped, they would begin to feel very bad very quickly. Devices need to be able to withstand such drops. Also, where possible, devices should minimize the number of parts that could break when accidents do happen.

Home devices involve more than just providing therapy. For example, the patient or caregiver may need to maintain the device, which might include calibration. Where possible design devices that do not require calibration. If this is not practical, make the calibration procedure as simple as possible, providing step by step instructions. Also, make sure that the user can always determine if the device is currently calibrated, when it was calibrated last, and when it needs to be calibrated next (FDA, 2014). In a recent study for a diabetes technology center, one of the authors conducted home-based interviews to understand how people were using their "artificial pancreas" system. When he visited people in their home, he realized that the majority of patients had abandoned the system. One of the main reasons was the need to constantly calibrate the sensors.

Maintenance also means cleaning, disinfection, and potentially reprocessing. Often, however, home users do not have access to cleaning, disinfecting, and sterilization supplies available in clinical environments. With this in mind, attempt to enable devices to be able to be cleaned, disinfected, or sterilized with readily available supplies and use simple methods.

Disposal of the supplies or device should also be taken into consideration. IFU should describe how to properly dispose of the device and its accessories. Be sure to provide proper warnings and precautions about safely disposing of the device and supplies.

Finally, it is important to design the home device to reduce its clinical appearance, choosing an aesthetic that is more consumer oriented.

- Resources National Research Council. (2010). The Role of Human Factors in Home Health Care: Workshop Summary. Steve Olson, Rapporteur. Committee on the Role of Human Factors in Home Health Care, Committee on Human-Systems Integration. Division of Behavioral and Social Sciences and Education. Washington, DC: The National Academies Press.
- United States FDA. (2014). Design considerations for devices intended for home use: Guidance for Industry and Food and Drug Administration Staff.
- Weick-Brady, M., & Singh, S. (2014). The Food and Drug Administration's initiative for safe design and effective use of home medical equipment. *Home Healthcare Now, 32*(6), 343–348.
- Winters, J.M. & Story, M. F. (Eds.). (2007). Medical instrumentation: Accessibility and usability considerations. Boca Raton, FL: CRC Press.

References

Abel, S. M., Giguère, C., Consoli, A., & Papsin, B. C. (2000). The effect of aging on horizontal plane sound localization. *The Journal of the Acoustical Society of America, 108*(2), 743–752.

Agrawal, Y., Niparko, J. K., & Dobie, R. A. (2010). Estimating the effect of occupational noise exposure on hearing thresholds: The importance of adjusting for confounding variables. *Ear and Hearing, 31*(2), 234–237.

Benito-Leon, J., & Louis, E. D. (2006). Essential tremor: Emerging views of a common disorder. *Nature Clinical Practice Neurology, 2*(12), 666–678.

Boot, W. R., Nichols, T. A., Rogers, W. A., & Fisk, A. D. (2012). Design for aging. *Handbook of Human Factors and Ergonomics, 4*, 1442–1471.

Branaghan, R. J. (2017). Human factors for medical devices—The home is not the same as the hospital. Medical Device and Diagnostic Industry (July, 2017). Retrieved September 1, 2020, from https://www.mddionline.com/design-engineering/human-factors-home-use-medical-devicesthe-home-not-same-hospital.

Brand, N., Hanson, E., & Godaert, G. (2000). Chronic stress affects blood pressure and speed of short-term memory. *Perceptual and Motor Skills, 91*(1), 291–298.

Charness, N. (2010). The healthcare challenge: Matching care to people in their home environ-
    ments. In National Research Council. The Role of Human Factors in Home Health Care:
    Workshop Summary. Steve Olson, Rapporteur. Committee on the Role of Human Factors in
    Home Health Care, Committee on Human-Systems Integration. Division of Behavioral and
    Social Sciences and Education. Washington, DC: The National Academies Press.
Charness, N. (2014). Utilizing technology to improve older adult health. *Occupational Therapy in
    Health Care, 28*(1), 21–30.
Charness, N., Kelley, C. L., Bosman, E. A., & Mottram, M. (2001). Word-processing training and
    retraining: Effects of adult age, experience, and interface. *Psychology and Aging, 16*(1), 110.
Cruickshanks, K. J., Wiley, T. L., Tweed, T. S., Klein, B. E., Klein, R., Mares-Perlman, J. A., et al.
    (1998). Prevalence of hearing loss in older adults in Beaver Dam, Wisconsin: The epidemiol-
    ogy of hearing loss study. *American Journal of Epidemiology, 148*(9), 879–886.
Cruttenden, A. (2014). *Gimson's pronunciation of English.* Abingdon: Routledge.
Druss, B. G., Hwang, I., Petukhova, M., Sampson, N. A., Wang, P. S., & Kessler, R. C. (2009).
    Impairment in role functioning in mental and chronic medical disorders in the United States:
    Results from the National Comorbidity Survey Replication. *Molecular Psychiatry, 14*(7),
    728–737.
Fisk, A. D., Rogers, W. A., Charness, N., Czaja, S. J., & Sharit, J. (2004). *Designing for older
    adults: Principles and creative human factors approaches.* Boca Raton, FL: CRC Press.
Gates, G. A., & Mills, J. H. (2005). Presbycusis. *The Lancet, 366*(9491), 1111–1120.
Humphrey, C. J., & Milone-Nuzzo, P. (1996). *Orientation to home care nursing.* Burlington, MA:
    Jones & Bartlett Learning.
Jastrzembski, T. S., & Charness, N. (2007). The Model Human Processor and the older adult:
    Parameter estimation and validation within a mobile phone task. *Journal of Experimental
    Psychology: Applied, 13*(4), 224.
Kaye, R., & Crowley, J. (2000). Medical device use-safety: Incorporating human factors engineer-
    ing into risk management. *Food and Drug Administration.*
Levine, C., Hunt, G. G., Halper, D., Hart, A. Y., Lautz, J., & Gould, D. A. (2005). Young adult
    caregivers: A first look at an unstudied population. *American Journal of Public Health, 95*(11),
    2071–2075.
National Institute of Health. (2015). *Facts about age-related macular degeneration.* Retrieved
    from: https://nei.nih.gov/health/maculardegen/armd_facts.
National Research Council. The Role of Human Factors in Home Health Care: Workshop
    Summary. Steve Olson, Rapporteur. Committee on the Role of Human Factors in Home
    Health Care, Committee on Human-Systems Integration. Division of Behavioral and Social
    Sciences and Education. Washington, DC: The National Academies Press. National Alliance
    of Caregiving (2020).
Noble, W., Byrne, D., & Lepage, B. (1994). Effects on sound localization of configuration and type
    of hearing impairment. *The Journal of the Acoustical Society of America, 95*(2), 992–1005.
Pratt, S. R., Kuller, L., Talbott, E. O., McHugh-Pemu, K., Buhari, A. M., & Xu, X. (2009).
    Prevalence of hearing loss in black and white elders: Results of the Cardiovascular Health
    Study. *Journal of Speech, Language, and Hearing Research, 52*, 973.
Rivner, M. H., Swift, T. R., & Malik, K. (2001). Influence of age and height on nerve conduction.
    Muscle & Nerve, 24(9), 1134–1141.
Salthouse, T. A. (2010). Selective review of cognitive aging. *Journal of the International
    Neuropsychological Society: JINS, 16*(5), 754.
Schulz, R. & Tomkins, C. A. (2010) Informal caregivers in the United States: Prevalence, caregiver
    characteristics, and ability to provide czare. In National Research Council. The Role of Human
    Factors in Home Health Care: Workshop Summary. Steve Olson, rapporteur. Committee on
    the Role of Human Factors in Home Health Care, Committee on Human-Systems Integration.
    Division of Behavioral and Social Sciences and Education. Washington, DC: The National
    Academies Press.

Seidler, R. D., Bernard, J. A., Burutolu, T. B., Fling, B. W., Gordon, M. T., Gwin, J. T., et al. (2010). Motor control and aging: Links to age-related brain structural, functional, and biochemical effects. *Neuroscience & Biobehavioral Reviews, 34*(5), 721–733.

Staal, M. A. (2004). *Stress, cognition, and human performance: A literature review and conceptual framework*. CA: NASA Ames Research Center.

Story, M. F. (2010). Medical devices in home health care. In National Research Council. The Role of Human Factors in Home Health Care: Workshop Summary. Steve Olson, Rapporteur. Committee on the Role of Human Factors in Home Health Care, Committee on Human-Systems Integration. Division of Behavioral and Social Sciences and Education. Washington, DC: The National Academies Press.

Strickland, B. B., Singh, G. K., Kogan, M. D., Mann, M. Y., van Dyck, P. C., & Newacheck, P. W. (2009). Access to the medical home: New findings from the 2005–2006 National Survey of Children with Special Health Care Needs. *Pediatrics, 123*(6), 996–1004.

Talley, R. C., & Crews, J. E. (2007). Framing the public health of caregiving. *American Journal of Public Health, 97*(2), 224–228.

The National Alliance for Caregiving Mission. (2020). Retrieved August 15, 2020, from Caregiving.org.

United States FDA. (2014). Design considerations for devices intended for home use: Guidance for Industry and Food and Drug Administration Staff.

Verdú, E., Ceballos, D., Vilches, J. J., & Navarro, X. (2000). Influence of aging on peripheral nerve function and regeneration. *Journal of the Peripheral Nervous System, 5*(4), 191–208.

Visser, M., Goodpaster, B. H., Kritchevsky, S. B., Newman, A. B., Nevitt, M., Rubin, S. M., et al. (2005). Muscle mass, muscle strength, and muscle fat infiltration as predictors of incident mobility limitations in well-functioning older persons. *The Journals of Gerontology Series A: Biological Sciences and Medical Sciences, 60*(3), 324–333.

Wingfield, A., Tun, P. A., & McCoy, S. L. (2005). Hearing loss in older adulthood: What it is and how it interacts with cognitive performance. *Current Directions in Psychological Science, 14*(3), 144–148.

Zayas-Cabán, T., & Valdez, R. S. (2012). Human factors and ergonomics in home care. In P. Carayon (Ed.), *Handbook of human factors and ergonomics in health care and patient safety* (pp. 772–791). Boca Raton, FL: CRC Press.

Index

© Springer Nature Switzerland AG 2021
R. J. Branaghan et al., *Humanizing Healthcare – Human Factors for Medical
Device Design*, https://doi.org/10.1007/978-3-030-64433-8

Printed in the United States
by Baker & Taylor Publisher Services